Analysis of Metals: Properties and Techniques

Analysis of Metals:
Properties and Techniques

Edited by **Peggy Rusk**

C WILLFORD PRESS

New York

Published by Willford Press,
118-35 Queens Blvd., Suite 400,
Forest Hills, NY 11375, USA
www.willfordpress.com

Analysis of Metals: Properties and Techniques
Edited by Peggy Rusk

International Standard Book Number: 978-1-68285-045-9 (Hardback)

The publisher's policy is to use permanent paper from mills that operate a sustainable forestry policy. Furthermore, the publisher ensures that the text paper and cover boards used have met acceptable environmental accreditation standards.

Trademark Notice: Registered trademark of products or corporate names are used only for explanation and identification without intent to infringe.

Printed in the United States of America.

Contents

Preface

Over the recent decade, advancements and applications have progressed exponentially. This has led to the increased interest in this field and projects are being conducted to enhance knowledge. The main objective of this book is to present some of the critical challenges and provide insights into possible solutions. This book will answer the varied questions that arise in the field and also provide an increased scope for furthering studies.

Each metal is a unique material and displays a distinct set of properties. This book unravels the recent researches in the study of metals. It covers in detail some existent theories and innovative concepts revolving around topics like metallurgy, noble metals, transition metals, alloys, metalloids, etc. The extensive content of this text provides the readers with a thorough understanding of metals. It aims to equip students and experts with the advanced theories and upcoming concepts related to metals. This book is ideal for students and research scholars pursuing chemistry, metallurgy and similar disciplines.

Over the recent decade, advancements and applications have progressed exponentially. This has led to the increased interest in this field and projects are being conducted to enhance knowledge. The main objective of this book is to present some of the critical challenges and provide insights into possible solutions. This book will answer the varied questions that arise in the field and also provide an increased scope for furthering studies.

Editor

Experimental Investigation of the Mg-Nd-Zn Isothermal Section at 300 °C

Ahmad Mostafa [1] and Mamoun Medraj [1,2,*]

[1] Department of Mechanical and Industrial Engineering, Concordia University,
1455 de Maisonneuve Blvd. West, Montreal, QC H3G 1M8, Canada;
E-Mail: ah_mosta@encs.concordia.ca

[2] Department of Mechanical and Materials Engineering, Masdar Institute, Masdar City,
P. O. Box 54224, Abu Dhabi, UAE

* Author to whom correspondence should be addressed; E-Mail: mmedraj@encs.concordia.ca

Academic Editor: Xing-Qiu Chen

Abstract: The Mg-Nd-Zn isothermal section at 300 °C was established in the full composition range using diffusion couples and equilibrated key alloys. Microstructural characterization was carried out using WDS, XRD, and metallographic methods. The homogeneity ranges of the binary and ternary compounds were determined by WDS analysis. Six ternary compounds were observed in the Mg-Nd-Zn system at 300 °C. These are: τ_1 ($Nd_5Mg_{21+x}Zn_{45-x}$; $0 \leq x \leq 4$), τ_2 ($Nd_5Mg_{3+y}Zn_{25-y}$; $0 \leq y \leq 1$), τ_3 ($NdMg_{1+z}Zn_{2-z}$; $0 \leq z \leq 0.44$), τ_4 ($Mg_{40}Nd_5Zn_{55}$), τ_5 ($Mg_{22-23.5}Nd_{15.5-17.5}Zn_{59.1-61.8}$), and τ_6 ($Nd_2(Mg,Zn)_{23}$). τ_5 was found to have a homogeneity range of 22.0–23.5 atom % Mg, 15.5–17.6 atom % Nd and 59.1–61.8 atom % Zn and τ_6 was found to have 54.1–61.3 atom % Mg at a constant Nd of 8.0 atom %. The ternary solubility of Zn in Mg-Nd compounds was found to increase with the decrease in Mg concentration. Accordingly, ($Mg_{41}Nd_5$) was found to have an extended solubility of 3.1 atom % Zn, whereas (Mg_3Nd) was found to have 30.0 atom % Zn. MgNd was found to have a complete substitution of Mg by Zn. The maximum solid solubility of Zn in α-Mg was measured as 4.8 atom % Zn.

Keywords: Mg-Nd-Zn system; diffusion couples; experimental investigation; ternary phase diagram

1. Introduction

Magnesium alloys are widely used in the automotive and aerospace industries, because of their high strength to weight ratio. The reason for this interest is the reduction of fuel consumption, because magnesium alloys are lighter than the traditional iron- and aluminum-based alloys [1]. The use of magnesium alloys is restricted to low stress level and low temperature components due to their poor mechanical properties and creep resistance at elevated temperatures [2,3]. Addition of zinc is known for improving the mechanical properties and corrosion resistance of magnesium-based alloys at room temperature [4]. However, the melting temperature of Mg-Zn alloys is very low, because of the deep eutectic which occurs at 342 °C [5,6]. For this reason, rare earth elements are added to improve the mechanical properties and creep resistance of magnesium alloys at elevated temperatures [7]. Addition of neodymium to magnesium-based alloys results in improving their hardness and strength after an age hardening treatment. This can be attributed to the formation of fine-scaled solute-rich regions, Guinier-Preston zones (GP zones), in the Mg matrix [8]. On the other hand, the Nd-Zn binary phase diagram [9] shows the formation of a series of stable intermetallic compounds with high melting temperatures, which may result in improving the creep behavior of Mg alloys.

The main objective of this work is to establish the Mg-Nd-Zn isothermal section at 300 °C in the full composition range by means of diffusion couples and key alloy experiments using X-ray diffraction (XRD), a scanning electron microscope (SEM), energy dispersive X-ray spectrometer (EDS), and wavelength dispersive X-ray spectrometer (WDS).

2. Literature Review

The Mg-Nd-Zn partial isothermal section (50–100 atom % Mg, 0–35 atom % Nd, and 0–55 atom % Zn) was studied by Drits et $al.$ [10] at 300 °C using X-ray diffraction and metallography. They reported several binary and ternary compounds that are in equilibrium with the α-Mg solid solution. These compounds are: $Mg_{12}Nd$ (denoted in their work as Mg_9Nd), Mg_7Zn_3, $MgZn$, $MgNd_4Zn_5$, $Mg_6Nd_2Zn_7$, and $Mg_2Nd_2Zn_9$. In addition, three vertical sections (section 1: from 80 atom % Mg and 20 atom % Nd to 70 atom % Mg and 30 atom % Zn; section 2: constant 10 atom % Nd and 0–60 atom % Zn; section 3: constant 20 atom % Zn and 0–35 atom % Nd) were determined using thermal analysis. Accordingly, two eutectic reactions: $L \leftrightarrow α\text{-}Mg + Mg_{12}Nd + MgNd_4Zn_5$ and $L \leftrightarrow α\text{-}Mg + Mg_2Nd_2Zn_9 + Mg_7Zn_3$, and two quasi-peritectic reactions: $L + MgNd_4Zn_5 \leftrightarrow α\text{-}Mg + Mg_6Nd_2Zn_7$ and $L + Mg_6Nd_2Zn_7 \leftrightarrow α\text{-}Mg + Mg_2Nd_2Zn_9$ were observed. Drits et $al.$ [10] found that Mg_7Zn_3 is stable within a narrow temperature range between 312 and 340 °C. Later, Drits et $al.$ [11] studied the combined solubility of Nd and Zn in an Mg solid solution at 200, 250, 300, 400, and 500 °C using X-ray diffraction, electrical resistivity, microhardness, and microstructural examination. Based on these results, the phase equilibria between $Mg_{12}Nd$, Mg_7Zn_3, $MgNd_4Zn_5$, $Mg_6Nd_2Zn_7$, and $Mg_2Nd_2Zn_9$ and the α-Mg solid solution were suggested. In his review paper, Raynor [12] presented the Mg-Nd-Zn ternary phase diagram based on the results obtained by Drits et $al.$ [10,11]. Kinzhibalo et $al.$ [13] reported different phase relations in the Mg-rich corner at 297 °C and ternary compounds than those reported by Drits et $al.$ [11] at 300 °C. Kinzhibalo et $al.$ [13] proposed that the α-Mg solid solution is in equilibrium with $Mg_{12}Nd$, Mg_6NdZn_3, $Mg_8Nd_3Zn_{11}$, Mg_7NdZn_{12}, and $MgZn$. Huang et $al.$ [14] investigated the structure, composition range, and phase equilibria of the ternary compound τ_1 ($Mg_{27-33.4}Nd_{6.1-7.4}Zn_{60.2-66.4}$) using SEM/EPMA, XRD, and TEM. The existence of τ_1 was confirmed in the low Nd side in the temperature range of 300–400 °C. Also, they reported the existence of three three-phase regions $\tau_1+\alpha$-Mg+MgZn, τ_1+MgZn+L, and $\tau_1+Mg_2Zn_3$+L at 300, 350, and 400 °C, respectively. Zhang *et al.* [15,16] reported the existence of a simple icosahedral quasi-crystal phase with a stoichiometric composition of $Mg_{40}Nd_5Zn_{55}$ using SEM, EDS, and TEM experiments. The quasi-lattice parameter a_R was determined based on X-ray diffraction as 0.525 nm. The phase was detected in the Mg-Nd-Zn as-cast alloys, which were made under conventional casting conditions [15,16]. However, the presence of the $Mg_{40}Nd_5Zn_{55}$ phase was not confirmed under equilibrium conditions.

Recently, Huang *et al.* [1] performed a systematic study on the Mg-rich corner of the partial Mg-Nd-Zn system at 400 °C using XRD, SEM/EPMA, and TEM. They reported a ternary compound τ_2, having the chemical formula $Nd(Mg,Zn)_{11.5}$, which was found in equilibrium with the α-Mg solid solution, τ_1, τ_3 ($MgNdZn_2$), and τ_4 ($Mg_{25.2}Nd_{15.7}Zn_{59.1}$). In addition, five three-phase regions, $\tau_2+\tau_3+\alpha$-Mg, $\tau_2+\tau_1+\alpha$-Mg, $\tau_2+\tau_3+\tau_4$, $\tau_2+\tau_1+\tau_4$, and α-Mg+τ_1+L, were identified [1]. More recently, Qi *et al.* [8] provided a thermodynamic description of the Mg-Nd-Zn system using the CALPHAD approach. Four stable ternary compounds, taken from the literature, were considered in their model [8]. These compounds are τ_1 (Mg_7NdZn_{12}), τ_2 ($Mg_7Nd_2Zn_{11}$), τ_3 (Mg_6NdZn_3), and τ_4 ($Mg_6Nd_3Zn_{11}$). The calculated isothermal section of the Mg-Nd-Zn system at 300 °C showed different phase relations than those reported in the experimental isothermal section of Kinzhibalo *et al.* [13] at 300 °C. Later, Zhang *et al.* [17] evaluated the Mg-Nd-Zn system at 300 °C and 400 °C using four key alloys and thermodynamic modeling. In their work, three ternary compounds, τ_1 ($Mg_{35}Nd_5Zn_{60}$), τ_2 ($Nd_8(Mg,Zn)_{92}$), and τ_3 ($Mg_{30}Nd_{15}Zn_{55}$), reported by Kinzhibalo *et al.* [13], were confirmed experimentally using EPMA and XRD. The various ternary compounds in the Mg-Nd-Zn system reported in the literature [1,10,11,13–16] are presented graphically in Figure 1.

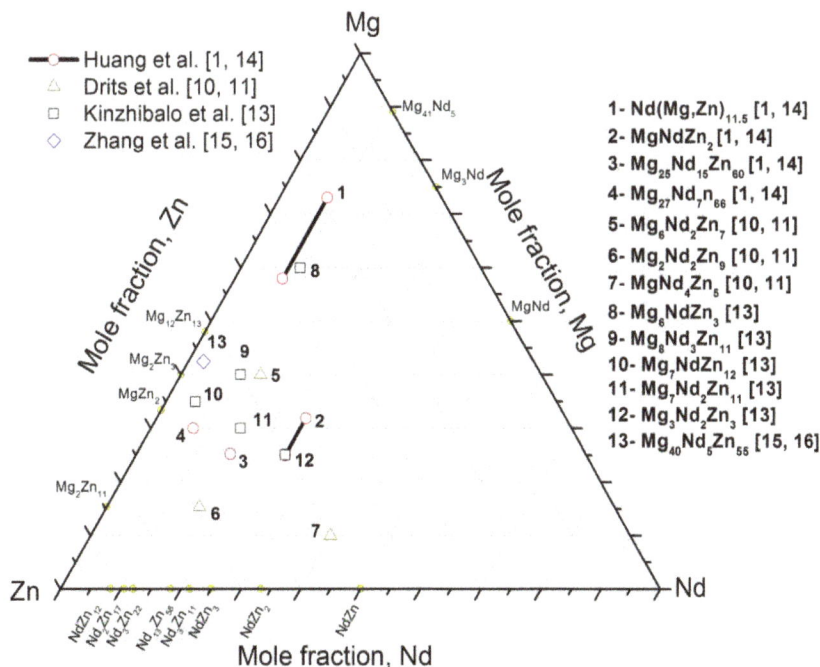

Figure 1. Graphical representation of the Mg-Nd-Zn ternary compounds reported in the literature [1,10,11,13–16] at different temperatures.

The available literature data focused on partial isothermal sections near the Mg-rich corner only and the phase relations near the Zn-rich corner were poorly studied. Also, many discrepancies on the composition of the ternary compounds were found. Thus, it is necessary to understand the phase relationships in the Mg-Nd-Zn system at 300 °C for the whole composition range by means of experimental techniques.

3. Results and Discussion

In the following section, ternary compounds are denoted by the Greek symbol τ and are numbered according to their order of appearance. Binary and ternary solid solutions are described using parentheses and superscripts. For example, the binary solid solubility of Zn in α-Mg is presented as $(\alpha\text{-Mg})^{Zn}$ and the ternary solid solubility of Zn in Mg_3Nd is presented as (Mg_3Nd).

3.1. Solid-Solid Diffusion Couples

The end-members of diffusion couple #1 were made from pure zinc and a two-phase binary alloy containing Mg_3Nd and $MgNd$, as shown in Figure 2d. The diffusion couple was annealed at 300 °C for 21 days. The SEM micrographs of diffusion couple #1 are shown in Figure 2a–c. Nine diffusion zones in this diffusion couple were identified using WDS spot analysis (±1 atom %). The compositions of the constituents of these nine zones are given in Table 1.

From Figure 2 and Table 1, diffusion zone #1 represents the pure-Zn end-member. Diffusion zone #2 represents the binary compound layer Mg_2Zn_{11}. Diffusion zones #3–5 represent two-phase layers. These are: $(Nd_2Zn_{17})+Mg_2Zn_{11}$, $(Nd_2Zn_{17})+MgZn_2$, and τ_1+MgZn_2, respectively. Three-phase equilibrium could be determined at every interface of these zones. For instance, the three-phase equilibria $(Nd_2Zn_{17})+Mg_2Zn_{11}+MgZn_2$ was determined at the interface between zones #3 and #4. The three-phase equilibria $(Nd_2Zn_{17})+MgZn_2+\tau_1$ was determined at the interface between zones #4 and #5. Each of the diffusion zones #6–8 represent a single phase layer as follows: $Mg_2Nd_2Zn_9$ (denoted as τ_2 in this work), $MgNdZn_2$ (denoted as τ_3 in this work), and $Nd(Mg,Zn)_3$, respectively. Drits et al. [11] reported the existence of $Mg_2Nd_2Zn_9$ at 200 and 300 °C and proposed that it is in equilibrium with α-Mg in both isothermal sections. Zones #7 and #8 show that Zn substitutes for Mg at a constant Nd concentration of 25 atom %. The composition of Mg varied from 25 to 31.5 atom % in the τ_3 ternary compound, and from 51.2 to 75 atom % in the (Mg_3Nd) ternary solid solution. Zone #9 is the $Mg_3Nd+MgNd$ end-member.

EDS elemental mapping was performed to determine the boundaries of the diffusion zones near the two-phase end-member (zone #9) as shown in Figure 3. These maps proved that the observed morphology in diffusion couple #1 has planar features with alternating phases and not cracks. The elemental maps of Mg, Nd, and Zn distribution are presented in Figure 3b–d, respectively.

According to EDS elemental mapping, the dark phase in Figure 3a, zones #4 and #5, represent the $MgZn_2$ compound, whereas the bright dendrites represent the ternary and the Nd-containing binary phases. These phases are Nd_2Zn_{17} in zone #4 and as τ_1 in zone #5. Zones #6 and #7 are τ_1 and τ_3 ternary compounds. Zone #8 is the (Mg_3Nd) ternary solid solution. These results were confirmed by WDS analysis, as listed in Table 1.

It is usually difficult to locate the original interface between the two end-members after diffusion takes place. However, knowing the location of the original interface provides information regarding atomic flux leading to diffusion layer formation [18]. The original interface of diffusion couple #1 was determined at the contacting interface between zones #2 and #3, as shown in Figure 2c. This means that the diffusion reaction took place based on Mg and Zn atom, exchange with less mobility for Nd atoms. As a result, a very wide matrix containing Mg_2Zn_{11} in zone #3, and $MgZn_2$ in zones #4 and #5, are formed, and the Nd-containing phases, the bright dendrites in Figure 2a, appeared within this matrix. Similar morphologies were observed at 300 °C in the Ce-Mg-Zn system studied by our research group [19]. The reason for having such periodic layer morphology could be due to the difference between the atomic sizes of Mg/Zn atoms that leads to lattice distortion, which imposes mechanical stresses that break the ternary compound layers. The atomic radius of Zn is about 133 pm, which is relatively smaller than the atomic radius of Mg (160 pm) [20]. Also, when a diffusion couple consists of species with very different mobilities, various phases are grown at different growth rates. Therefore, the slowly growing phase will be under tension as the other phase grows rapidly, and it finally splits off from the reaction front [21]. In this work, the mobility of Zn atoms is expected to be higher than that of Nd. Thus, the very mobile Zn atoms diffuse into the Mg-Nd end-member, leading to the formation of a Mg-Zn matrix and Mg-Nd-Zn ternary compounds.

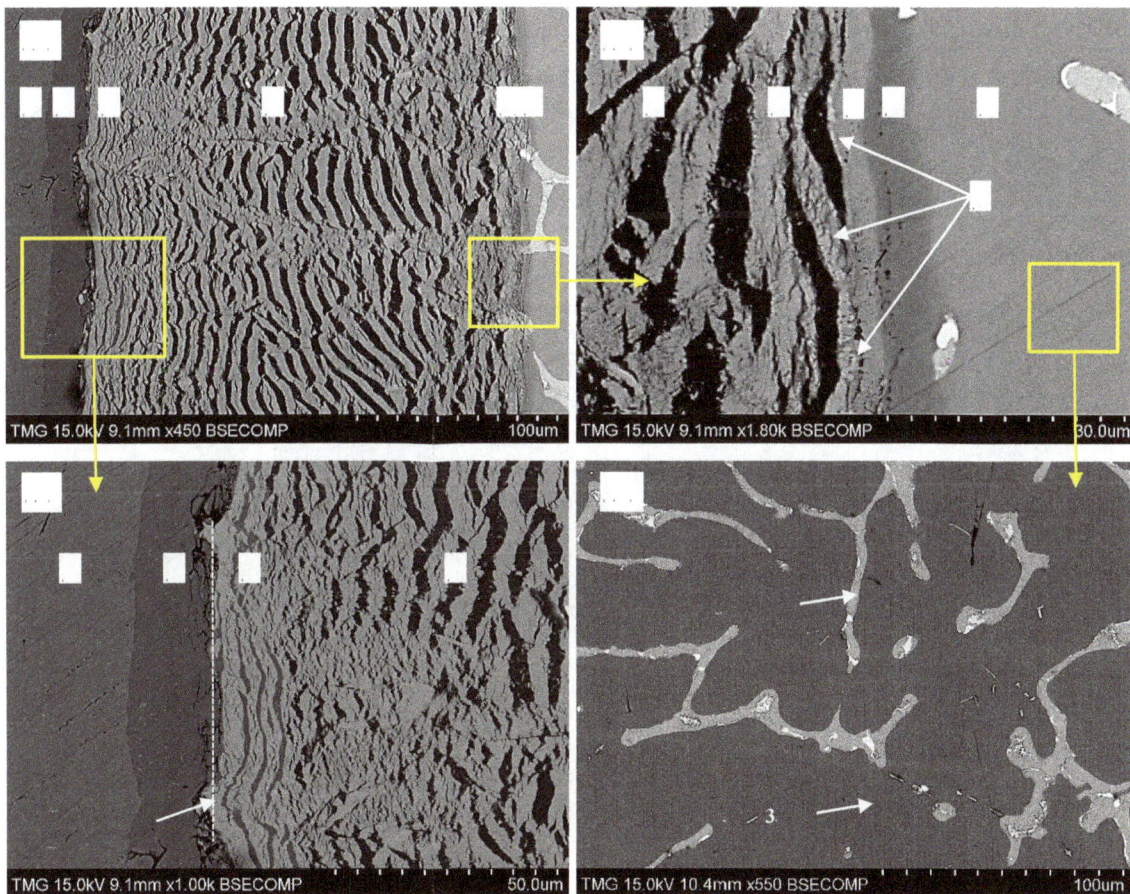

Figure 2. (**a–c**) SEM micrographs of diffusion couple #1 annealed at 300 °C for 21 days. The numbers represent the diffusion zones and correspond to those in Table 1 and Figure 4; (**d**) is the SEM micrograph of MgNd+Mg_3Nd end-member.

Table 1. WDS spot analysis of different diffusion zones of the diffusion couple #1.

Zone	Description	Composition (atom %)			Corresponding Phase
		Mg	Nd	Zn	
1	Pure Zn (end-member)	-	-	100	Zn
2	Single-phase layer	15.4	-	84.6	Mg_2Zn_{11}
3	Two-phase layer	10.8	1.9	87.3	(Nd_2Zn_{17})
		14.7	85.3	-	Mg_2Zn_{11}
4	Two-phase layer	2.0	8.9	89.1	(Nd_2Zn_{17})
		32.5	0	67.5	$MgZn_2$
5	Two-phase layer	6.5	29.5	64	τ_1
		32.1	-	67.9	$MgZn_2$
6	Single-phase layer	14.5	14.7	71.8	τ_2
7	Single-phase layer	25.0–31.5	24.5	44.0–51.5	τ_3
8	Single-phase layer	51.2–75.0	24.6	0–24.2	(Mg_3Nd)
9	Two-phase alloy	74.8	25.2	-	Mg_3Nd
	(end-member)	50.0	50.0	-	MgNd

Based on the phase equilibria determined from diffusion couple #1, the phase equilibrium and diffusion path can be depicted as follows: pure Zn (end-member)/Mg_2Zn_{11}/Mg_2Zn_{11}+(Nd_2Zn_{17})/(Nd_2Zn_{17})+$MgZn_2$+Mg_2Zn_{11}/(Nd_2Zn_{17})+$MgZn_2$/$MgZn_2$+(Nd_2Zn_{17})+τ_1/$MgZn_2$+τ_1/τ_2/τ_3/(Mg_3Nd)/Mg_3Nd+MgNd (end-member). Accordingly, the phase relations can be represented graphically, as shown in Figure 4. The two end-members are connected by a dashed line. The numbered boxes represent the corresponding diffusion zones in Figure 2 and Table 1.

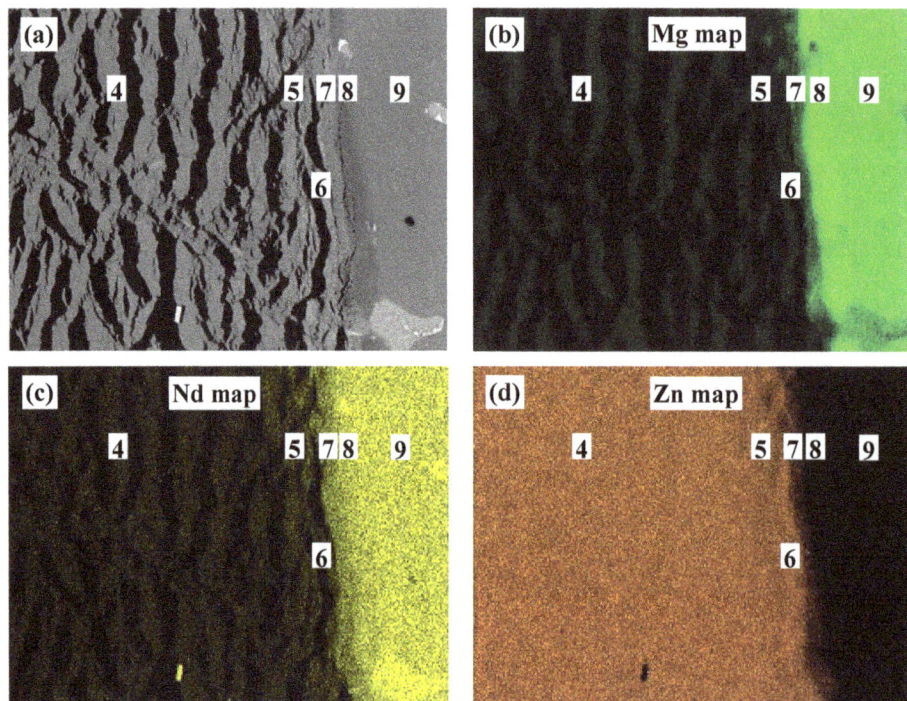

Figure 3. (**a**) SEM micrograph of diffusion couple #1 near zones #4–8; (**b–d**) EDS elemental maps for the elements Mg, Nd, and Zn, respectively. Readers are encouraged to see the online version for the colored maps.

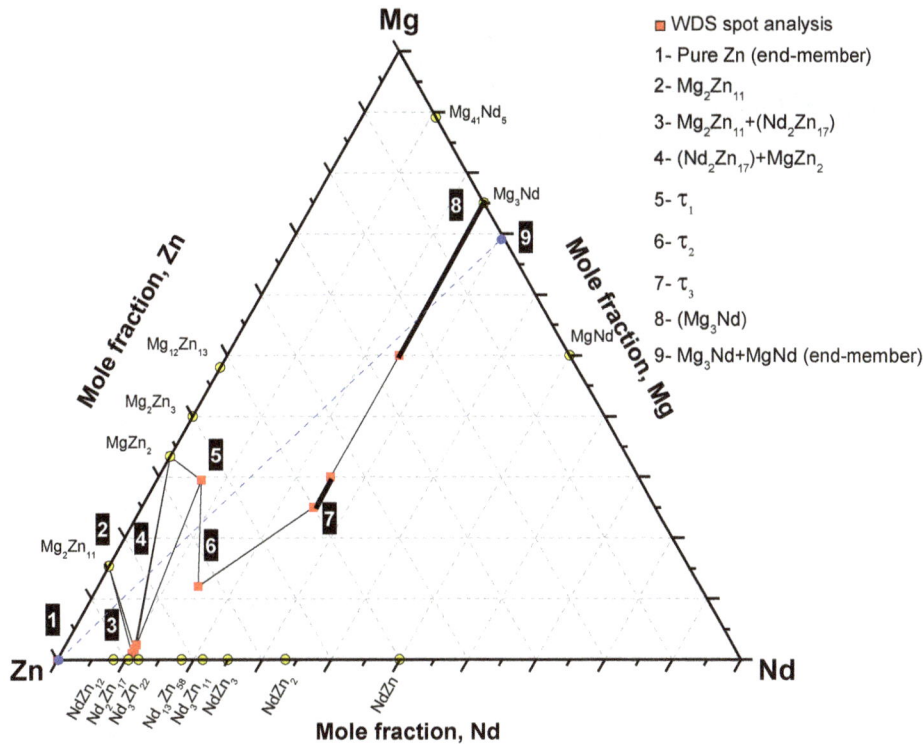

Figure 4. The phase equilibria depicted from diffusion couple #1.

SEM micrographs of diffusion couple #2, annealed at 300 °C for 40 days, are shown in Figure 5a,b. A long annealing time (40 days) was required, because no successful diffusion occurred when a shorter annealing time (21 days) was used. This can be attributed to the composition of the end-members, because the diffusion process is governed by the affinity of different species to one another. For instance, the affinity of Nd to Zn is reduced in the presence of Mg. This can be attributed to the high diffusivity of Mg in both Zn and Nd [22]. Evidence can be seen in the thickness of diffusion zones in diffusion couple #1, where very wide diffusion layers of Mg-Zn compounds were formed at 300 °C within 21 days only.

Both end-members of diffusion couple #2 were composed of two-phase binary alloys. The actual composition of the first end-member, Figure 5c, is 63Mg-37Zn atom % and it contains $Mg_{12}Zn_{13}$ and $(\alpha\text{-}Mg)^{Zn}$. The actual composition of the second end-member, Figure 5d, is 55Mg-45Nd atom % and it contains Mg_3Nd and $MgNd$. WDS spot analysis, summarized in Table 2, was used to determine the composition of the diffusion zones and their constituents. A WDS line-scan, shown in Figure 6, was performed across the diffusion zones of diffusion couple #2. The original interface location was determined at the junction between zones #2 and #3, as indicated in Figure 5a,b.

Based on WDS spot analysis and line-scan, seven diffusion zones were determined. Diffusion zones #1 and #7 are the end-members. The SEM micrographs and WDS results revealed that the other diffusion zones are single-phase layers, unlike diffusion couple #1, where most of the layers contained two phases. Zone #2 is the $Mg_{40}Nd_5Zn_{55}$ ternary compound (denoted as τ_4 in this work), that conforms with that reported by [15,16]. Diffusion zones #3–6 represent single-phase diffusion layers as follows: τ_1, Mg_2NdZn_4 (denoted as τ_5 in this work), τ_3, and $Nd(Mg,Zn)$, respectively.

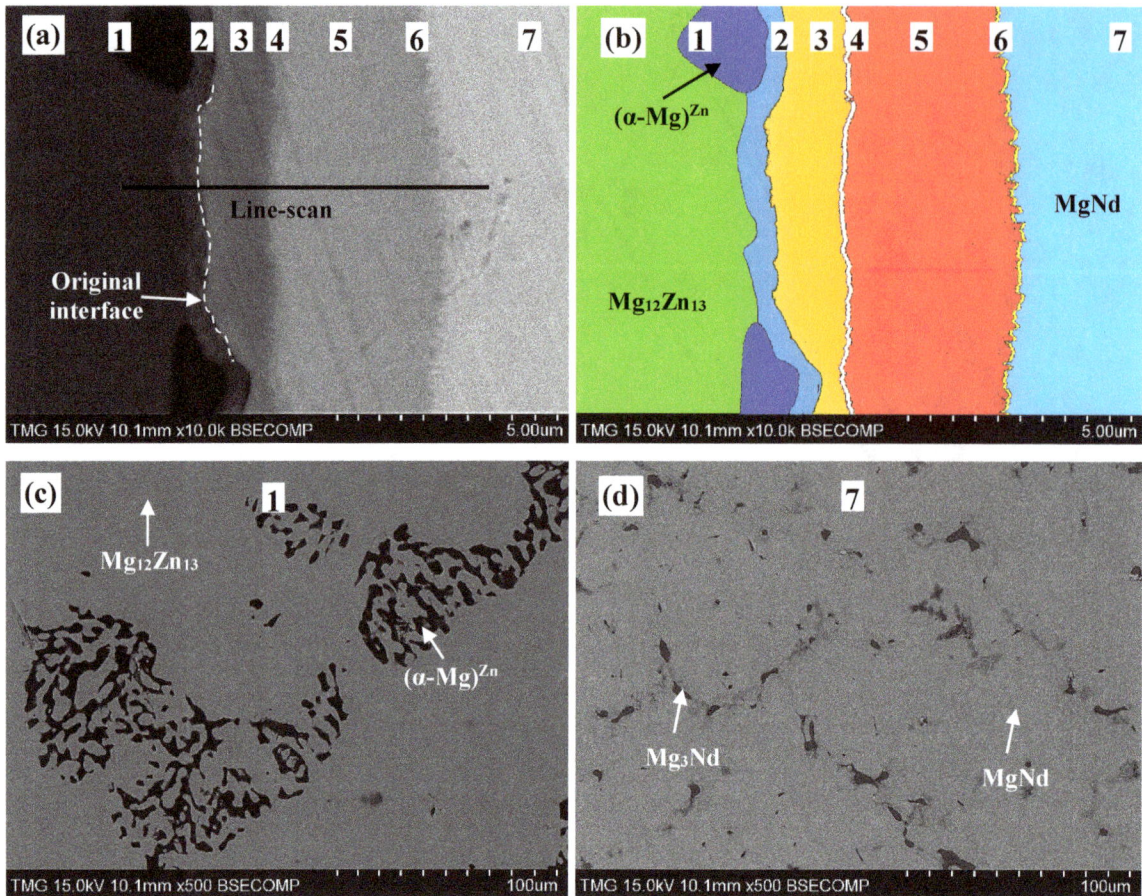

Figure 5. (**a**) A SEM micrograph of diffusion couple #2 annealed at 300 °C for 40 days; (**b**) the colored version of the same diffusion couple (readers are encouraged to see the online version for the colored image); (**c**) is the (α-Mg)Zn+Mg$_{12}$Zn$_{13}$ end-member; and (**d**) is the MgNd+Mg$_3$Nd end-member. The numbers represent the diffusion zones and correspond to those in Table 2 and Figure 7.

From the WDS results in Figure 6 and Table 2, τ_1 was recognized as a ternary solid solution extending from 26.6 to 29.5 atom % Mg at 6.5 atom % Nd, τ_3 extended from 24.6 to 38.0 atom % Mg at 25.0 atom % Nd, and Nd(Mg,Zn) extended from 43.3 to 48.1 atom % Mg at 50.0 atom % Nd. Diffusion couples do not provide the solubility limits very accurately. However, they are useful in detecting whether solubility exists or not. The exact solubility limits can be obtained using equilibrated key alloys. Besides, diffusion zones #4 and #6 are relatively thinner than other zones, which can lead to large errors in WDS results. Therefore, the composition of the corresponding compounds must be confirmed by the key alloys method.

The SEM micrograph in Figure 5a and the colored version in Figure 5b showed that (α-Mg)^{Zn} and Mg$_{12}$Zn$_{13}$ are in equilibrium with τ_4 from one side, and τ_4 is in equilibrium with τ_1 from the other side of the original interface of diffusion couple #2. Based on the phase equilibria determined from diffusion couple #2, the diffusion path can be depicted as follows: α-Mg+Mg$_{12}$Nd$_{13}$ (end-member)/τ_4/τ_1/τ_5/τ_3/(Mg$_3$Nd)/Nd(Mg,Zn)/Mg$_3$Nd+MgNd (end-member). Accordingly, the phase relations can be represented graphically as shown in Figure 7. The two end-members are connected by a dashed line.

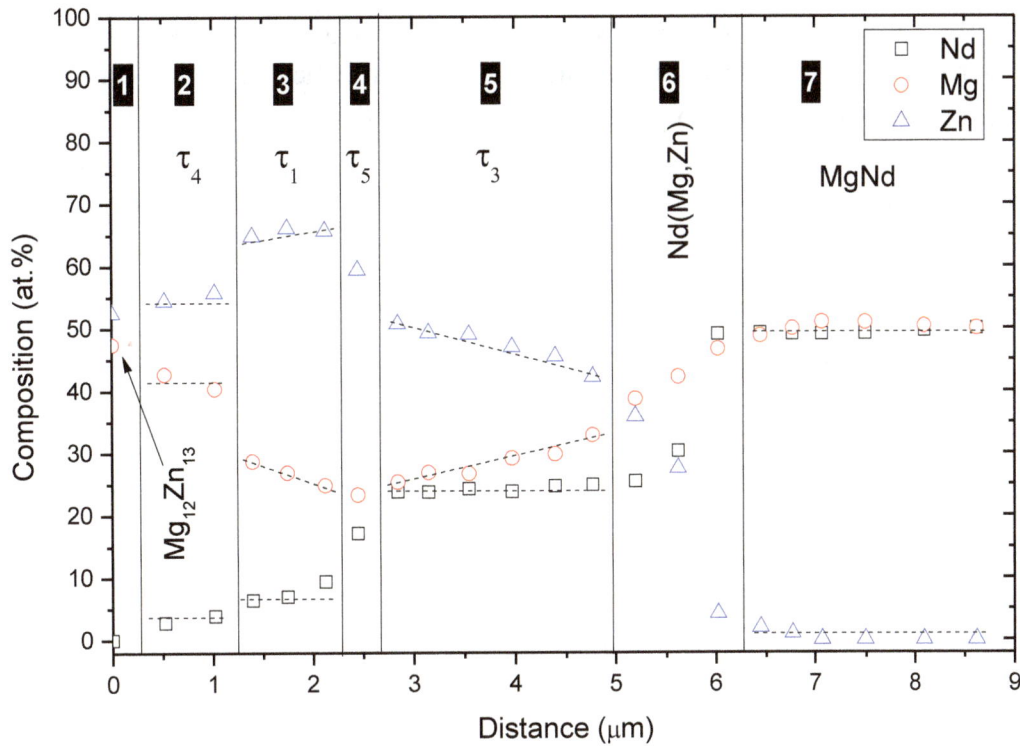

Figure 6. WDS line-scan across diffusion couple #2.

Table 2. WDS spot analysis of diffusion zones of diffusion couple #2.

Zone	Description	Composition (atom %)			Corresponding Phase
		Mg	**Nd**	**Zn**	
1	Two-phase alloy (end-member)	95.6	-	4.4	$(\alpha\text{-Mg})^{Zn}$
		47.3	-	52.7	$Mg_{12}Zn_{13}$
2	Single-phase layer	45.6	1.8	52.6	τ_4
3	Single-phase layer	26.6–29.5	6.5	64.0–66.9	τ_1
4	Single-phase layer	28.8	13.1	58.1	τ_5
5	Single-phase layer	24.6–38.0	25.0	37.0–50.4	τ_3
6	Single-phase layer	43.3–48.1	50.0	1.9–6.7	$Nd(Mg,Zn)$
7	End-member	50.0	50.0	-	$MgNd$

Figure 7 shows that the diffusion path is mainly passing across the lower side of the line connecting the two end-members. However, the mass balance principle must be taken into account in any diffusion couple. To understand the mass balance in this diffusion couple, the diffusion reaction starting from zone #7 to zone #1 can be described as follows. In the case of binary diffusion couples, the growth of diffusion layers is sequential rather than simultaneous [23]. However, the case could be different for ternary diffusion couples. Due to the presence of more than two species having different mobilities, layers can form simultaneously. For instance, if two of the diffusing species react to form a binary compound, the third element could be simultaneously consumed adjacent to the interface to form another layer. Evidence could be observed through the layers around the original interface, where the less mobile Nd atoms form τ_4 on one side, and the Zn atoms with high mobility form many layers on the other side. More specifically, Nd(Mg,Zn) firstly formed adjacent to MgNd (from zone #7) by dissolving diffusing Zn form zone #1. Simultaneously, Zn depletion from $Mg_{12}Zn_{13}$ produced a layer of τ_4, with low Nd concentration, in zone #2. The concentration of Zn increased gradually with the decrease in Mg and Nd concentrations, starting from zone #6 until the interface between zones #2 and #3, as illustrated in Figure 6. As a result, the diffusion path, in Figure 7, was pulled down towards the Zn-rich side. This could be due to the mobility of Zn atoms. In zone #3, there is Zn/Mg exchange at around 6.5 atom % Nd to form τ_1. At the end of zone #3, the diffusion path changed its direction towards the Mg-Zn side.

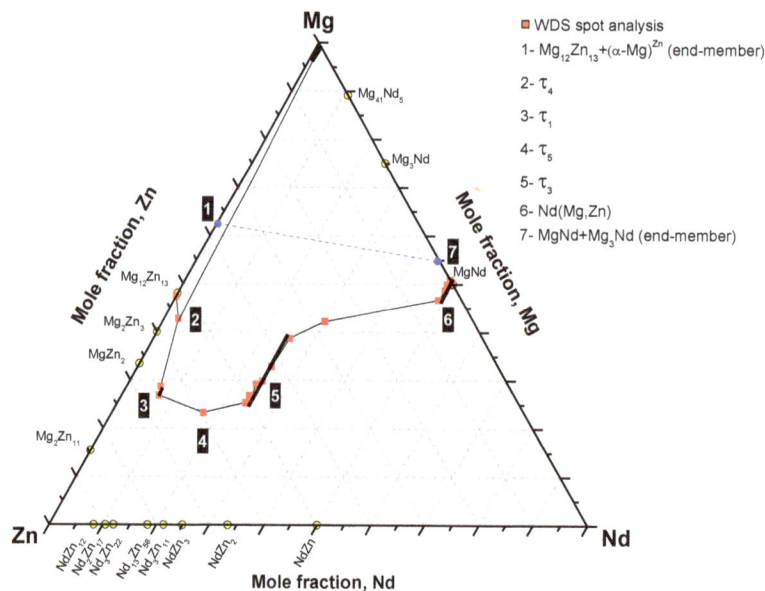

Figure 7. The phase equilibria depicted from diffusion couple #2.

Diffusion couple experiments revealed the existence of five ternary compounds in the system at 300 °C. These are: τ_1 (Mg_7NdZn_{12}), τ_2 ($Mg_2Nd_2Zn_9$), τ_3 ($MgNdZn_2$), τ_4 ($Mg_{40}Nd_5Zn_{55}$), and τ_5 (Mg_2NdZn_4). The preliminary results obtained from this work showed that $(\alpha\text{-}Mg)^{Zn}$ is in equilibrium with $Mg_{12}Zn_{13}$, τ_1, and τ_4. Also, the ternary compounds τ_1 and τ_3 were recognized with extended solid solubility. However, these results must be confirmed using equilibrated key alloys. The combined results of diffusion couples and key alloys improve the accuracy of the obtained phase equilibrium information. In the following section, the results obtained from equilibrated key alloys are discussed.

3.2. Equilibrated Key Alloys

Twenty-five equilibrated key alloys, selected at different compositions of the Mg-Nd-Zn system, were annealed at different time intervals. Figure 8 shows the distribution of the selected key alloys based on their actual composition obtained by SEM/EDS area analysis.

Because annealing was performed at 300 °C, a long time was required to achieve the equilibrium structure. For some compositions, annealing for about 45 days was not sufficient to homogenize the structure. In such cases, the equilibrium was inferred by comparing the as-cast microstructures and XRD patterns with the annealed ones. Tables 3 and 4 list the experimental results obtained from equilibrated key alloys containing three-phase and two-phase equilibria, respectively. The question marks in the XRD results in Table 3 denote that the crystal structure of the compound is unknown. The crystal structure of τ_1 was taken from [14] and of τ_3 and τ_6 was taken from [1].

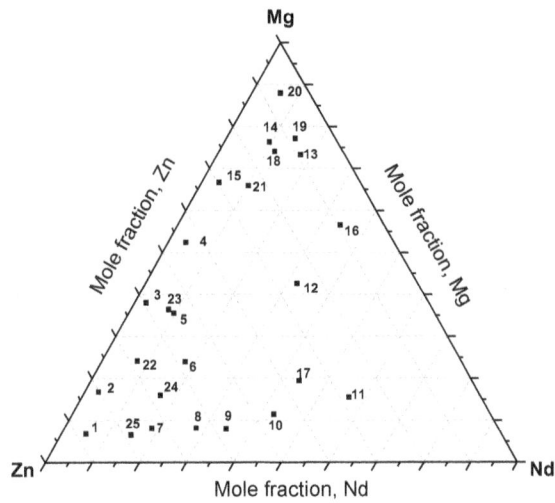

Figure 8. The actual composition of the Mg-Nd-Zn key alloys.

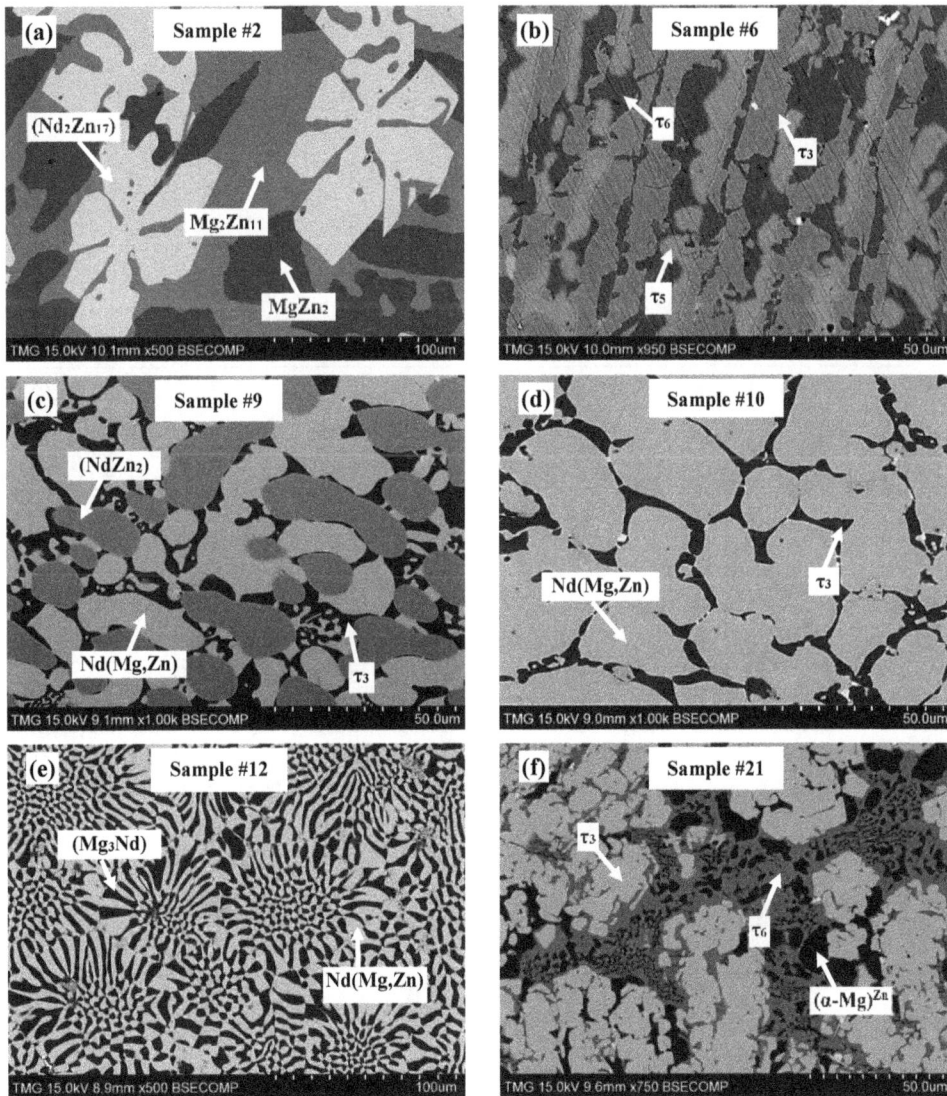

Figure 9. SEM micrographs of alloys annealed at 300 °C selected from different regions in the Mg-Nd-Zn system: (**a**) sample #2, (**b**) sample #6, (**c**) sample #9, (**d**) sample #10, (**e**) sample #12, and (**f**) sample #21.

Table 3. Actual composition of key alloys containing three-phase equilibria and their experimental results.

Sample	Actual alloy Composition (atom %)			Phase Identification XRD	Phase Composition (atom %)			Corresponding Phase
	Mg	Nd	Zn		Mg	Nd	Zn	
1	6.8	5.3	87.9	Zn	-	-	100.0	Zn
				Mg_2Zn_{11}	15.4	-	84.6	Mg_2Zn_{11}
				(Nd_2Zn_{17})	<1.0	10.1	89.1	(Nd_2Zn_{17})
2	16.8	3	80.2	Mg_2Zn_{11}	15.4	-	84.6	Mg_2Zn_{11}
				$MgZn_2$	33.3	-	66.7	$MgZn_2$
				(Nd_2Zn_{17})	1.1	10.1	88.9	(Nd_2Zn_{17})
3	38.0	2.5	59.5	$Mg_{12}Zn_{13}$	48.0	-	58.0	$Mg_{12}Zn_{13}$
				Mg_2Zn_3	40.0	-	60.0	Mg_2Zn_3
				τ_1	32.0	7.0	61.0	τ_1
4	52.4	3.8	43.8	$(\alpha\text{-}Mg)^{Zn}$	95.2	-	4.8	$(\alpha\text{-}Mg)^{Zn}$
15	66.7	3.7	29.6	$Mg_{12}Zn_{13}$	48.1	-	51.9	$Mg_{12}Zn_{13}$
				?	34.9	7.0	58.1	τ_1
5	35.5	9.7	54.8	$(\alpha\text{-}Mg)^{Zn}$	95.3	-	4.7	$(\alpha\text{-}Mg)^{Zn}$
				τ_6	49.0	7.2	43.8	τ_6
				τ_1	30.0	6.5	63.5	τ_1
6	24	17.8	58.2	τ_6	54.8	8.3	36.9	τ_6
				?	22.7	17.6	59.7	τ_5
				τ_3	26.5	24.9	48.6	τ_3
7	8.12	18.6	73.28	(Nd_3Zn_{11})	3.5	21.4	75.1	(Nd_3Zn_{11})
				τ_3	25.1	25.0	49.9	τ_3
				?	11.0	15.0	74.0	τ_2
8	8.2	28	63.8	(Nd_3Zn_{11})	3.5	21.4	75.1	(Nd_3Zn_{11})
				$(NdZn_2)$	2.5	33.4	64.1	$(NdZn_2)$
				τ_3	26.2	25.0	48.8	τ_3
9	8.0	34.4	57.6	$(NdZn_2)$	2.5	33.4	64.1	$(NdZn_2)$
				$Nd(Mg,Zn)$	2.0	50.0	48.0	$Nd(Mg,Zn)$
				τ_3	28.3	25.0	46.7	τ_3
20	88.0	6.0	6.0	$(\alpha\text{-}Mg)^{Zn}$	99.4	-	<1.0	$(\alpha\text{-}Mg)^{Zn}$
				$(Mg_{41}Nd_5)$	87.0	10.0	3.0	$(Mg_{41}Nd_5)$
				(Mg_3Nd)	55.0	25.0	20.0	(Mg_3Nd)
21	65.8	10.4	23.8	$(\alpha\text{-}Mg)^{Zn}$	98.2	-	1.8	$(\alpha\text{-}Mg)^{Zn}$
				τ_6	63.5	7.5	29	τ_6
				τ_3	30.0	25.0	45.0	τ_3
24	16.0	16.5	67.5	?	21.0	16.0	63.0	τ_5
				τ_3	25.2	25.0	49.8	τ_3
				?	12.5	15.5	72.0	τ_2

"?" is used to indicate unknown crystal structure.

Table 4. Actual composition of key alloys containing two-phase equilibria and their experimental results.

Sample Number	Actual alloy Composition (atom %)			Phase Identification XRD	Phase Composition (atom %)			Corresponding Phase
	Mg	Nd	Zn		Mg	Nd	Zn	
10	11.5	42.7	45.8	Nd(Mg,Zn)	3.5	50.0	46.5	Nd(Mg,Zn)
				τ_3	30.5	25.0	44.5	τ_3
11	15.5	56.4	28.1	Nd(Mg,Zn)	19.0	50.0	31.0	Nd(Mg,Zn)
				Nd	100.0	-	-	Nd
12	42.6	32.1	25.3	Nd(Mg,Zn)	28.3	50.0	21.7	Nd(Mg,Zn)
				(Mg_3Nd)	60.0	25.0	15.0	(Mg_3Nd)
13	73.3	17.5	9.2	$(Mg_{41}Nd_5)$	89.0	10.1	<1.0	$(Mg_{41}Nd_5)$
				(Mg_3Nd)	70.0	25.0	5.0	(Mg_3Nd)
14	76.3	9.5	14.2	$(\alpha\text{-Mg})^{Zn}$	98.8	-	1.2	$(\alpha\text{-Mg})^{Zn}$
				(Mg_3Nd)	48.1	25.0	26.9	(Mg_3Nd)
16	56.5	34.1	9.4	Nd(Mg,Zn)	43.0	50.0	7.0	Nd(Mg,Zn)
				(Mg_3Nd)	70.0	25.0	5.0	(Mg_3Nd)
17	19.5	44	36.5	Nd(Mg,Zn)	8.6	50.0	41.4	Nd(Mg,Zn)
				(Mg_3Nd)	47.0	25.0	28.0	(Mg_3Nd)
18	74.1	11.7	14.2	$(\alpha\text{-Mg})^{Zn}$	97.8	-	2.2	$(\alpha\text{-Mg})^{Zn}$
				(Mg_3Nd)	47.0	25.0	28.0	(Mg_3Nd)
19	77.2	14.5	8.3	$(Mg_{41}Nd_5)$	87.0	10.0	3.0	$(Mg_{41}Nd_5)$
				(Mg_3Nd)	63.0	25.0	12.0	(Mg_3Nd)
22	24.2	7.5	68.3	(Nd_2Zn_{17})	7.5	10.5	82.0	(Nd_2Zn_{17})
				τ_1	29.0	6.5	64.5	τ_1
23	36.4	8.1	55.5	$(\alpha\text{-Mg})^{Zn}$	94.3	-	5.7	$(\alpha\text{-Mg})^{Zn}$
				τ_1	31.5	7.2	61.3	τ_1
25	6.5	15	78.5	(Nd_2Zn_{17})	7.0	10.5	82.5	(Nd_2Zn_{17})
				τ_1	25.5	7.5	67.0	τ_1

The microstructures of equilibrated key alloys selected from different regions of the Mg-Nd-Zn system are shown in Figure 9. The equilibrated phases are labeled on the same micrographs. WDS analysis of all the equilibrated key alloys is shown graphically in Figure 10, where the arrows point to the detected phases.

The existence of ternary compounds τ_1, τ_2, τ_3, and τ_5, detected by diffusion couple experiments, was confirmed using the equilibrated key alloys. Furthermore, key alloys revealed the presence of a ternary compound equilibrating with $(\alpha\text{-Mg})^{Zn}$. This compound was given the formula $Nd(Mg,Zn)_{11.5}$ in the work of Huang *et al.* [1] at 400 °C. It is denoted as τ_6 in the current work. We assume that τ_4 was not detected by key alloys due to the low Nd concentration, which falls within the WDS detection limits. Thus, the phase might be read as $Mg_{12}Zn_{13}$ for negligible Nd concentration. However, it was clearly recognized in diffusion couple #2, where τ_4 is in equilibrium with $(\alpha\text{-Mg})^{Zn}$, $Mg_{12}Zn_{13}$, and τ_1. More data and discussion regarding the ternary compounds and solid solutions are provided in the following sections.

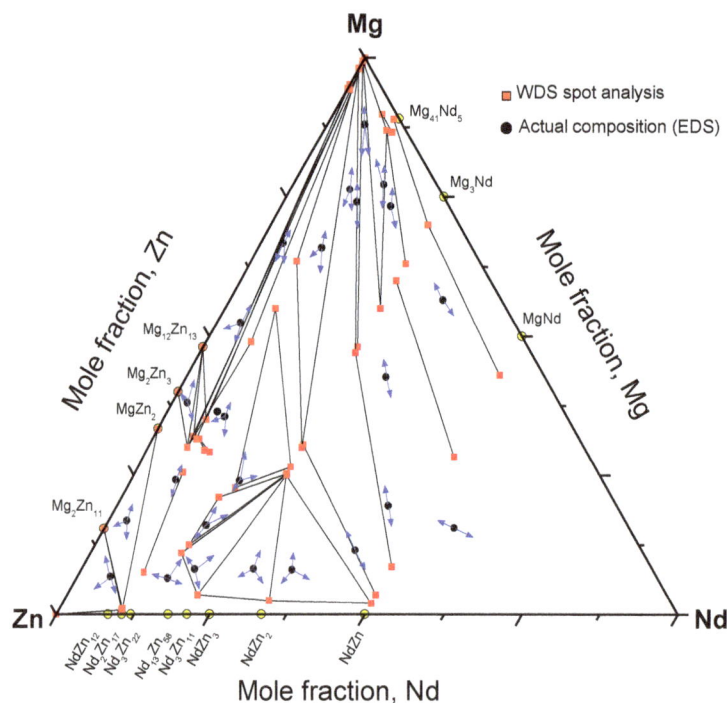

Figure 10. Phase equilibrium results obtained from WDS spot analysis of equilibrated key alloys, where the arrows point to the detected phases.

3.2.1. Ternary Compounds

Six ternary compounds in the Mg-Nd-Zn system were confirmed experimentally at 300 °C by diffusion couples and key alloys. The homogeneity range of these compounds was determined by WDS analysis. τ_1 extends from 29.0 to 35.0 atom % Mg at a constant Nd concentration of ~6.8 atom % and has a hexagonal crystal structure with lattice parameters of $a = b = 1.5$ nm and $c = 0.87$ nm [14]. τ_2 extends from 10.0 to 12.0 atom % Mg at a constant Nd concentration of 15.1 atom %. τ_3 extends from 25.0 to 36.0 atom % Mg at a constant Nd concentration of 25.0 atom % and has a face-centered cubic crystal structure with lattice parameters of $a = b = c = 0.68$ nm [1]. τ_5 was found with a homogeneity range of 22.0–23.5 atom % Mg, 15.5–17.6 atom % Nd and 59.1–61.8 atom % Zn. τ_6 extends from 54.1 to 61.3 atom % Mg at a constant Nd of 8.0 atom % and has a C-centered orthorhombic crystal structure with lattice parameters of $a = 0.965$–0.984 nm, $b = 1.18$–1.135 nm and $c = 0.946$–0.963 nm [1].

3.2.2. Solid Solutions

Although the binary solid solubility of Nd in Mg is negligible [10,11], Kinzhibalo *et al.* [13] showed that (α-Mg) contains ~5 atom % Nd in the Mg-Nd-Zn isothermal section at 300 °C. In this work, the solubility of Nd in α-Mg is negligible and the maximum solid solubility of Zn in $(\alpha\text{-Mg})^{Zn}$ was measured as 4.8 atom % Zn at 300 °C.

The ternary solubility of Zn in Mg-Nd compounds was found to increase with a decrease in Mg concentration. For instance, ($Mg_{41}Nd_5$) was found to have an extended solubility of 3.1 atom % Zn, whereas (Mg_3Nd) was found to have 30.0 atom % Zn. MgNd was found to have a complete substitution of Mg by Zn. Thus, the complete solid solubility line Nd(Mg,Zn) was established between MgZn and NdZn. This was also reported in the literature for this system at 300 °C [13] and confirmed in the current work.

3.3. Mg-Nd-Zn Isothermal Section at 300 °C

The Mg-Nd-Zn isothermal section at 300 °C, established using the diffusion couples and key alloy results, is shown in Figure 11. The existence of six ternary compounds was confirmed by WDS analysis. The ternary solubility ranges were determined using key alloy experiments. Because of many wide two-phase fields, many key alloys are required to determine the boundary tie-lines. In this work, tie-lines were interpolated based on the mass balance principle as described in [19]. In this method, each two-phase field must be determined experimentally by at least three tie-lines. Two lines, parallel to the boundary composition lines, are extended from the two ends of each experimental tie-line until they intersect. For each region, three intersection points are assigned. Then, these points are connected together to give the balance curve, which represents the amount of the equilibrated phases *versus* composition. Each point on the produced curve represents an intersection of a tie-line that occurs in that field with the extended composition axis. Once the curve is determined, the process can be reversed and additional tie-lines can be interpolated by extending two lines parallel to the phase boundary composition lines. Using this method, many regions in the Mg-Nd-Zn isothermal section could be determined.

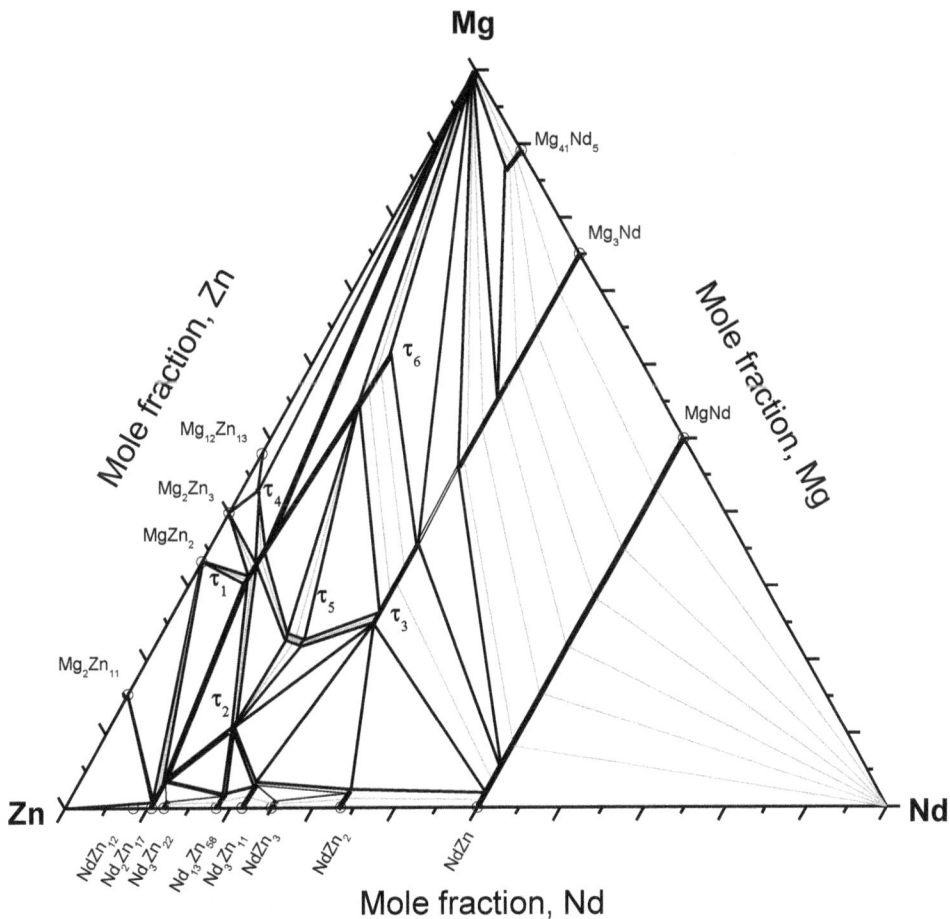

Figure 11. The Mg-Nd-Zn isothermal section at 300 °C.

The current study showed different phase relations than those reported by Qi *et al.* [8], Drits *et al.* [11], and Kinzhibalo *et al.* [13] at 300 °C. However, the phase relations near the Mg-rich corner are consistent with Huang *et al.* [1] except for the presence of the liquid phase, and consistent with Zhang *et al.* [17] except for the presence of τ4 at 300 °C.

4. Experimental Procedures

In order to study the phase relationships in the Mg-Nd-Zn isothermal section at 300 °C, solid-solid diffusion couples along with key alloys were prepared and analyzed using a PANalytical (Almelo, The Netherlands) X-ray diffractometer (Cu K-α radiations) and a Hitachi (Tokyo, Japan) S-3400N SEM equipped with Oxford® (Abingdon, UK) EDS/WDS detectors. EDS and WDS analyses were used to determine the phases and to quantify their compositions. Pure metals (Mg, Nd, and Zn) were used as standards for WDS calibration. X-ray scans were performed on powder-form samples, in the range from 20 to 120 degree 2θ with 0.02° step size, to identify and confirm the phases obtained by WDS measurements. X-ray phase analysis was carried out using X'pert Highscore Plus software (Almelo, The Netherlands). The standard intensity data were taken from Pearson's Crystal Data software (Crystal Impact, Materials Park, OH, USA). Silicon was added to the powder samples and used as a calibration standard to correct for the zero shift and specimen displacement of the obtained diffraction patterns.

4.1. Key Alloy Preparation

Pure elements were used for sample preparation and diffusion couple end-members. Mg ingots with purity of 99.8% were supplied by CANMET Materials Technology Laboratory (Ottawa, ON, Canada). Nd ingots with purity of 99.9% were supplied by STREM Chemicals Incorporated (Newburyport, MA, USA). Zn rods with purity of 99.9% were supplied by Alfa Aesar Company (Haverhill, MA, USA). The key alloys, 3–4 gm each, were prepared in a tantalum crucible under argon atmosphere using induction-melting furnace. The melts were left to solidify in the crucible.

Alloys near the Zn-rich corner were prepared by adding pure Zn to previously prepared Mg-Nd binary alloys to avoid the exothermic reaction between Zn and Nd. The other compositions were prepared by melting the three pure metals together. Excess amounts of Mg and Zn (around 10%) were added to compensate for losses due to evaporation. The actual global composition was determined using EDS/SEM area analysis. An average of three readings from three different locations was taken as the actual composition. The WDS error of measurements was estimated as ±1 atom %.

4.2. Solid-Solid Diffusion Couples

The end-members of the solid-solid diffusion couples were prepared from pure metals and/or alloys. The contacting surfaces were ground gradually up to 1200 SiC paper using 99% pure ethanol to reduce friction and to prevent oxidation. The ground surfaces were polished down to 1 μm using an alcohol diamond suspension. The end-members were attached together strongly using stainless steel clamping rings.

For annealing purposes, samples and diffusion couples were wrapped in tantalum foil and encapsulated inside an argon-purged quartz tube with an internal pressure of about 8×10^{-1} torr. To reach equilibrium, diffusion couples and key alloys were heated up to 300 °C for a predefined period of time. Annealing time was chosen based on the composition of the selected end-members. After annealing, the quartz tubes, containing samples and diffusion couples, were rapidly quenched in cold water in order to maintain the high temperature structure. The equilibrated samples and the diffusion couples were analyzed using SEM/WDS spot analysis and line-scans. The isothermal section of the Mg-Nd-Zn phase diagram at 300 °C was constructed based on these results.

5. Conclusions

The Mg-Nd-Zn isothermal section at 300 °C was established for the whole composition range by solid-solid diffusion couples and equilibrated key alloys. The microstructural characterization was performed using XRD, SEM/EDS/WDS, and metallography. An annealing temperature of 300 °C was chosen to avoid the liquid phase formation upon annealing and to resolve the inconsistent phase relations found in the literature [1,8,10,11,13,14,17]. The current study showed different phase relations from those reported by Qi *et al.* [8], Drits *et al.* [11], and Kinzhibalo *et al.* [13] at 300 °C. However, the phase relations near the Mg-rich corner are consistent with Huang *et al.* [1] and Zhang *et al.* [17].

Six ternary compounds were detected in the Mg-Nd-Zn system at 300 °C. These are: τ_1 ($Nd_5Mg_{21+x}Zn_{45-x}$; $0 \leq x \leq 4$), τ_2 ($Nd_5Mg_{3+y}Zn_{25-y}$; $0 \leq y \leq 1$), τ_3 ($NdMg_{1+z}Zn_{2-z}$; $0 \leq z \leq 0.44$), τ_4 ($Mg_{40}Nd_5Zn_{55}$), τ_5 ($Mg_{22-23.5}Nd_{15.5-17.5}Zn_{59.1-61.8}$), and τ_6 ($Nd_2(Mg,Zn)_{23}$). The ternary solubility of Zn in Mg-Nd compounds was found to increase with decrease in Mg concentration so that ($Mg_{41}Nd_5$) and (Mg_3Nd) were found to have an extended solubility of 3.1 and 30.0 atom % Zn, respectively. MgNd was found to have a complete substitution of Mg by Zn.

Acknowledgments

This research was supported by funding from the Magnesium Strategic Research Network (MagNET), www.MagNET.ubc.ca.

Author Contributions

Ahmad Mostafa carried out the experiments and analysis of the results. Ahmad Mostafa and Mamoun Medraj prepared and revised the manuscript. Mamoun Medraj initiated and directed the project. He helped in the interpretation of the results and followed up on the progress step by step.

Conflicts of Interest

The authors declare no conflict of interest.

References

1. Huang, M.; Li, H.; Ding, H.; Tang, Z.; Mei, R.; Zhou, H.; Ren, R.; Hao, S. A ternary linear compound T2 and its phase equilibrium relationships in Mg-Zn-Nd system at 400 °C. *J. Alloy. Compounds* **2010**, *489*, 620–625.
2. Mordike, B.; Ebert, T. Magnesium: Properties—applications—potential. *Mater. Sci. Eng.* **2001**, *302*, 37–45.
3. Mordike, B.L. Creep-resistant magnesium alloys. *Mater. Sci. Eng. A* **2002**, *324*, 103–112.
4. Ghosh, P.; Mezbahul-Islam, M.; Medraj, M. Critical assessment and thermodynamic modeling of Mg-Zn, Mg-Sn, Sn-Zn and Mg-Sn-Zn systems. *Calphad* **2012**, *36*, 28–43.

5.   Okamoto, H. Supplemental literature review of binary phase diagrams: Cs-In, Cs-K, Cs-Rb, Eu-In, Ho-Mn, K-Rb, Li-Mg, Mg-Nd, Mg-Zn, Mn-Sm, O-Sb, and Si-Sr. *J. Phase Equilib. Diff.* **2013**, *34*, 251–263.

6.   Mezbahul-Islam, M.; Mostafa, A.O.; Medraj, M. Essential magnesium alloys binary phase diagrams and their thermochemical data. *J. Mater.* **2014**, *2014*, doi:10.1155/2014/704283.

7.   Ferro, R.; Saccone, A.; Borzone, G. Rare earth metals in light alloys. *J. Rare Earth* 1997, *15*, 45–61.

8.   Qi, H.-Y.; Huang, G.-X.; Bo, H.; Xu, G.-L.; Liu, L.-B.; Jin, Z.-P. Thermodynamic description of the Mg-Nd-Zn ternary system. *J. Alloy. Compounds* **2011**, 3274–3281.

9.   Okamoto, H. Nd-Zn (neodymium-zinc). *J. Phase Equilib. Diff.* **2012**, *33*, doi:10.1007/s11669-011-9969-8.

10. Drits, M.; Padezhnova, E.; Miklina, N. Phase diagram of the magnesium-neodymium-zinc system in the magnesium-rich phase. *Izvestiya Vysshikh Uchebnykh Zavedenii Tsvetn. Met.* **1971**, *14*, 104–107.

11. Drits, M.; Padezhnova, E.; Miklina, N. The combined solubility of noedymium and zinc in solid magnesium. *Rus. Metallurgy* **1974**, *3*, 143–146.

12. Raynor, G. Constitution of ternary and some complex alloys of magnesium. *Int. Mater. Rev.* **1977**, *22*, 65–96.

13. Kinzhibalo, V.; Tyvanchuk, A.; Melnik, E. Stable and metastable phase equilibria in metallic systems. *Nauka Mosc. USSR* **1985**, 70–74.

14. Huang, M.; Li, H.; Yang, J.; Ren, Y.; Ding, H.; Hao, S. Research on a ternary compound T1 at the low Nd side in the Mg-Zn-Nd alloy. *ACTA Metall Sin.* **2008**, *44*, 385–390,

15. Zhang, J.; Yan, J.; Liang, W.; Xu, C.; Zhou, C. Icosahedral quasicrystal phase in Mg-Zn-Nd ternary system. *Mater. Lett* **2008**, *62*, 4489–4491.

16. Zhang, J.; Yan, J.; Liang, W.; Du, E.; Xu, C. Microstructures of Mg-Zn-Nd alloy including small quasicrystalline grains. *J. Non-Cryst. Solids* **2009**, *355*, 836–839.

17. Zhang, C.; Luo, A.A.; Peng, L.; Stone, D.S.; Chang, Y.A. Thermodynamic modeling and experimental investigation of the magnesium-neodymium-zinc alloys. *Intermetallics* **2011**, *19*, 1720–1726.

18. Kirkendall, E.O. Diffusion of zinc in alpha brass. *Trans. AIME Metall Pet. Eng.* **1942**, *147*, 104–110.

19. Mostafa, A.; Medraj, M. Phase equilibria of the Ce-Mg-Zn ternary system at 300 °C. *Metals* **2014**, *4*, 168–195.

20. Callister, W.; Rethwisch, D. *Materials Science and Engineering: An Introduction*, 7th ed.; John Wiley & Sons, Inc.: New York, NY, USA, 2007.

21. Chen, Y.C.; Zhang, Y.G.; Chen, C.Q. Quantitative descriptions of periodic layer formation during solid state reactions. *Mater. Sci. Eng. A* **2003**, *362*, 135–144.

22. Mostafa, A.; Medraj, M. On the atomic interdiffusion in Mg-{Ce, Nd, Zn} and Zn-{Ce, Nd} binary systems. *J. Mater. Res.* **2014**, *29*, 1463–1479.

23. Dybkov, V.I. Solid state growth kinetics of the same chemical compound layer in various diffusion couples. *J. Phys. Chem. Solids* **1986**, *47*, 735–740.

Effects of Silicon on Mechanical Properties and Fracture Toughness of Heavy-Section Ductile Cast Iron

Liang Song [1,2,*], Erjun Guo [2], Liping Wang [2] and Dongrong Liu [2]

[1] School of Materials Science and Engineering, Heilongjiang University of Science and Technology, Harbin 150022, China

[2] School of Materials Science and Engineering, Harbin University of Science and Technology, Harbin 150080, China; E-Mails: guoerjun@126.com (E.G.); Wangliping@126.com (L.W.); dong-rong.liu@im2np.fr (D.L.)

* Author to whom correspondence should be addressed; E-Mail: songliang16888@163.com

Academic Editor: Hugo F. Lopez

Abstract: The effects of silicon (Si) on the mechanical properties and fracture toughness of heavy-section ductile cast iron were investigated to develop material for spent-nuclear-fuel containers. Two castings with different Si contents of 1.78 wt.% and 2.74 wt.% were prepared. Four positions in the castings from the edge to the center, with different solidification cooling rates, were chosen for microstructure observation and mechanical properties' testing. Results show that the tensile strength, elongation, impact toughness and fracture toughness at different positions of the two castings decrease with the decrease in cooling rate. With an increase in Si content, the graphite morphology and the mechanical properties at the same position deteriorate. Decreasing cooling rate changes the impact fracture morphology from a mixed ductile-brittle fracture to a brittle fracture. The fracture morphology of fracture toughness is changed from ductile to brittle fracture. When the Si content exceeds 1.78 wt.%, the impact and fracture toughness fracture morphology transforms from ductile to brittle fracture. The *in-situ* scanning electronic microscope (SEM) tensile experiments were first used to observe the dynamic tensile process. The influence of the vermicular and temper graphite on fracture formation of heavy section ductile iron was investigated.

Keywords: heavy-section ductile cast iron; silicon content; *in situ* SEM tensile; fracture toughness

1. Introduction

Silicon is an extremely sensitive element in heavy section ductile iron, which affects the formation of chunky graphite [1–3]. S.I. Karsay and E. Campomanes [4–9] studied the influence of silicon content on ductile iron, and found that lower silicon content can reduce the performance of chunky graphite iron, especially in its capacity to demonstrate good impact performance at low temperatures. The proper content of silicon can increase the nodularity and improve the mechanical properties of heavy section ductile iron, and inhibit graphite floatation and chunky graphite formation [10–13].

Although many researchers have studied the influences of Si on the microstructures and mechanical properties of ductile cast iron [14], systematic research on effects of Si on large-scale heavy section ductile cast iron is scarce.

The aim of this paper is to investigate the effects of Si on microstructures and mechanical properties as well as fracture toughness of heavy section ductile cast iron. Two cubic-shaped castings with different Si contents of 1.78 wt.% and 2.74 wt.% were prepared. Specimens were taken at four positions from the edge to the center of the castings, which representthe different typical cooling rates of heavy section ductile cast iron. The influence of Si content on morphology and distribution of graphite, and on the impact toughness, tensile strength, elongation and fracture toughness of the specimens at different positions of castings were studied. The *in-situ* scanning electronic microscope (SEM) tensile experiment was used to observe the dynamic tensile process. Fracture analysis was carried out to clarify how the vermicular graphite and temper graphite affect the fracture process of heavy section ductile iron.

2. Experimental Procedure

The heavy section ductile cast iron castings were obtained by melting pig iron, 45 steel and graphite in a medium frequency induction furnace. The Ce-Mg-Si and 75 wt.% Si-Fe alloys were used as nodularizer and inoculant, respectively. The composition of nodularizer used in the experiments is given in Table 1. The molten iron was poured into the furan resin sand mould to obtain cubic-shaped block casting, with dimension of 400 mm × 400 mm × 400 mm. Two castings with Si contents of 1.78 wt.% and 2.74 wt.% were prepared, and they were denoted as Casting-A and Casting-B, respectively. The compositions of castings are given in Table 2. Four positions (P1, P2, P3 and P4) were chosen from the edge to the core of the as-casts (Figure 1). As shown in Figure 1, specimens were fabricated at four positions from the edge to the center of the castings, and they were denoted as A1 (B1), A2 (B2), A3 (B3) and A4 (B4), respectively. The cooling curves of the specimens from the four positions were measured by using a Kingview temperature monitoring system.

Table 1. Composition of nodularizer (wt.%).

Elements of Nodularizer/Content				
Ce	Mg	Si	Mn	Ca
6.49	7.88	43.04	2.0	<3

Table 2. Compositions of castings (wt.%).

Elements	C	Si	Mn	S	P	Mg
Casting-A	3.61	1.78	<0.2	<0.02	<0.05	0.037
Casting-B	3.57	2.74	<0.2	<0.02	<0.05	0.046

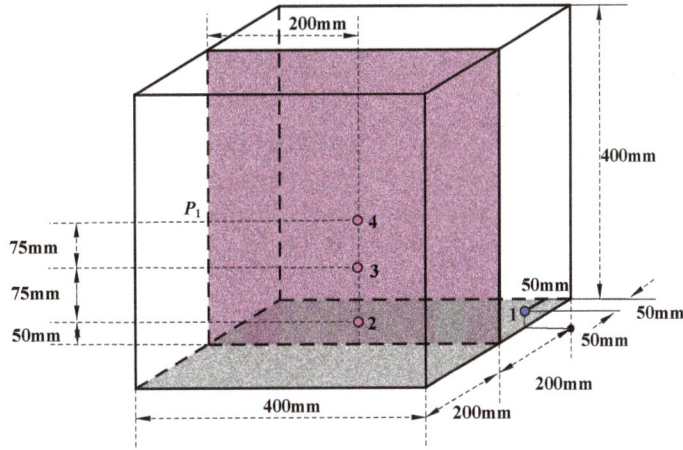

Figure 1. Four positions (P1, P2, P3 and P4) in castings chosen for measuring temperature and preparing specimens.

The morphology of graphite in specimens was observed by using an optical microscope (GX71, Olympus, Shanghai, China). The tensile tests were conducted at room temperature on the tensile tester (1186, Instron, Norwood, MA, USA) according to the standard GB228-2002. The gauge length of the sample is 50 mm. The tensile rate is 0.8 mm·min^{-1}. Impact toughness specimens (50 mm × 10 mm × 10 mm) were tested in a standard impact testing machine (JBN-300B, Chenda, Jinan, China) at room temperature according to the standard GB229-2007.

The fracture toughness test was carried out at room temperature on electro-hydraulic servo testing machine (809, MTS, Minneapolis, MN, USA) according to the standard GB/T4161-2007. The sample size is shown in Figure 2. The *in situ* SEM tensile experiment was performed at room temperature on SEM (S-570, Hitachi, Tokyo, Japan) loaded by manual. The maximum tensile load and the maximum stretching distance is 10 kg and 5 mm, respectively. As shown in Figure 3, the thickness of *in situ* SEM tensile specimen is 0.1 mm, and the pre-crack is 0.5 mm in depth.

3. Results and Discussion

3.1. Cooling Curves and Microstructures of Ductile Cast Iron

Figure 4 shows the cooling curves at four positions of two castings during solidification. It can be seen that the cooling rate at position 1 is the highest and the subsequent sequence is positions 2, 3, and 4. At position 3 and position 4, the solidification time is more than 250 min. The cooling curves in Figure 4 represent the cooling rate of four positions in two castings owing to the micro-amount change of Si content. The influence of Si content on the cooling process of ductile iron (more than 500 kg) can be ignored [15].

Φ12 + 2.5 mm

50 ± 0.25 mm

45°

A

B

5 + 0.1 mm

30°

2.5 mm

30 ± 0.25 mm

13.75 ± 0.25 mm

30 ± 0.25 mm

35 ± 0.1 mm

25 + 0.5 mm

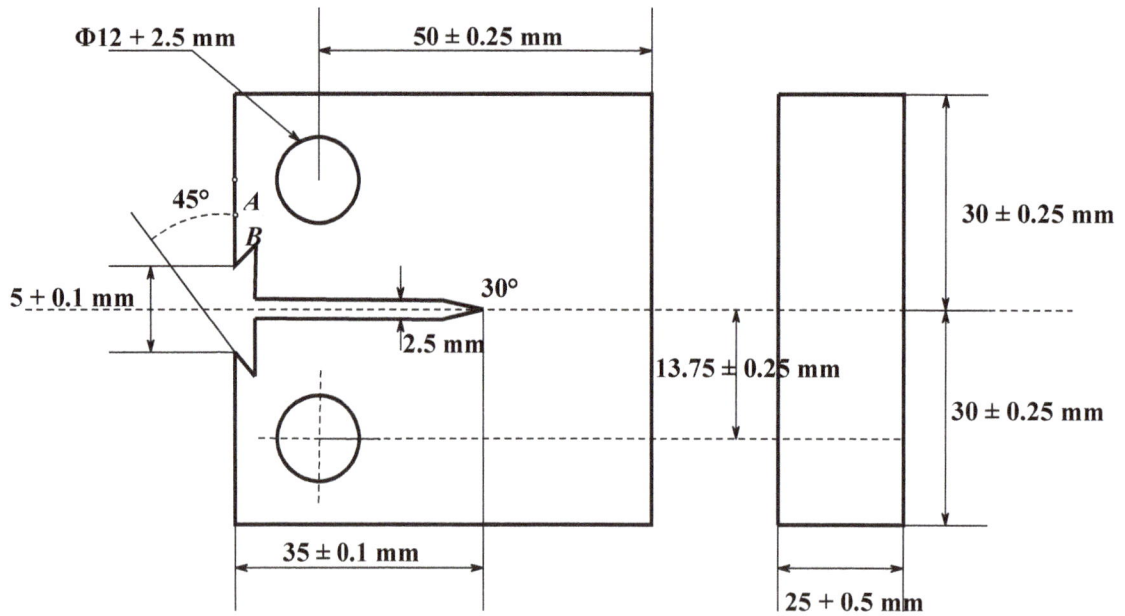

Figure 2. Dimensions of the fracture toughness specimen.

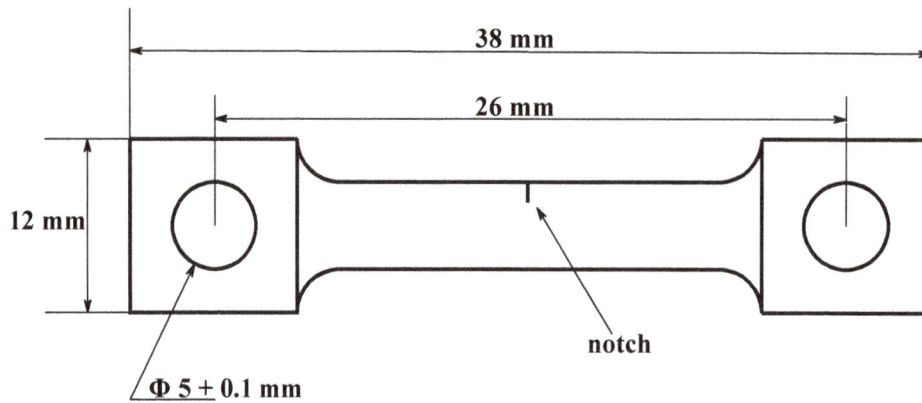

38 mm

26 mm

12 mm

notch

Φ 5 + 0.1 mm

Figure 3. Dimensions of the *in situ* SEM tensile specimen.

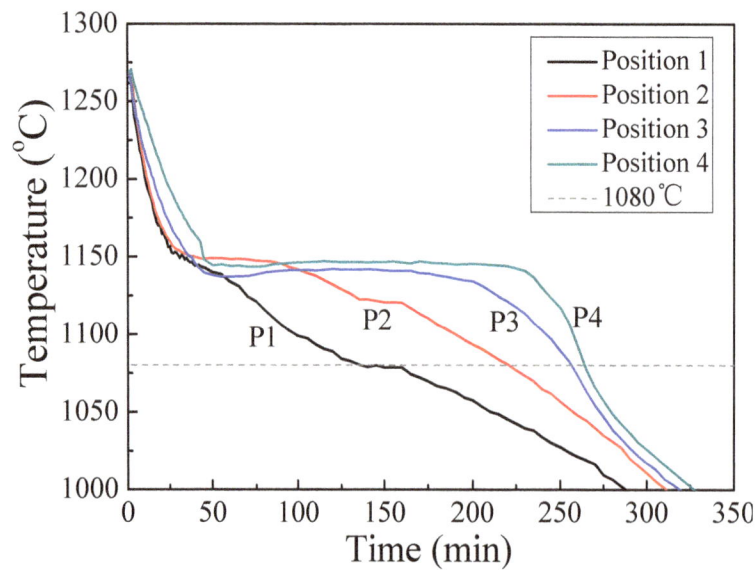

Figure 4. Cooling curves at four positions of castings during solidification.

The OM observation results show that the matrix structure of all specimens contains ferrite without pearlite. Figure 5 shows the graphite morphology of specimens at four positions of Casting-A (Si content of 1.78 wt.%) and Casting-B (Si content of 2.74 wt.%), respectively. As shown in Figure 5a1, the graphite nodules are almost spheroidal and a small amount of quasi-spheroidal temper graphite can be observed at position 1 of casting-A. With the increase of solidification time, the quasi-spheroidal temper graphite nodule and temper graphite nodule increases, as shown in Figure 5a2. With the further increase of solidification time, some vermicular graphite nodules occur, as shown in Figure 5a3. It shows that the spheroidalization decaying happens in Casting-A. The graphite nodules are almost chunky and the graphite is tempered at position 4 of casting-A, as shown in Figure 5a4. It can be inferred that with the decrease of cooling rate, the spheroidalization decaying becomes more significant and the graphite morphology gradually deteriorates.

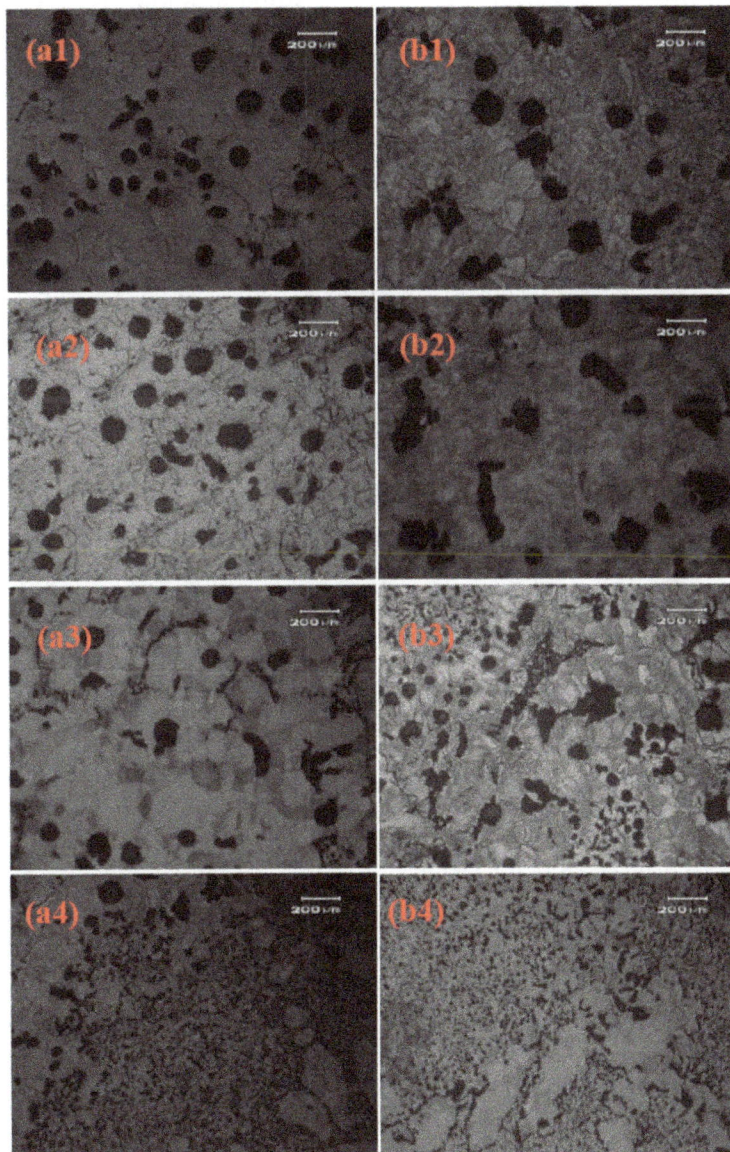

Figure 5. Graphite morphology of at four positions of two castings (**a1**) specimen A1; (**a2**) specimen A2; (**a3**) specimen A3; (**a4**) specimen A4; (**b1**) specimen B1; (**b2**) specimen B2; (**b3**) specimen B3; and (**b4**) specimen B4.

As shown in Figure 5b1, the graphite nodules at position 1 of casting-B are almost that of spheroidal and quasi-spheroidal temper graphite, and some temper graphite can be observed. With the decrease of cooling rate, the number and diameter of graphite nodules of specimen B2 decreases. Some vermicular graphite can be observed at position 2, as shown in Figure 5b2. The graphite morphology of specimens B3 and B4 is almost that of chunky, vermicular graphite. It can be inferred that significant spheroidalization decaying occurs, as shown in Figure 5b3,b4. The morphology of graphite, graphite sphere grade and nodularity of the specimens in Casting-A and Casting-B are listed in Table 3. It can be seen that the graphite sphere grade and nodularity of Casting-A are higher than that of Casting-B at the same position.

The silicon, like cerium, calcium and nickel, can cause melting point of austenitic shell to drop and promote the formation of chunky graphite. Therefore, the content of silicon must be controlled in the production of heavy-section ductile cast iron. Generally, a suitable addition of silicon can obviously improve the graphite sphere grade and nodularity of heavy section ductile iron [9]. However, when the content is 2.74 wt.%, the aggregation of chunky graphite occurs, as shown in Figure 5b4. This is because the excess addition of Si will result in too much heterogeneous nucleation, causing an increase in the amount of chunky graphite and the aggregation of chunky graphite. Areas of aggregation become weak points during tensile and impact tests [16,17].

Table 3. Morphology of graphite, graphite sphere grade and nodularity in Casting-A (Si content = 1.78 wt.%) and Casting-B (Si content = 2.74 wt.%).

Specimens	Graphite Morphology	Graphite-sphere Grade	Nodularity%
A1	Spheroidal + Quasi-spheroidal temper	3	80
A2	Spheroidal + Quasi-spheroidal temper + temper	5	60
A3	Spheroidal + Temper + Vermicular	6	50
A4	Spheroidal + Temper + chunky	-	-
B1	Spheroidal+Quasi-spheroidal temper + temper	5	60
B2	Spheroidal + Temper + Vermicular	6	50
B3	Temper + Vermicular + chunky	-	-
B4	Vermicular + Temper + chunky	-	-

3.2. Mechanical Properties of Ductile Cast Iron

Figure 6 shows the tensile strength, impact toughness, elongation and fracture toughness of the specimens at the four positions of Casting-A and Casting-B. With the decrease of the cooling rate, the mechanical properties of ductile iron decrease. The mechanical properties of the specimens at position 4 of the two castings are the poorest. Compared with the specimen at position 1 of Casting-A, the tensile strength, impact toughness, elongation and fracture toughness of the specimens at position 4 are decreased by 6.7%, 75.8%, 81.3% and 28.3%, respectively. As for Casting-B, compared with the specimen at position 1, the tensile strength, impact toughness, elongation and fracture toughness of the specimens at position 4 are decreased by 5.8%, 62.5%, 66.7% and 29.8%, respectively.

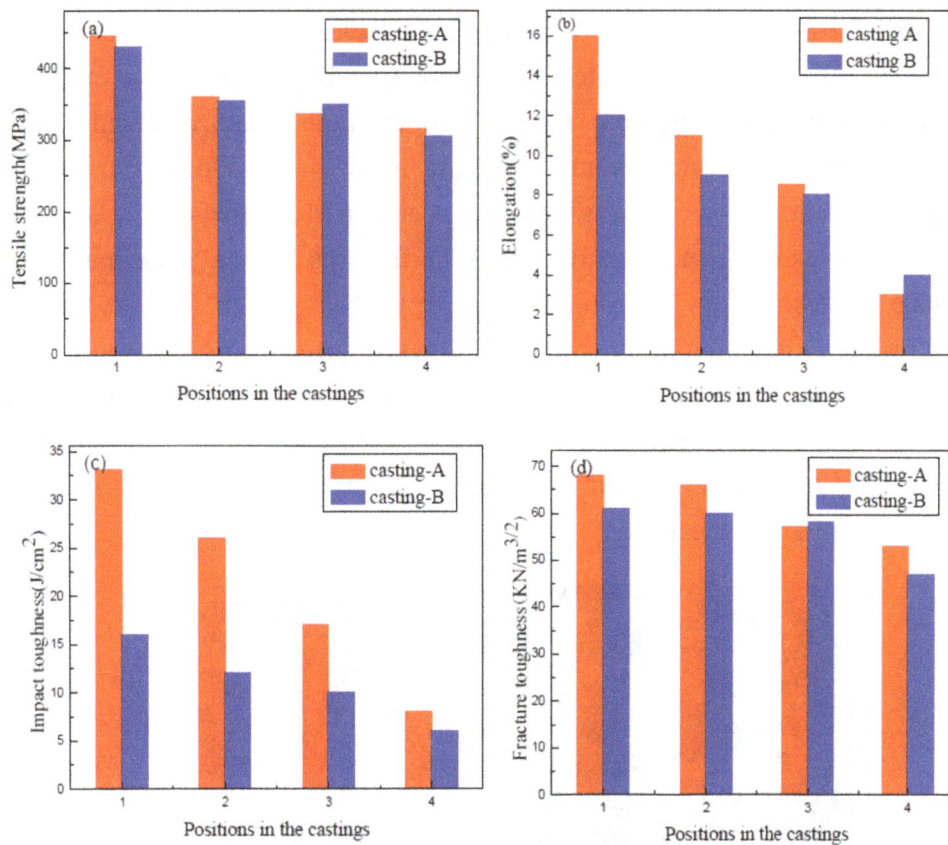

Figure 6. The tensile strength, elongation, impact toughness and fracture toughness at four positions of two castings (**a**) Tensile strength; (**b**) Elongation; (**c**) Impact toughness; and (**d**) Fracture toughness.

As shown in Figure 6, it can be seen that the mechanical properties of the Casting-A is higher than that of Casting-B at the same position. With the increase of Si content, the mechanical properties of heavy section ductile iron decrease. This is because Si can increase the eutectic temperature, reduce the carbon content of the eutectic ferrite and increase the amount of ferrite. However, Si is an extremely sensitive element in heavy section ductile iron, and the excess of it can result in the generation of chunky graphite [17].

3.3. Fracture Microstructure

Figure 7 shows the fracture toughness fracture microstructures of the specimens in Casting-A and Casting-B. As shown in Figure 7a, there are many dimples in the matrix at position 1 of casting-A. The graphite nodules are surrounded by dense dimples, forming tearing ridges. Most of the tearing ridges are closed. The closed tearing ridges can retard the spread of micro-cracks and the connections between the cracks. Therefore, the specimen A1 shows better ductility. The graphite nodules being peeled off from the ferrite under applied force creates the fracture mechanism. The cavities are formed and grow under the effect of slip. Eventually, the cavities connect with each other and cause macroscopic fracture. A relatively larger plastic deformation of ferrite is generated during this process. The fracture morphology of specimen A1 is a typical ductile fracture.

As shown in Figure 7a2,b1, there exists a small amount of dimples in the matrix. Some tearing ridges are connected with each other. A few graphite nodules are split by the spread of micro-cracks that originate from the adjacent abnormal graphite, such as temper graphite and vermicular graphite. The fracture morphology of specimen A2 and specimen B1 is the mixed cleavage-dimple fracture. There are some abnormal graphite nodules in specimen A3 and B2, as shown in Figure 7a3,b2. When the force is applied on the specimens, stress concentration forms at the tips of the abnormal graphite, and micro-cracks originate at these places. The micro-cracks quickly reach an unstable spreading stage without experiencing a cavity aggregation growth stage due to high stress concentration at the tips of abnormal graphite and inconsistency in the deformation performance of ferrite. Most of the graphite nodules are fractured because they cannot be protected by the dimples. Cracks spread quickly in the ferrite, causing cleavage fracture with flat cleavage planes. The fracture morphology of specimens A3 and B2 shows brittle fracture of the cleavage.

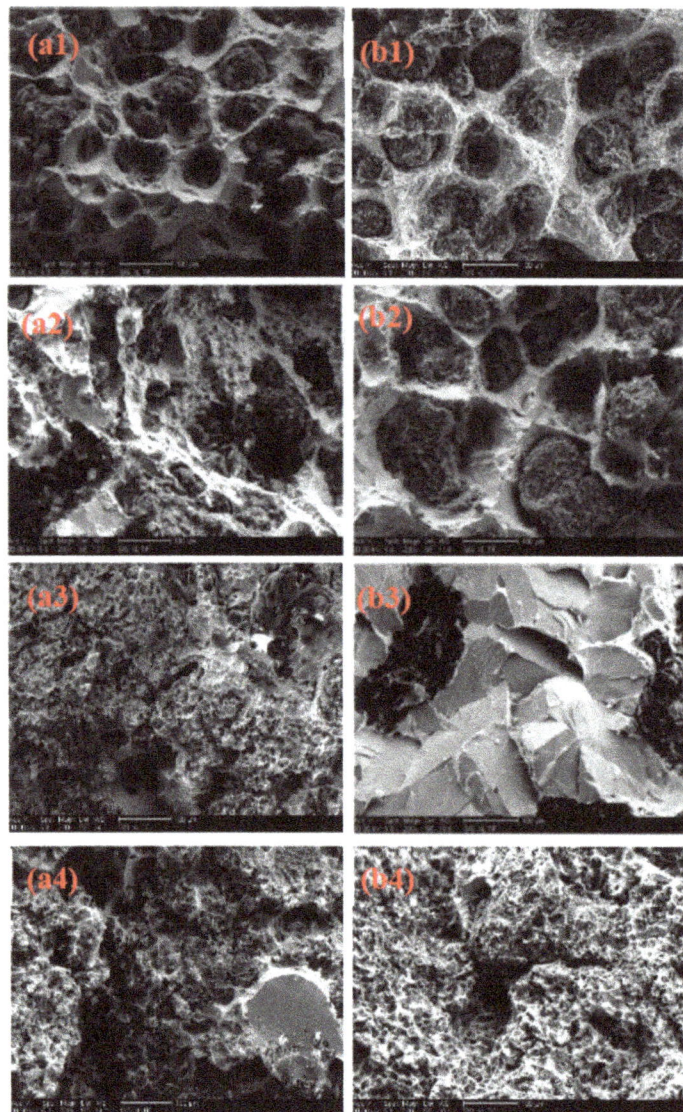

Figure 7. The fracture toughness fracture morphology at four positions of two castings (**a1**) specimen A1; (**a2**) specimen A2; (**a3**) specimen A3; (**a4**) specimen A4; (**b1**) specimen B1; (**b2**) specimen B2; (**b3**) specimen B3; and (**b4**) specimen B4.

As shown in Figure 7a4,b3,b4, there is a large amount of small cleavage planes. The aggregation of chunky graphite occurs. The stress concentration forms easily at the areas of aggregation during the test, while micro-cracks originate and connect with the other cracks surrounding the areas of aggregation. Therefore, the fracture morphology of specimens A4, B3 and B4 is that of brittle fracture.

3.4. The Microscopic Fracture Process of Vermicular Graphite of Heavy Section Ductile Iron

Figure 8 shows the microscopic fracture process of vermicular graphite of heavy-section ductile cast iron. Several typical locations were selected to observe the fracture process of graphite. In Figure 8a1, VG1 is vermicular graphite. When the stretch of specimen is 2.1 mm, micro-crack CVG1 occurs within VG1 (Figure 8a2). The micro-crack initiates at the graphite-matrix(G-M) interface and extends along the G-M interface. The micro-crack CVG1 tears the ferrite matrix at the sharp corners of vermicular graphite. Some local plastic deformation and a small amount of slip bands can be observed in front of the micro-cracks (Figure 8a3). It can be inferred that the stress concentration is significant at the location of sharp cracks in CVG1.

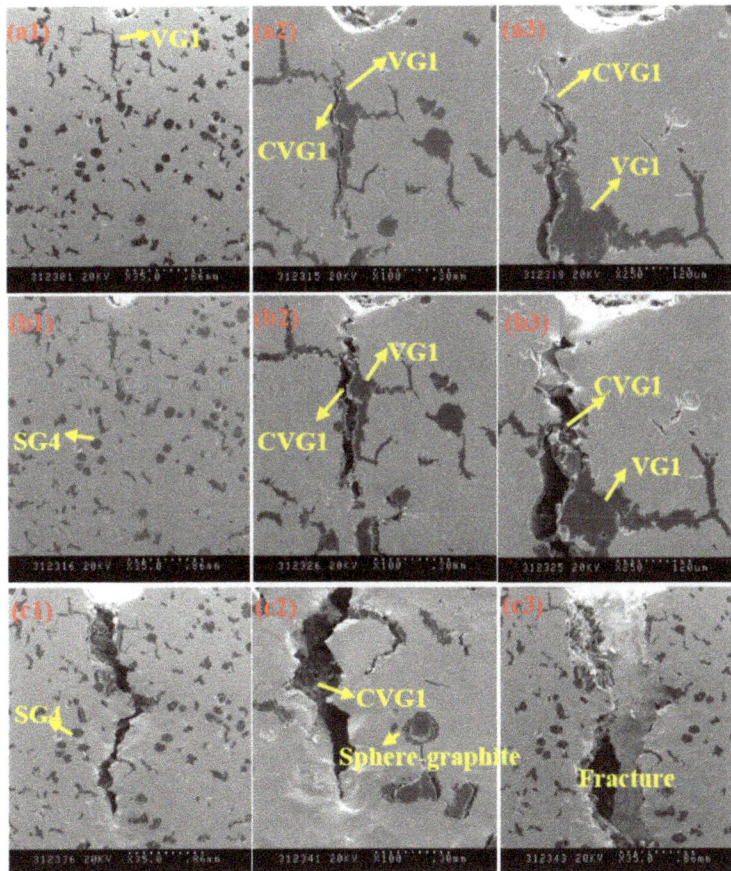

Figure 8. The crack process of vermicular graphite of heavy section ductile iron. (**a1**), (**a2**) and (**a3**): the stretch of specimen is 2.1 mm; (**b1**), (**b2**) and (**b3**): the stretch of specimen is 2.25 mm; (**c1**), (**c2**) and (**c3**): the stretch of specimen is 2.45 mm.

As shown in Figure 8b1, SG4 is spheroidal graphite. When the stretch of specimen is 2.25 mm, the spheroidal graphite SG4 shows no micro-crack initiation. The micro-crack CVG1 extends and becomes wider, as shown in Figure 8b2. The micro-crack CVG1 tears the ferrite matrix and extends to the pre-crack notch. Obvious plastic deformation can be observed in front of the micro-crack in the ferrite matrix, as shown in Figure 8b3.

When the stretch of specimen is 2.45 mm, the micro-crack CVG1 further widens and extends, as shown in Figure 8c1. CVG1 tears the ferrite matrix and connects with the pre-crack notch. A main crack in the matrix is formed. At the tip of the main crack, severe plastic deformation can be observed in the matrix, as shown in Figure 8 (c2). The spherical graphite around the main crack is stripped from the ferrite matrix and plastic deformation is generated, which causes a further concentration of stress on the main crack tip. The main crack propagates unstably and eventually causes fracture, as shown in Figure 8c3.

From the discussion above, it is clear that the vermicular graphite most likely initiates cracks along the G-M interface under the external force. The micro-cracks cause the ferrite matrix to be easily torn and extend unstably because the stress cannot be released by the plastic deformation of the matrix. The vermicular graphite can significantly reduce the conventional mechanical properties and fracture toughness of heavy section ductile iron.

3.5. The Microscopic Fracture Process of Temper Graphite of Heavy Section Ductile Iron

Figure 9 shows the microscopic fracture process of temper graphite of heavy section ductile iron. Several typical locations were selected to observe the fracture process of graphite. As shown in Figure 9a1, TG1, TG4 and VG2 is temper graphite, temper graphite and vermicular graphite, respectively. When the stretch of specimen is 2.1 mm, the micro-cracks can be observed in TG1, TG4 and VG2, as shown in Figure 9a2. When the stretch of specimen is 2.25 mm, two micro-cracks CTG1 and CTG2 initiate at the tangential direction of the spheroidal section of temper graphite TG1, as shown in Figure 9b1. Obvious plastic deformation and slip bands can be observed in front of micro-cracks in the ferrite matrix. It can be inferred that stress concentration occurs at the corners of TG1 and the ferrite matrix is torn. When the stretch of specimen is 2.45 mm, the micro-cracks in TG1 further extend, as shown in Figure 9c1. The micro-crack CTG1 in temper graphite TG1 tears the ferrite matrix and connects with the micro-crack in vermicular graphite VG2. The micro-crack CTG2 extends along with G-M interface of dendritic branching part of TG1, and eventually tears the ferrite matrix. When the stretch of specimen is 2.79 mm, the micro-cracks in TG1 and VG2 further extend and connect with each other, forming a major crack and causing final fracture, as shown in Figure 9d1.

From the discussion above, it is clear that the temper graphite most likely initiates cracks throughout the graphite or along the G-M interface under the external force. High local stress concentration occurs due to the presence of sharp corners of temper graphite. The matrix at the sharp corners is torn up and the micro-cracks extend unstably. The temper graphite can significantly reduce the conventional mechanical properties and fracture toughness of heavy section ductile iron.

4. Conclusions

Under the present conditions, with the increase of Si content, the mechanical properties of ductile iron decreased. When the Si content exceeds 1.78 wt.% and reaches 2.74 wt.%, the impact and fracture toughness of the fracture morphology transforms from ductile to brittle fracture.

The vermicular graphite most likely initiates micro-cracks along the G-M interface under the external force. The micro-cracks cause the ferrite matrix to easily torn, and extend unstably because the stress cannot be released by the plastic deformation of the matrix. The vermicular graphite can significantly reduce the conventional mechanical properties and fracture toughness of heavy section ductile iron.

The temper graphite most likely initiates cracks throughout the graphite or along the G-M interface under the external force. High local stress concentration occurs due to the presence of sharp corners of temper graphite, which causes the unstable extension of micro-cracks. The temper graphite can significantly reduce the mechanical properties and fracture toughness of heavy section ductile iron.

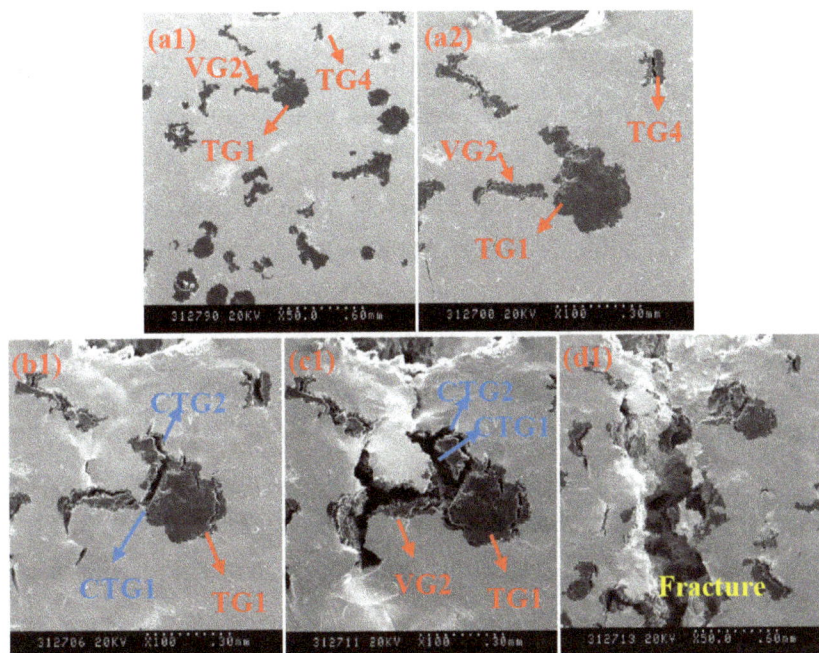

Figure 9. The crack process of temper graphite in heavy section ductile iron. (**a1**) and (**a2**): the stretch of specimen is 2.1 mm; (**b1**): the stretch of specimen is 2.25 mm; (**c1**): the stretch of specimen is 2.45 mm; (**d1**): the stretch of specimen is 2.79 mm.

Acknowledgments

This study was financially supported by National Natural Science Foundation of China (No. 51174068 and 51374086).

Author Contributions

The authors would like to acknowledge Liang Song for carrying out the experimental work and mechanism analyzing work at "Heilongjiang University of Science and Technology", China. Erjun Guo, Liping Wang, and Dongrong Liu would like to acknowledge very much, for the experimental support.

Conflicts of Interest

The authors declare no conflict of interest.

References

1   Cai, Q.H.; Wei, B.K. Recent Development of ductile cast iron production technology in China. *China Foundry* **2008**, *5*, 82–91.

2   Dong, M.J.; Berdin, C.; Beranger, A.S. Damage effect in the fracture toughness of nodular cast iron. *J. Phys. IV France* **1996**, *6*, 65–74.

3   König, M. Literature review of microstructure formation in compacted graphite iron. *Int. J. Cast Met. Res.* **2010**, *23*, 185–192.

4   Riposan, I.; Chisamera, M.; Stanper, S. Peformance of heavy ductile iron castings for windmills. *China Foundary* **2010**, *3*, 163–170.

5   Larrañaga, P.; Asenjo, I.; Sertucha, J. Effect of antimony and cerium on the formation of chunky graphite during solidification of heavy-section castings of near-eutectic spheroidal graphite irons. *Metall. Mater. Trans. A* **2009**, *40*, 65–74.

6   Toktaş, G.; Toktaş, A.; Tayanç, M. Influence of matrix structure on the fatigue properties of an alloyed ductile iron. *Mater. Des.* **2008**, *29*, 1600–1608.

7   Fredriksson, H.; Stjerndahl, J.; Tinoco, J. On the solidification of nodular iron and its relation to the expansion and contraction. *Mater. Sci. Eng. A* **2005**, *413*, 363–372.

8   Cho, G.S.; Choe, K.H.; Lee, K.W. Effects of alloying elements on the microstructures and mechanical properties of heavy section ductile cast iron. *J. Mater. Sci. Technol.* **2007**, *23*, 97–101.

9   Yeung, C.F.; Zhao, H.; Lee, W.B. Effect of homogenisation treatment on segregation of silicon in ferritic ductile irons: A colour metallographic study. *Mater. Sci. Technol.* **1999**, *15*, 733–737.

10 Wang, L.; Guo, E.; Jiang, W. Physical modeling of spent-nuclear fuel container. *China Foundry* **2012**, *9*, 366–369.

11 Baer, W.; Wossidlo, P.; Abbasi, B. Large scale tesing and statistical analysis of dynamic fracture toughness of ductile cast iron. *Eng. Fract. Mechanics* **2009**, *76*, 1024–1036.

12 Minnebo, P.; Nilsson, K.F.; Blagoeva, D. Tensile, compression and fracture properties of thick-walled ductile cast iron components. *J. Mater. Eng. Perform.* **2007**, *16*, 35–45.

13 Iacoviello, F.; Di Bartolomeo, O.; Di Cocco, V. Damaging micromechanisms in ferritic-pearlitic ductile cast irons. *Mater. Sci. Eng. A* **2008**, *478*, 181–186.

14 Nakae, H.; Jung, S.; Kitazawa, T. Eutectic solidification mode of spheroidal graphite cast iron and graphitization. *China Foundry* **2007**, *4*, 34–37.

15 Song, L.; Guo, E.; Tan, C. Effect of Bi on graphite morphology and mechanical properties of heavy section ductile cast iron. *China Foundry* **2014**, *2*, 125–131.

16 Mourujärvi, A.; Widell, K.; Saukkonen, T. Influence of chunky graphite on mechanical and fatigue properties of heavy-section cast iron. *Fatigue Fract. Eng. Mat. Str.* **2009**, *32*, 379–390.

17 Asenjo, I.; Larranaga, P.; Sertucha, J. Effect of mould inoculation on formation of chunky graphite in heavy section spheroidal graphite cast iron parts. *Int. J. Cast. Metal. Res.* **2007**, *20*, 319–324.

3

Wetting by Liquid Metals—Application in Materials Processing: The Contribution of the Grenoble Group

Nicolas Eustathopoulos

SIMAP, University Grenoble Alpes-CNRS, F-38000 Grenoble, France;
E-Mail: nikos@simap.grenoble-inp.fr

Academic Editor: Enrique Louis

Abstract: The wettability of ceramics by liquid metals is discussed from both the fundamental point of view and the point of view of applications. The role of interfacial reactions (simple dissolution of the solid in the liquid or formation of a layer of a new compound) is illustrated and analysed. Several results are presented in order to illustrate the role of wettability in materials processing, namely infiltration processing, joining dissimilar materials by brazing and selecting crucibles for crystallising liquid metals and semiconductors. The review includes results obtained during the last 15 years mainly, but not only, by the Grenoble group.

Keywords: wetting; reactivity; metals; ceramics; composites; brazing; infiltration; crucible

1. Introduction

The intrinsic aptitude of a non-reactive liquid to wet a flat, smooth and chemically homogeneous solid surface is quantified by the value of Young's contact angle θ_Y, a unique characteristic of a solid S–liquid L–vapour V system (Figure 1). θ_Y enters into all model equations describing the wetting of liquids on real solid surfaces, $i.e.$, surfaces with a certain roughness and degree of heterogeneity, as well as into equations modelling wetting in reactive liquid–solid systems. Moreover, by measuring θ_Y and the surface energy (or surface tension) σ_{LV} of the liquid Dupré's adhesion energy W_a can be evaluated. This quantity characterises the thermodynamic stability of interfaces between dissimilar materials and is widely used in practice for predicting their potential bonding properties.

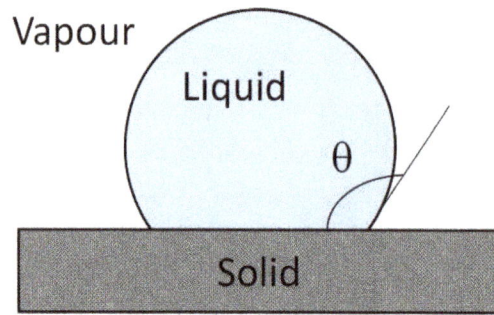

Figure 1. Definition of the equilibrium contact angle θ. For a flat, smooth and chemically homogeneous solid surface, θ is the Young contact angle θ_Y.

During the last decades, significant improvements have been made in the measurement of contact angles of high temperature systems [1]. In this period, wetting studies have benefited from high resolution techniques for characterising the topological and chemical features of surfaces at nanometric scale. Another reason for this improvement has been the use of monocrystalline or vitreous solids to prepare the high-quality surfaces required for Young's contact angle determinations. Over the past 15 years, further improvements have been made as an increasing number of laboratories are now using more sophisticated versions of the sessile drop method (dispensed drop, transferred drop) enabling *in situ* cleaning of surfaces. Finally, automatic systems for data acquisition and image analysis leading to the simultaneous measurement of contact angle and surface tension have been developed and used widely.

Data on the wetting of ceramics by liquid metals published until the end of the last century have been extensively reviewed in [2]. Another review with special emphasis on alloys used for soldering in microelectronics was published in 2007 [3]. After giving a brief overview of the fundamental equations of wetting and adhesion for both smooth and rough solid surfaces, this article aims to review results obtained during the last 15 years for non-reactive and reactive liquid–solid systems. Although this review focuses on liquid metal/ceramic systems, a limited number of results concerning liquid metal–solid metal systems are also given because they are useful in understanding bonding at metal/ceramic interfaces. Several experimental results are presented in the paper in order to illustrate the role of wettability in different applications, namely materials' infiltration processing, joining dissimilar materials by brazing and selecting crucibles for solidifying liquid metals and semiconductors. The results are mainly, but not only, from studies performed by the Grenoble group.

2. Non-Reactive Wetting

2.1. Thermodynamics

The intrinsic contact angle θ_Y in a non-reactive solid-liquid system is given by the classical equations of Young (Equation (1a)) and Young-Dupré (Equation (1b)):

$$\cos\theta_Y = \frac{\sigma_{SV} - \sigma_{SL}}{\sigma_{LV}}$$

(1a)

$$\cos\theta_Y = \frac{W_a}{\sigma_{LV}} - 1 \tag{1b}$$

where the quantities σ_{SV} and σ_{LV} define the surface energy of the solid and liquid, respectively, and σ_{SL} the solid/liquid interface energy. W_a is the adhesion energy of the system defined as the energy required to separate reversibly a solid and a liquid having a common interface of unit area, creating two free surfaces, one solid-vapour and one liquid-vapour. Therefore, W_a is related to the surface energies of the system by $W_a = \sigma_{SV} + \sigma_{LV} - \sigma_{SL}$.

According to Equation (1b), the intrinsic contact angle θ_Y in a non-reactive liquid/solid system results from two types of competing forces: (i) adhesion forces that develop between the liquid and the solid phases, expressed by the quantity of adhesion energy which promotes wetting, and (ii) cohesion forces of the liquid taken into account by the surface energy of the liquid σ_{LV} acting in the opposite direction (the cohesion energy of the liquid is equal to $2\sigma_{LV}$).

Usual liquid metals are high surface energy liquids. Their surface energy σ_{LV} lies between 0.5 J m^{-2} for low melting point (m.p.) metals such as Pb and Sn and 2 J m^{-2} for high m.p. metals such as Fe and Mo [2]. These values, reflecting the high cohesion of metals due to their metallic (*i.e.*, chemical) bonding, are one to two orders of magnitude greater than the surface energies of room temperature liquids in which bonding is achieved by weak, intermolecular interactions (*i.e.*, physical interactions). Then, according to Equation (1b), good wetting (*i.e.*, a contact angle of a few degrees or tens of degrees) of a liquid metal on a solid substrate can be observed if the adhesion energy is close to the cohesion energy of the liquid $2\sigma_{LV}$. This is possible only if the interfacial bond is strong, *i.e.*, chemical in nature. This condition is fulfilled for liquid metals on solid metals regardless of the miscibility between the liquid and the solid, because in this type of system the interfacial bond is metallic. For instance, good wetting is observed for liquid Cu on solid Mo despite the absence of any miscibility in this system (Table 1). Liquid metals also wet semiconductors such as Si, Ge or SiC because these solids, that are covalent in the bulk, are metallic in nature near the surface. Finally, liquid metals also wet ceramics such as carbides, nitrides or borides of transition metals because a significant part of the cohesion of these materials is provided by metallic bonds. Among the solids that are not wetted by non-reactive liquid metals are the different forms of carbon, the ionocovalent oxides and the predominantly covalent ceramics with a high band gap like BN. In these non-wetting systems' adhesion is provided by weak van der Waals interactions.

Table 1. Wetting of different types of solids by non-reactive liquid metals at temperatures close to the metal melting point. The contact angle values are from the review [2].

Type of substrate	Type of interaction	θ (degrees)	Examples
Solid metals			Cu/Mo: 10°–30°
Semiconductors	Strong (chemical)	$\theta \ll 90°$	Sn/Ge: 40°; Si/SiC: 35°–45°
Ceramics with a partially metallic character			Cu/WC: 20°; Au/ZrB$_2$: 25°
Carbon materials	Weak (physical)	$\theta \gg 90°$	Au/C: 120°–135°
Ionocovalent ceramics			Ag/Al$_2$O$_3$, Cu/SiO$_2$: 120°–140°; Au/BN: 135°–150°
Ionocovalent oxides	Moderate (chemical)	$\theta \approx 90°$	(Ag+O)/Al$_2$O$_3$; Al/Al$_2$O$_3$

Between systems with $\theta_Y \gg 90°$, corresponding to weak, physical solid–liquid interactions, and systems with $\theta_Y \ll 90°$ (strong, chemical interactions), there are some liquid/solid combinations where θ_Y is in the range 80°–100°. These values correspond to an adhesion energy that is 2–2.5 times higher than the adhesion energy of non-reactive metals on oxides for the same σ_{LV} value and reflect the development of moderate interactions of a chemical nature through the interface. In a given metal/oxide couple, when the mole fraction of oxygen dissolved in the metal x_{ox} increases progressively (the source of oxygen being an oxygen-rich gas), the contact angle starts to decrease at a threshold value $x_{ox}*$ around 10^{-5}, by chemisorption of oxygen at the metal/oxide interface [4]. Silver has a very limited affinity for oxygen (silver oxides are in fact unstable), but it can dissolve large amounts of this element resulting in a strong decrease in contact angle on alumina, from a value close to 130° for pure Ag to 90° for Ag with $x_{ox} > 10^{-5}$. Contrary to Ag, Al has a very high affinity for oxygen while the solubility of this element in liquid Al is extremely low. Indeed, the maximum value of x_{ox}, that is the value of this quantity at the Al/Al_2O_3 interface at 700 °C, is only 10^{-9}, *i.e.*, several orders of magnitude lower than $x_{ox}* \approx 10^{-5}$. Therefore, oxygen chemisorption cannot occur in this case. However, due to its high affinity for oxygen, Al can modify the surface chemistry of the oxide substrate by forming a two-dimensional layer with a lower O/metal ratio than in the bulk oxide, thus improving wetting.

Equations (1a,b) give the "Young contact angle" θ_Y or intrinsic contact angle of a liquid on a perfectly smooth and chemically homogeneous solid surface. In most applications, chemical heterogeneities (such as, oxide inclusions on a metallic surface) and the roughness of real solid surfaces lead to deviations in the observed contact angles from θ_y that can attain several degrees and in some cases tens of degrees [1].

The roughness of the solid surface affects wetting through two different effects: the first is the increase in the actual surface area and the second is pinning of the triple line by sharp defects.

The first effect is expressed by Wenzel's equation:

$$\cos\theta_w = s_r \cos\theta_Y \tag{2}$$

where s_r denotes the ratio of the actual area to the planar area ($s_r > 1$). According to this equation, if $\theta_Y < 90°$, θ_w will be lower than θ_Y and, for $s_r > 1/\cos\theta_Y$, perfect wetting will be observed.

Sharp defects can pin the triple line at positions far from stable equilibrium, *i.e.*, at contact angles markedly different from θ_Y. This effect is illustrated schematically in Figure 2 where θ is the macroscopic (or observed) contact angle, α is the inclination of the defect and θ_Y the microscopic contact angle. This configuration corresponds to a metastable equilibrium state (*i.e.*, to a local minimum of the total energy of the system) where:

$$\theta = \theta_Y + \alpha \tag{3}$$

In systems with good wetting ($\theta_Y \ll 90°$), the predominant effect of surface roughness is that given by Wenzel's equation. For a θ_Y value close to or higher than 90°, the pinning effect predominates.

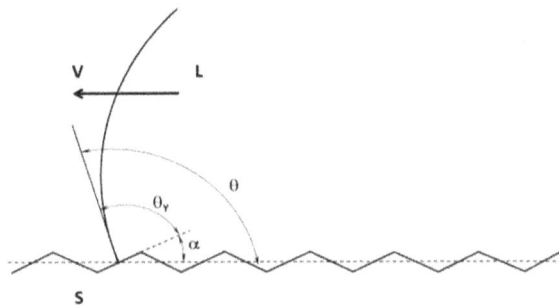

Figure 2. Pinning of the triple line during wetting on a rough surface. Pinning corresponds to a metastable state where the macroscopic contact angle θ satisfies the relation $\theta = \theta_Y + \alpha$.

When $\theta_Y \gg 90°$, wetting on high-roughness solids leads to the formation of "composite interfaces", partly solid–liquid and partly solid–vapour, (Figure 3a) resulting in contact angles θ well above θ_Y. In this case, even limited stress produced during cooling leads to detachment of the solidified metal from the substrate by a purely adhesive rupture. Finally, when the goal of an experiment is to attain θ_Y, solid surfaces with an average roughness less than 100 nm must be used [1]. The results given in Table 1 have been obtained with this type of substrate.

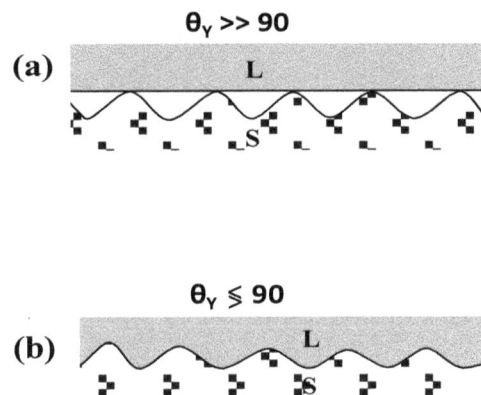

Figure 3. Microscopic configuration at solid/liquid interfaces: (**a**) For $\theta_Y \gg 90°$, at microscopic scale, the liquid contacts the rough surface of the solid only at a few points. During cooling the solidified liquid detaches spontaneously from the solid; (**b**) For θ_Y values lower than 90° or higher but close to this value, an intimate contact exists at any point of the interface.

When a solid α contains a dispersion of particles β, the equilibrium contact angle θ_C on the heterogeneous surface is given by Cassie's equation:

$$\cos\theta_C = f_\alpha \cos\theta_\alpha + f_\beta \cos\theta_\beta \qquad (4)$$

where f_α is the surface area fraction of the matrix. This equation is also applicable to a porous solid taken for the pore $\theta_\beta = 180°$.

2.2. Kinetics

As the viscosity of liquid metals and alloys is very low—a few mPas [5]—spreading of this type of liquid is a very fast process. As a general rule, for $\theta_Y > 20°$, the "spreading time" t_{spr} (defined as the time needed for a millimetre sized droplet to attain the equilibrium contact angle) is around 10 ms [6–8] (see an example in Figure 9). Significantly higher spreading times can be observed in two cases (i) in systems with equilibrium contact angles close to zero. In this case, instead of a drop, the liquid rapidly forms a film in which the viscous friction during further spreading is no longer negligible (Figure 4) and (ii) when wetting is assisted by the modification of the 2-d interface caused by an element contained in the liquid alloy.

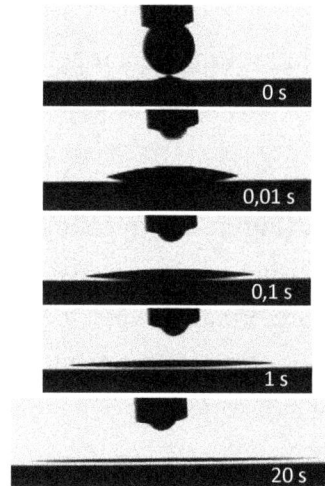

Figure 4. Spreading of a CuAg alloy saturated in Cu on monocrystalline Cu studied by the dispensed drop technique. The diameter of the droplet before wetting is 0.9 mm [9].

An example is silicon in Cu on silica substrates (Figure 5). The addition of 25 at% of Si in Cu leads to a decrease in the contact angle from 135° for pure copper, to about 105°. Silicon has two effects: first it decreases the surface energy σ_{LV} (from 1300 mJ/m^2–1075 mJ/m^2), and second it increases the adhesion energy W_a (from 380 mJ/m^2–780 mJ/m^2) due to the formation of a two-dimensional layer on the silica surface with a lower O/metal ratio than in the bulk oxide, as in the case of the Al/Al$_2$O$_3$ system. As can be easily seen from Equation (1b), both silicon effects act in the same direction, *i.e.*, to a decreasing contact angle. The interface modification is probably the reason why the spreading time in this system is several minutes, which is orders of magnitude higher than the spreading time in systems where the wetting process takes place without modification of the interfacial chemistry.

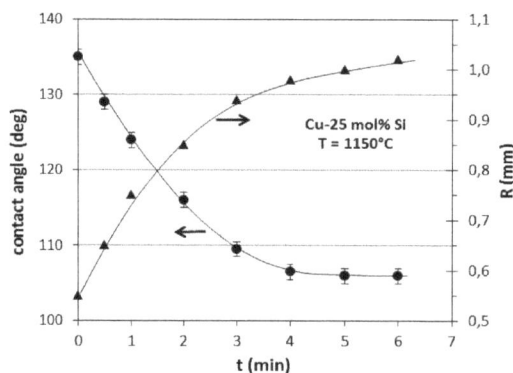

Figure 5. Contact angle and drop base radius as a function of time for a Cu-Si alloy on silica under high vacuum. Dispensed drop technique. Results from [10].

3. Reactive Wetting

Wetting in metal/metal and metal/ceramic systems is often accompanied by reactions at the solid–liquid interface, namely simple dissolution of the solid into the liquid or formation of a 3-d layer of a new compound.

3.1. Wetting with Formation of a New Compound at the Interface

The thermodynamics and kinetics of reactive wetting with formation of a new compound at the interface are given by the Reaction Product Control (RPC) model which can be summarised as follows: [2]. The contact angle in such a system varies between two characteristic contact angles, the initial contact angle θ_0, which is the contact angle on the unreacted substrate, and the final contact angle θ_F, which is the contact angle on the reaction product (Figure 6):

$$\theta_0 = \theta_S \tag{5a}$$

$$\theta_F = \theta_P \tag{5b}$$

This is true not only when the reaction product is better wetted than the initial substrate, as in the case of Si/C couple depicted in Figure 6, but also when the opposite situation occurs, for instance in the couples Ag/SiC, Cu/SiC (see Sections 3 and 4) and Au/TiC. Pure Au does not wet TiC with a contact angle of about 130°, a value that is similar to that obtained for Au on carbon substrates [11] (Figure 7). Because of the strong interaction between Au and Ti a slight dissolution of Ti from the substrate to the liquid occurs. However, given that the solubility of C is much smaller than for Ti, graphite precipitates at the surface and the measured contact angle of 130° is characteristic of the reacted interface: Au on graphite. Since the solubility of C in molten Ni and Fe (but not in Cu) is several orders of magnitude higher than in Au, small additions of these elements in Au increase the solubility of C significantly and, thereby, prevent the formation of the graphite interface layer on the TiC substrate. As a result, the experimental results show a dramatic decrease in contact angle with a few at.% of Ni or Fe without any reaction product at the interface. In this system, the action of Ni and Fe is to remove, by dissolution, the wetting barrier that inhibits spreading.

The spreading time t_{spr} in metal/metal and metal/ceramic reactive systems is in the range 10–10^4 s, i.e., several orders of magnitude higher than $t_{spr} \approx 10^{-2}$ s observed for non-reactive metals. Therefore, in a given reactive system, $t(\theta_F) >> t(\theta_0)$, which implies that the spreading rate of the reactive stage is limited by the interfacial reaction itself. In [12], the RPC model was greatly improved thanks to an analytical approach used to describe spreading in systems where the growth rate of the reaction layer parallel to the interface is controlled by the chemical reaction localized at the triple line.

Note that in 1998, Saiz et al. [13] proposed an approach in which the reactive wetting is assumed to be caused mainly by adsorption, whereas spreading kinetics is controlled by the migration of a ridge formed at the solid-liquid-vapour triple line. However, detailed comparisons between predictions made by this model and experimental data did not confirm the Saiz et al. approach [14,15].

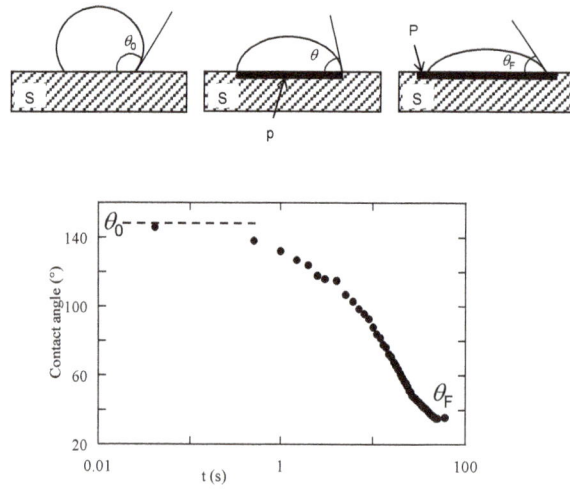

Figure 6. Top: Schematic representation of the "Reaction Product Control" model [2]. The initial contact angle θ_0 is the contact angle on the surface of the unreacted ceramic substrate S. After a transient stage, a quasi-state configuration is established at the triple line where the advance of the liquid is hindered by the presence of a non-wettable substrate in front of the triple line. Thus, the only way to move ahead is by lateral growth of the wettable reaction product layer P until the macroscopic contact angle equals the equilibrium contact angle θ_F of the liquid on the reaction product. **Bottom**: Contact angle *vs*. logt for Si on vitreous carbon (where P = SiC) according to [16].

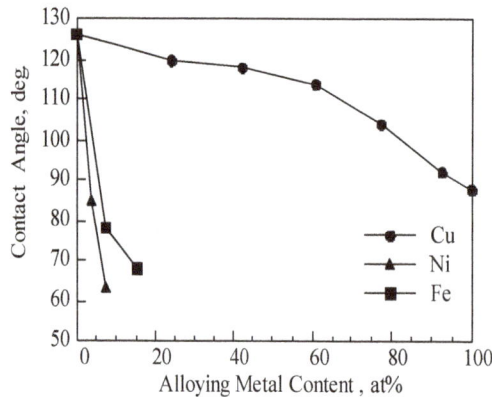

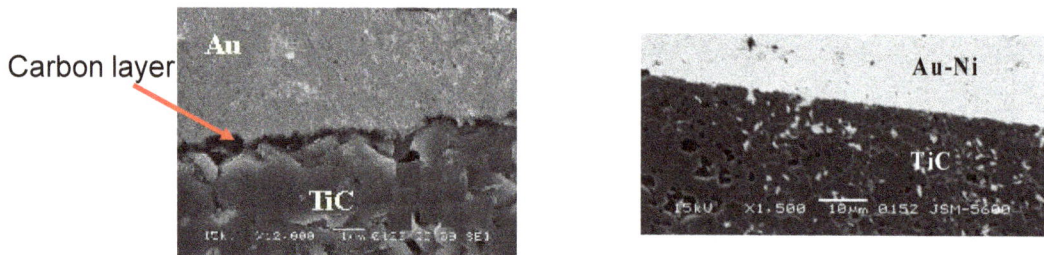

Figure 7. Top: Contact angle between TiC and Au alloys at 1150 °C. **Bottom**: Metal/TiC interfaces for pure Au and an Au-Ni alloy [11].

3.2. Dissolutive Wetting

Extensive dissolution of a solid in a liquid is a phenomenon occurring in many liquid metal/solid metal systems as well as in some metal/ceramic ones, such as Ni/C and Ni/SiC. Dissolution can reduce the observed (or visible) contact angle through two effects: first by decreasing the surface tension of the liquid, which occurs when the surface tension of the solid is much lower than that of the liquid; second by forming under the droplet, a crater such that the observed contact angle θ_{ap} is lower than the true angle θ_F formed at the solid/liquid/vapour junction (Figure 8).

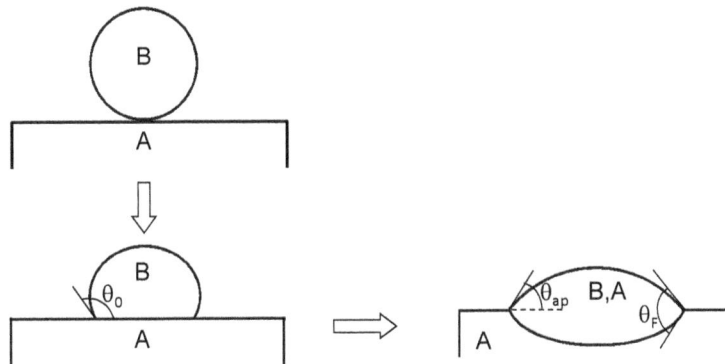

Figure 8. Dissolutive wetting: Wetting of a liquid metal B leads rapidly, within a few ms, to a first contact angle θ_0 corresponding to the intrinsic contact angle of pure B in metastable equilibrium with A. Then, the wetting process continues with the formation of a crater until saturation of B in A [8].

Protsenko *et al.* proposed that dissolution results in further spreading of the liquid on the substrate surface in order to maintain the capillary equilibrium at the three phase junction [8]. Accordingly, θ_F should be the intrinsic contact angle of a B liquid saturated in A on solid A measured on an inclined solid surface. Conversely, other authors consider that this angle does not result from the equilibrium of surface energies but by diffusion of dissolved species from the interface to the liquid bulk [17]. As for the spreading time, experimental values of this quantity are higher than for non-reactive spreading, varying between seconds and hundreds of seconds, depending on the solubility of the solid in the liquid, droplet size, contact angle and temperature [8,17]. An example is given in Figure 9 for pure Cu on solid Si compared to copper presaturated in Si.

A new effect of dissolution on wettability was recently highlighted by Lai *et al.* [18] in wetting experiments of two-phase composite Cu-Fe substrates by molten Sn. It was found that the curve of contact angle *versus* the surface fraction of composite components passes through a minimum, behaviour that cannot be interpreted by Equation (4). It was shown that the enhanced wetting observed on the composite substrate (Figure 10) can be explained by the dissolution contrast of Cu and Fe phases, leading to increasing interfacial roughness, thus providing an additional driving force for wetting.

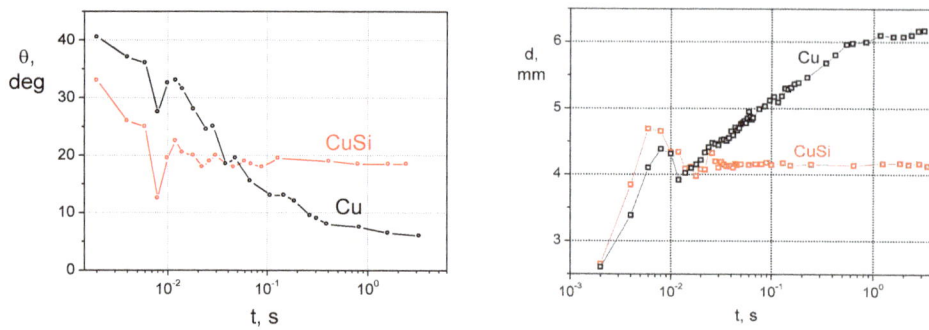

Figure 9. Visible contact angle (**left**) and drop base diameter (**right**) as a function of time plotted on a logarithmic scale for Cu and Cu presaturated in Si on a (001) Si surface at 1100 °C. Dispensed drop technique [8].

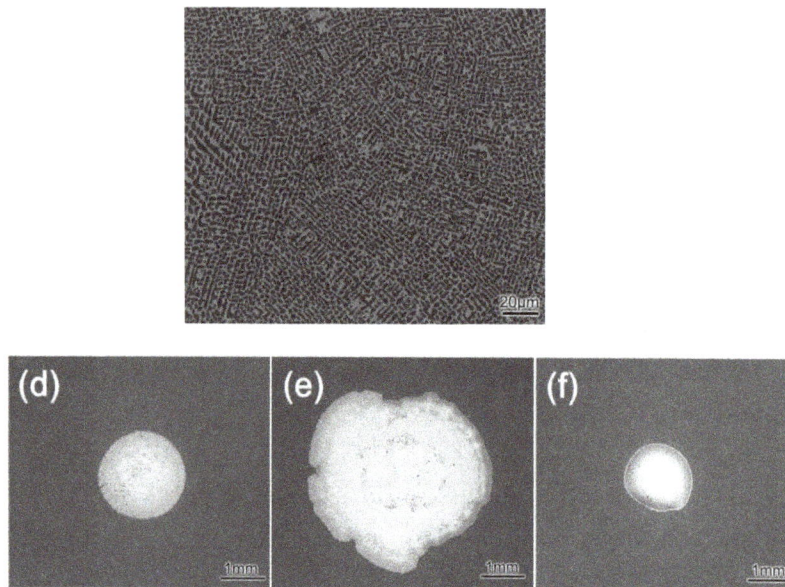

Figure 10. Top: Microstructure of a Cu-Fe alloy with 50 wt% Cu. Iron is dark, copper is grey. **Bottom**: top-view of areas wetted by Sn on Cu, $\theta = 25°$ (**d**), 50Cu-50Fe, $\theta = 5°$ (**e**) and Fe $\theta = 47°$ (**f**) [18].

3.3. Dissolution vs. Formation of a New Compound

In many liquid metal–solid systems, the formation of a new compound is preceded by the dissolution of the solid in the liquid. The initial dissolution rate is high; however, once a layer of the new compound is formed at the interface, the dissolution rate decreases rapidly. An example which illustrates the two types of reactive wetting is given by results obtained by the dispensed drop technique with an Au-40 at% Ni alloy on ZrB_2 substrates [19]. The experiments were performed at two temperatures, 980 °C and 1170 °C. A nickel boride is formed at the alloy/ZrB_2 interface at 980 °C while dissolution of ZrB_2 into the liquid alloy occurs only at 1170 °C. Thus, by varying the temperature, it is possible to change the type of metal–ceramic reactivity by keeping the composition of the metallic phase constant. Figure 11 gives the time-dependent change of drop base diameter in an experiment at 1170 °C, as well as during the subsequent cooling. Due to the metal-like character of ZrB_2, the initial contact angle is low (28°) and decreases slightly to 25° thanks to substrate dissolution.

Further spreading was observed when the temperature decreased to values where formation of the new compound (Ni_2B) becomes possible.

The micrograph of the region near the triple line (Figure 11) shows the position of the triple line before cooling, separating the dissolution cavity formed at 1170 °C from the interface zone produced at lower temperature. The reaction product layer covering the interface, a few microns thick, can also be distinguished. From these observations it is easy to understand the unusual result depicted in Figure 11, namely that a decrease in temperature improves wettability.

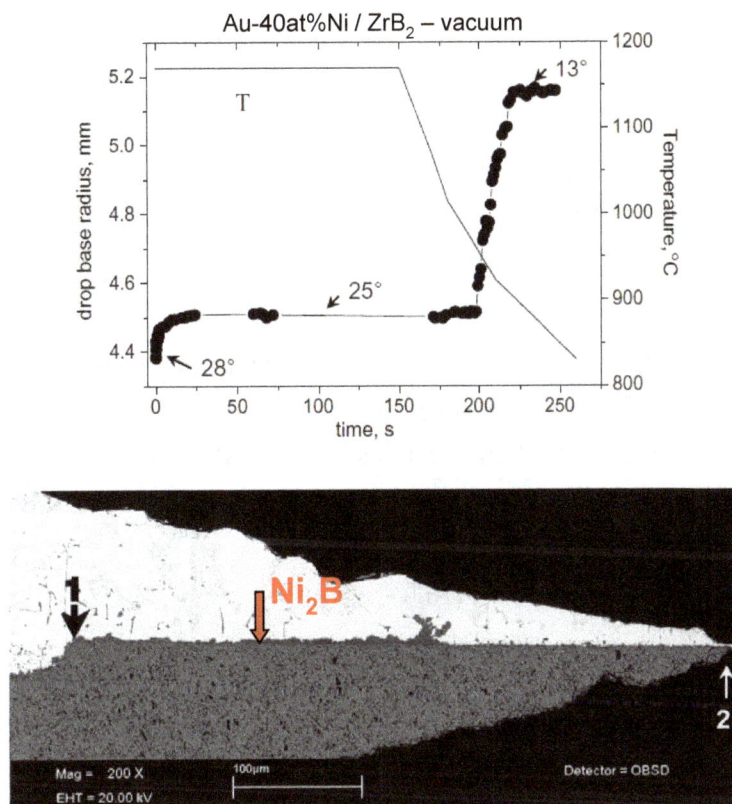

Figure 11. Au-40 at% Ni/ZrB_2 system. **Top**: Drop base radius and temperature as a function of time. The numbers on the drop base radius curve represent the corresponding values of the contact angle. **Bottom**: Micrograph of the interface close to the triple line formed during cooling. 1 and 2 indicate the positions of the triple line before and after cooling [19].

In the previous example, the formation of a new compound at the interface leads to a net, but limited, improvement in wetting. This is because both the initial solid (ZrB_2) and the reaction product (nickel boride) are metallic in character and thus wettable by liquid metals. A much more pronounced improvement is observed in the example depicted in Figure 12 for two Ni-Si alloys on graphite at 1270 °C [20]. For the Ni-21 at% Si alloy, only a limited dissolution of graphite in the liquid takes place. The steady contact angle attained a few minutes after melting is close to 120°. A completely different behaviour is exhibited by a Ni-47 at% Si alloy, leading to the formation of wettable SiC at the interface. The alloy was shown to wet the graphite with a steady contact angle of 40°. As will be discussed in the following section, this wetting results in liquid alloy infiltrating the porous graphite.

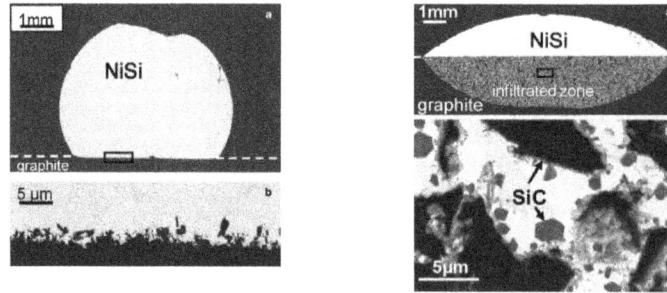

Figure 12. Micrographs of two NiSi alloy/graphite sectioned samples, **Left**: Ni-21 at% Si, (**a**) general view; (**b**) alloy/carbon interface. Only a limited dissolution of graphite is observed. **Right**: Ni-47 at% Si, (**a**) general view; (**b**) an area of the infiltrated zone [20]. $T = 1270$ °C.

4. Application in Processing of Materials

4.1. Wetting in Infiltration

Metal/ceramic composites are often processed by infiltration of porous ceramics by liquid metals [21,22]. Under the temperature and atmosphere conditions normally used in practice, liquid metals (Cu, Al, *etc.*) do not wet ionocovalent ceramics such as alumina, silicon carbide or graphite and, for this reason, infiltration is achieved by applying a sufficiently high pressure P_0 (Figure 13a) to overcome the capillary pressure

$$P_C = -(2\sigma_{LV}/r_{eff}) \cos\theta \tag{6}$$

where r_{eff} is an effective pore radius characteristic of the preform and θ the contact angle on pore walls. In numerous studies published since the 1990s (see for instance [23,24]), it has been found that the infiltration distance h increases parabolically with both time t and excess pressure $\Delta P = P_0 - P_C$. This agrees with the following equation established by Washburn [25] assuming that infiltration is limited by viscous friction:

$$h^2 = r_{eff}^2 \frac{\Delta P}{4\eta} t \tag{7}$$

For wetting liquids (Figure 13b) pressureless infiltration becomes possible, described by Equation (7) which, introducing the expression of capillary pressure P_C and taking $P_0 = 0$, becomes:

$$h^2 = r_{eff} \frac{\sigma_{LV} \cos\theta}{2\eta} t \tag{8}$$

According to Equations (6) and (8), pore infiltration is possible if the equilibrium contact angle θ is lower than a threshold contact angle $\theta^* = 90°$. This value is only valid for strictly parallel pore walls having smooth surfaces. For real porous media, the inequality $\theta < 90°$ for a liquid with an intrinsic contact angle θ does not guarantee that infiltration will occur. In the first place, pore walls are rough and if θ is too close to the threshold contact angle, this roughness can hinder infiltration. Secondly, the pore walls are never parallel. Consequently, if during infiltration the liquid–vapour surface area increases, as in the V-shaped pore depicted in Figure 14, θ^* is no longer equal to 90° but to 90° − φ.

In practice, contact angles as low as 60° seem to be necessary in order to ensure spontaneous infiltration [26,27].

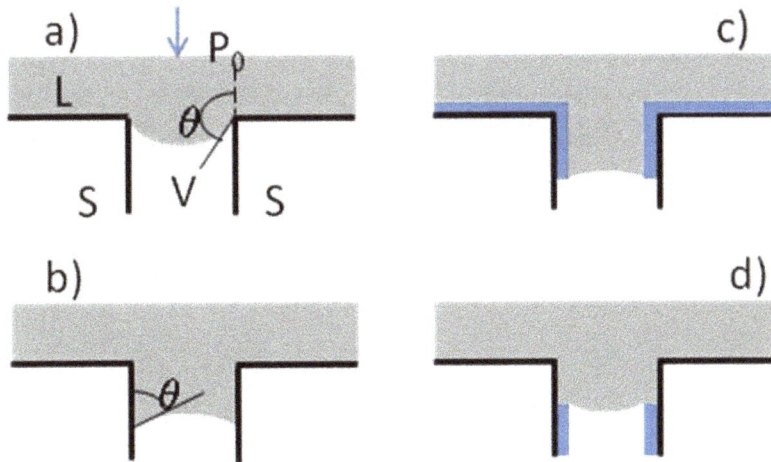

Figure 13. Schematic presentation of pore infiltration by a liquid in non-reactive (**a,b**) and reactive (**c,d**) systems. In case (**c**), the infiltration of a non-wettable solid S is assisted by the formation of a wettable reaction product layer (example: Si/porous graphite). In case (**d**), the infiltration of a wettable solid is hindered by the presence of a non-wettable layer covering the pore walls (example: Si/oxidised porous Si_3N_4).

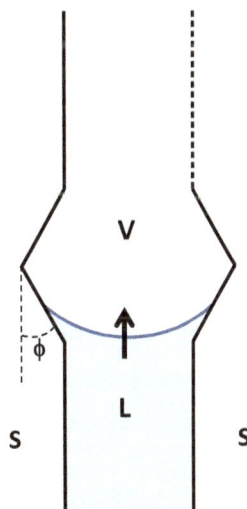

Figure 14. Infiltration in a pore locally exhibiting an increase in pore area perpendicular to the infiltration direction.

An example of spontaneous infiltration is silicon in porous carbon performs. Liquid silicon is known to wet carbon and it can therefore spontaneously infiltrate a carbon preform. Infiltrated Si reacts with carbon to form SiC. Reactive infiltration is used to process the so-called "reaction-bonded silicon carbide" or SiC composites [28,29].

For a long time, infiltration of Si and silicon rich alloys into porous carbon has been described as consisting of rapid, non-reactive, infiltration obeying Washburn's equation followed by the reaction between Si and C to form SiC [30–32]. Three experimental facts disagree with this model: (i) Infiltration of porous graphite by Si and NiSi alloys rich in silicon does not show a parabolic trend as predicted by Wasbhurn's equation, but is linear with time (Figure 15) (ii) Si does not wet unreacted carbon substrates as shown by the large obtuse θ_0 value in the curve of Figure 6. This is also indicated by the results obtained for the Ni-Si/C couple presented in the previous section and depicted in Figure 12. These results strongly suggest that silicon carbide formation at the Si/C interface is a necessary condition for wetting and infiltration. (iii) Further evidence is given by the experimental values of the infiltration rate dh/dt which, in the case of pure Si, is 5–10 μm/s, nearly equal to the spreading rate dR/dt [33] but several orders of magnitude lower than the values predicted by Washburn's equation which are in the mm/s range. It is concluded that the infiltration rate in this system is controlled by the rate of growth of SiC parallel to the pore walls at the infiltration front (Figure 13c).

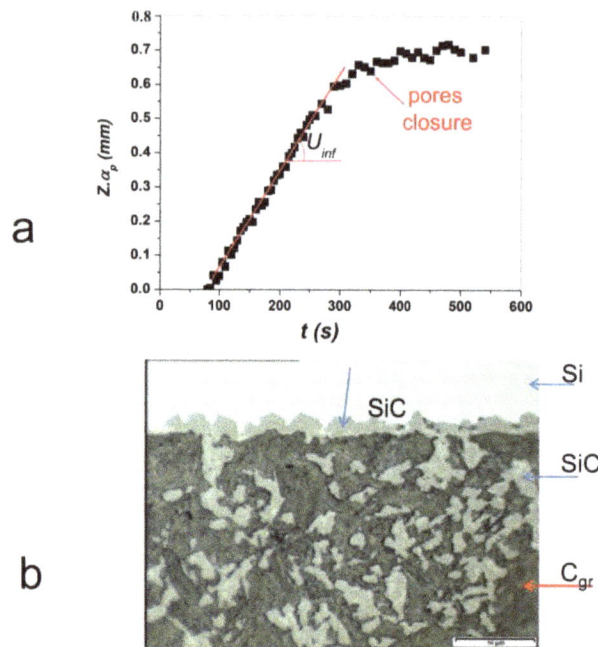

Figure 15. Infiltration of porous graphite (pore volume fraction α_p = 0.15) by molten silicon. (**a**) Variation with time of the quantity $Z\alpha_p$ where Z is the infiltration distance (Z = h in Equations (7) and (8)). (**b**) Reacted interface and infiltrated zone [33,34].

Another example of spontaneous infiltration is infiltration of silicon in porous silicon nitride coatings of silica crucibles used in the crystallisation of photovoltaic (PV) quality silicon. When Si-PV is processed by liquid state routes, solidified Si adheres to the crucible walls leading to thermo-mechanical stress and resulting in loss of Si electrical performance and crucible deterioration and even destruction.

In order to avoid adherence, large non-wetting contact angles are needed to obtain the "composite interface" configuration shown in Figure 3a. However, as a general rule the contact angles of silicon on ceramic materials are close to or lower than 90° [35], far below the values needed in order to obtain non-adherence. The only exception is hc-BN, a ceramic for which non-wetting behaviour of Si is observed. However, for this ceramic, boron contamination leads to overdoping of silicon (hundreds of ppm) incompatible with applications in PV-silicon [36]. In the absence of a satisfactory dense material crucible, photovoltaic silicon ingots are currently grown in SiO_2 crucibles coated with a silicon nitride powder, which acts as an interface releasing agent between silicon and the crucible. The phenomena occurring at the liquid Si/porous Si_3N_4 interface were studied and modelled in [37]. The coating is processed starting from a silicon nitride submicronic powder where all Si_3N_4 particles are covered by a silica layer a few nanometres thick. Figure 16 presents the results obtained in a sessile drop experiment performed under argon flow for uncoated dense Si_3N_4 used as a reference substrate. In order to simulate the process conditions, the substrate was preoxidised in air. When a droplet of silicon is placed on the surface of such a silicon nitride substrate the contact angle varies with time between two values: an initial contact angle close to 90 degrees which is the contact angle on the silica skin (the contact angle on the surface of bulk silica is indeed close to 90°), and a final contact angle of about 45° attained after about 500 s, which is the equilibrium contact angle on a clean, deoxidised silicon nitride surface. Therefore, the wetting kinetics in this system is controlled by the removal of the oxide film occurring by the dissolution of silica in molten silicon followed by the diffusion of dissolved oxygen towards the Si free surface where oxygen is evacuated in the form of gaseous silicon monoxide SiO. The overall reaction is:

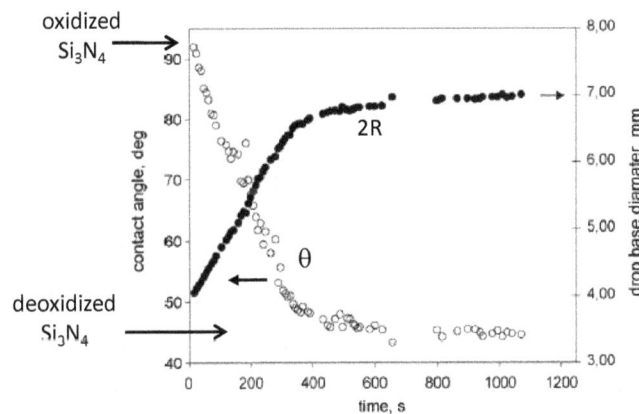

$$SiO_2(S) + Si(L) \rightarrow 2SiO(G) \qquad (9)$$

Figure 16. Wetting curves (contact angle and drop base diameter 2R *vs.* time) for liquid Si on Si_3N_4 covered by a silica layer a few nm thick. $T = 1430$ °C, in Ar [36].

As for infiltration, since the contact angle of silicon on oxidised silicon nitride is close to 90°, molten Si does not infiltrate the porous coating as long as the pore walls are covered by an oxide film (Figure 13d). The infiltration rate is in this case equal to the dissolution rate of oxide film in silicon at the infiltration front [37]. The resulting rates dh/dt are small enough to avoid the detrimental phenomenon of sticking during a melting/crystallisation cycle of Si-PV.

4.2. Wetting in Brazing: Effect of Interfacial Reactions

When a small piece of a metal such as Cu or Ni is placed on silicon carbide, a strong reaction between the metal and silicon carbide takes place [38]. The reaction consists of SiC dissolution in the metal while the excess carbon forms graphite precipitates close to the interface:

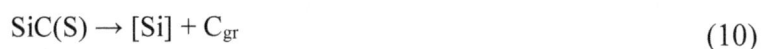

$$SiC(S) \rightarrow [Si] + C_{gr} \qquad (10)$$

where the parenthesis means silicon dissolved in the metal. While liquid metals wet clean SiC surfaces, they do not wet graphite. Therefore, this reaction inhibits wetting and additionally leads to the formation of brittle graphite particles that are detrimental to mechanical properties. To avoid these effects, metal-silicon alloys are used instead of pure metal. For a given system, the silicon concentration is chosen such that the formation of graphite precipitates is suppressed. For instance, a non-reactive Si-Pr alloy/silicon carbide interface is obtained (Figure 17) together with good wetting (the contact angle in this system is around 40 degrees) [39]. These findings are the basis of new alloys developed in Grenoble for joining SiC and other carbonaceous materials by brazing.

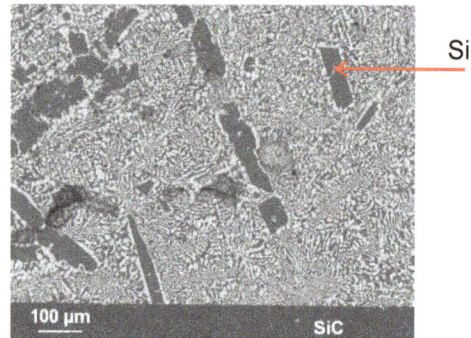

Figure 17. Cross-section of a Si-Pr alloy on sintered α-SiC obtained after 15 min at 1250 °C in high vacuum. The primary Si crystals are surrounded by the eutectic $PrSi_2$-Si. The reactivity is negligible and the interface is mechanically strong [39].

In the above example, the interfacial reaction is detrimental for wetting and brazing. In many other situations, the interfacial reaction can promote wetting and brazeability. This is the case depicted in Figure 6 for the Si/carbon system where SiC formed at the interface is better wetted than the initial graphite. Another example is wetting of alumina by CuAg alloys containing a few per cent of Ti. The effect of Ti on wetting is illustrated by the sessile drop experiment, depicted in Figure 18, in which a CuAg-1.75wt% Ti alloy was processed *in situ*, by placing a small quantity of Ti on top of a piece of AgCu on an alumina substrate [40].

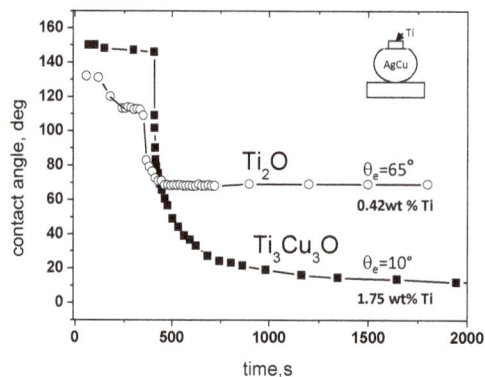

Figure 18. Contact angle as a function of time for two CuAg-Ti droplets of different composition processed *in situ* on monocrystalline alumina in He at 900 °C. According to [40].

The large, non-wetting, contact angle observed at melting, which is typical of a non-reactive noble metal on an oxide, is the contact angle of CuAg alloy on alumina. This angle remains constant for several minutes, the time needed for Ti to dissolve in the liquid and reach the metal–oxide interface by diffusion. Afterwards, the contact angle decreases rapidly towards a value close to 10°, which corresponds to excellent wetting.

The micrograph in Figure 19 presents the interface formed between the alloy and alumina at the end of the wetting experiment. The reactive interface is composed of two layers. The layer formed in contact with the liquid alloy, was identified as the Cu_3Ti_3O compound. The low final contact angle observed in this system is precisely due to the metallic character of this compound. A thin, submicronic layer is also formed in contact with alumina and identified as the Ti_2O suboxide. By decreasing the Ti concentration in the alloy from 1.75–0.42 wt%, only the Ti suboxide was formed at the interface and, as a consequence, a very significant increase in the equilibrium contact angle was observed. Even higher contact angles can be observed with further reduction in Ti concentration or, more generally, of titanium *activity* in the alloy, which is in reality the relevant thermodynamic quantity involved in the interfacial reactivity. Information obtained by the above sessile drop experiments provided useful information concerning brazing of alumina by CuAgTi alloys [41].

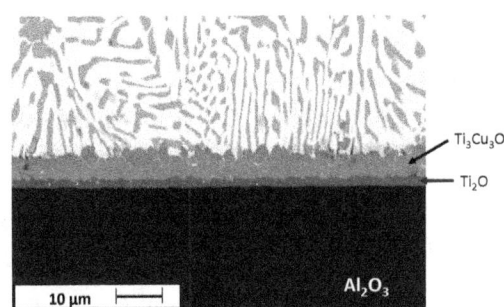

Figure 19. Cross-section of a CuAg-3 at% Ti/Al₂O₃ interface, after 30 min at 900 °C in He. According to [40].

When alumina (or another ceramic) is brazed to a metallic partner Me, the interactions between Me and the brazing alloy can affect the joint microstructure and the interfacial chemistry. Indeed, as the diffusion time in a liquid gap of 100 microns is only a few seconds, Me dissolves in the liquid braze and attains rapidly its equilibrium concentration. For instance, when one of the alumina pieces is replaced by a copper-nickel plate, the dissolution of Ni and Cu from the plate in the liquid braze leads to a dramatic change in joint microstructure and also a net change in reactivity at the alumina/braze interface (Figure 20) consisting of a decrease in the thickness of the interfacial reaction layer by one order of magnitude (from ~2 μm to ~200 nm). Moreover, instead of the wettable Cu_3Ti_3O compound formed at the interfaces in the alumina to alumina joint, high oxidation-level Ti oxides are formed in the case of CuNi/alumina joint, resulting in a very significant increase in contact angle towards 90° and ultimately in the mechanical weakening of the interface and of the actual joint [42]. From these results, it appears that when ceramics are brazed to metallic parts, allowance must be made not just for the well-known problem of thermo-mechanical compatibility between these two types of solid but also for the interactions between the metallic part and the braze that can affect wetting as well as the joint composition and microstructure.

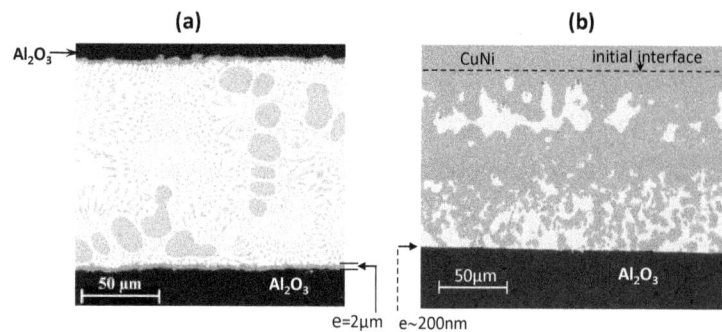

Figure 20. Cross-sections of (**a**) alumina to alumina and (**b**) alumina to CuNi plates. Assemblies brazed by a CuAg-Ti alloy (900 °C, 15 min). *e* denotes here the thickness of the interfacial reaction product. According to [42].

5. Conclusions and Perspectives

During the last 15 years, a large number of studies have been published on the fundamentals of reactive wetting. Some of them have focused on the dissolutive wetting studied using simple model systems. Although these studies have contributed significantly to improving our knowledge of dissolutive wetting, more work is needed to clarify points where diverging opinions persist, especially on the driving force of this type of wetting.

For reactive systems, where wetting is accompanied by the formation of a layer of a new compound at the interface, a first version of the Reaction Product Control (RCP) model was proposed in [43]. The model was greatly improved thanks to the analytical approach proposed in [12] for the case where spreading kinetics is limited by the local chemical process at the triple line. This approach completes the previously published, similar study by Mortensen *et al.* on diffusion controlled reactive spreading [44].

The RCP model has been applied with success in the analysis of experimental data obtained with various ceramic substrates, namely the different types of carbon [15,16], oxides [40], nitrides [45] or borides [19]. A main assumption in the model is that the new compound growth process takes place inside a zone of submicronic size localised around the solid–liquid–vapour junction, where the reactive element contained in the liquid has direct access to the solid substrate. This seems to be a reasonable assumption when experiments are performed in neutral gas or in vacuum but at low or moderate temperatures. However, experiments performed for the Si/graphite couple in high vacuum at 1430 °C, showed different wetting kinetics than in neutral gas [33]. In high vacuum, the transport of a reactive solute through the vapour can modify the surface chemistry of the substrate in front of the triple line and thus enhance spreading. A similar effect is expected to occur under an inert gas but at higher temperatures. A general description of reactive wetting taking into account both the localised and delocalised reaction is still lacking.

In this review, several examples have been given showing how simple wetting experiments can provide very useful information on the basic mechanisms involved in materials processing by infiltration or in joining similar or dissimilar materials by brazing alloys.

As for the experimental methods used in wettability studies, great improvements are expected to be made in the near future, including the development of new devices where sessile drop experiments can be coupled in the same chamber with high temperature surface analysis techniques (by Auger or XPS spectroscopy). The execution of surface analysis at high temperatures is a difficult but very exciting objective.

Conflicts of Interest

The author declares no conflict of interest.

References

1. Eustathopoulos, N.; Sobczak, N.; Passerone, A.; Nogi, K. Measurement of contact angle and work of adhesion at high temperature. *J. Mater. Sci.* **2005**, *40*, 2271–2280.
2. Eustathopoulos, N.; Drevet, B.; Nicholas, M.G. *Wettability at High Temperatures*; Pergamon: Oxford, UK, 1999; Volume 3.
3. Kumar, G.; Prabhu, K.N. Review of non-reactive and reactive wetting of liquids on surfaces. *Adv. Colloid Interface Sci.* **2007**, *133*, 61–89.
4. Eustathopoulos, N.; Drevet, B.; Muolo, M.L. The oxygen-wetting transition in metal/oxide systems. *Mater. Sci. Eng. A* **2001**, *300*, 34–40.
5. Battezzati, L.; Greer, L.A. The viscosity of liquid metals and alloys. *Acta Metall.* **1989**, *37*, 1791–1802.
6. Saiz, E.; Tomsia, A.P. Atomic dynamics and Marangoni films during liquid-metal spreading. *Nat. Mater.* **2004**, *3*, 903–909.
7. Saiz, E.; Tomsia, A.P.; Rauch, N.; Scheu, C.; Ruehle, M.; Benhassine, M.; Seveno, D.; de Coninck, J.; Lopez-Esteban, S. Nonreactive spreading at high temperature: Molten metals and oxides on molybdenum. *Phys. Rev. E* **2007**, *76*, 041602–041615.
8. Protsenko, P.; Garandet, J.-P.; Voytovych, R.; Eustathopoulos, N. Thermodynamics and kinetics of dissolutive wetting of Si by liquid Cu. *Acta Mater.* **2010**, *28*, 6565–6574.
9. Kozlova, O.; Voytovych, R.; Protsenko, P.; Eustathopoulos, N. Non-reactive versus dissolutive wetting of Ag-Cu alloys on Cu substrates. *J. Mater. Sci.* **2010**, *45*, 2099–2105.
10. Rado, C. Contribution à l'étude du mouillage et de l'adhésion thermodynamique des métaux et alliages sur le carbure de silicium. Ph.D. Thesis, Grenoble-INP, Grenoble, France, 1997.
11. Frage, N.; Froumin, N.; Dariel, M.P. Wetting of TiC by non-reactive liquid metals. *Acta Mater.* **2002**, *50*, 237–245.
12. Dezellus, O.; Hodaj, F.; Eustathopoulos, N. Progress in modeling of chemical-reaction limited wetting. *J. Eur. Ceram. Soc.* **2003**, *23*, 2797–2803.
13. Saiz, E.; Tomsia, A.P.; Cannon, R.M. Ridging effects on wetting and spreading of liquids on solids. *Acta Mater.* **1998**, *46*, 2349–2361.
14. Eustathopoulos, N. Progress in understanding and modelling reactive wetting of metals on ceramics. *Curr. Opin. Solid State Mater. Sci.* **2005**, *9*, 152–160.
15. Bougiouri, V.; Voytovych, R.; Dezellus, O.; Eustathopoulos, N. Wetting and reactivity in Ni-Si/C system: Experiments versus model predictions. *J. Mater. Sci.* **2007**, *42*, 2016–2023.
16. Dezellus, O.; Jacques, S.; Hodaj, F.; Eustathopoulos, N. Wetting and infiltration of carbon by liquid silicon. *J. Mater. Sci.* **2005**, *40*, 2307–2311.

17. Yin, L.; Murray, B.T.; Singler, T.J. Dissolutive wetting in the Bi-Sn system. *Acta Mater.* **2006**, *54*, 3561–3574.

18. Lai, Q.Q.; Zhang, L.; Eustathopoulos, N. Enhanced wetting of dual-phase metallic solids by liquid metals: A new effect of interfacial reaction. *Acta Mater.* **2013**, *61*, 4127–4134.

19. Voytovych, R.; Koltsov, A.; Hodaj, F.; Eustathopoulos, N. Reactive *vs.* non-reactive wetting of ZrB$_2$ by azeotropic Au-Ni. *Acta Mater.* **2007**, *55*, 6316–6321.

20. Bougiouri, V.; Voytovych, R.; Rojo-Calderon, N.; Narciso, J.; Eustathopoulos, N. The role of the chemical reaction in the infiltration of porous carbon by NiSi alloys. *Scr. Mater.* **2006**, *54*, 1875–1878.

21. Evans, A.; SanMarchi, C.; Mortensen, A. *Metal Matrix Composites in Industry: An Introduction and a Survey*; Kluwer Academic Publishers: Dordrecht, NL, USA, 2003.

22. Mortensen, A. Melt infiltration of metal matrix composites. In *Comprenhensive Composite Materials*; Kelly, A., Zweben, C., Eds.; Pergamon: Oxford, UK, 2000; Volume 3, pp. 521–554.

23. Michaud, V.J.; Compton, L.M.; Mortensen, A. Capillarity in isothermal infiltration of alumina fiber preforms with aluminum. *Metall. Mater. Trans. A* **1994**, *25A*, 2145–2152.

24. Alonso, A.; Pamies, A.; Narciso, J.; Garcia-Cordovilla, C.; Louis, E. Evaluation of the wettability of liquid aluminum with ceramic particulates (SiC, TiC, Al$_2$O$_3$) by means of pressure infiltration. *Metall. Trans. A* **1993**, *24A*, 1423–1432.

25. Washburn, E.W. The dynamics of capillary flow. *Phys. Rev.* **1921**, *17*, 273–283.

26. Trumble, K.P. Spontaneous infiltration of non-cylindrical porosity: Close-packed spheres. *Acta Mater.* **1998**, *46*, 2363–2367.

27. Kaptay, G.; Barczy, T. On the asymmetrical dependence of the threshold pressure of infiltration on the wettability of the porous solid by the infiltrating liquid. *J. Mater. Sci.* **2005**, *40*, 2531–2535.

28. Ness, J.N.; Page, T.F. Microstructural evolution in reaction-bonded silicon carbide. *J. Mater. Sci.* **1986**, *21*, 1377–1397.

29. Wang, Y.; Tan, S.; Jiang, D. The effect of porous carbon preform and the infiltration process on the properties of reaction-formed SiC. *Carbon* **2004**, *42*, 1833–1839.

30. Einset, E.O. Capillary infiltration rates into porous media with applications to Silcomp processing. *J. Am. Ceram. Soc.* **1996**, *79*, 333–338.

31. Sangsuwan, P.; Tewari, S.N.; Gatika, J.E.; Singh, M.; Dickerson, R. Reactive infiltration of silicon melt through microporous amorphous carbon performs. *Metall. Mater. Trans. B* **1999**, *30B*, 933–944.

32. Kumar, S.; Kumar, A.; Devi, R.; Shukla, A.; Gupta, A.K. Capillary infiltration studies of liquids into 3D-stitched C-C preforms Part B: Kinetics of silicon infiltration. *J. Eur. Ceram. Soc.* **2009**, *29*, 2651–2657.

33. Israel, R.; Voytovych, R.; Protsenko, P.; Drevet, B.; Camel, D.; Eustathopoulos, N. Capillary interactions between molten silicon and porous graphite. *J. Mater. Sci.* **2010**, *45*, 2210–2217.

34. Israel, R. Etude des interactions entre silicium liquide et graphite pour l'application à l'elaboration du silicium photovoltaique. Ph.D. Thesis, Grenoble-INP, Grenoble, France, 2009.

35. Drevet, B.; Eustathopoulos, N. Wetting of ceramics by molten silicon and silicon alloys: A review. *J. Mater. Sci.* **2012**, *47*, 8247–8260.

36. Drevet, B.; Voytovych, R.; Israel, R.; Eustathopoulos, N. Wetting and adhesion of Si on Si₃N₄ and BN substrates. *J. Eur. Ceram. Soc.* **2009**, *29*, 2363–2367.

37. Huguet, C.; Deschamps, C.; Voytovych, R.; Drevet, B.; Camel, D.; Eustathopoulos, N. Initial stages of silicon–crucible interactions in crystallisation of solar grade silicon: Kinetics of coating infiltration. *Acta Mater.* **2014**, *76*, 151–167.

38. Rado, C.; Drevet, B.; Eustathopoulos, N. The role of compound formation in reactive wetting: The Cu/SiC system. *Acta Mater.* **2000**, *48*, 4483–4491.

39. Koltsov, A.; Hodaj, F.; Eustathopoulos, N. Brazing of AlN to SiC by a Pr silicide: Physicochemical aspects. *Mater. Sci. Eng. A* **2008**, *495*, 259–264.

40. Voytovych, R.; Robaut, F.; Eustathopoulos, N. The relation between wetting and interfacial chemistry in the CuAgTi/alumina system. *Acta Mater.* **2006**, *54*, 2205–2214.

41. Kozlova, O.; Braccini, M.; Voytovych, R.; Eustathopoulos, N.; Martinetti, P.; Devismes, M.-F. Brazing copper to alumina using reactive CuAgTi alloys. *Acta Mater.* **2010**, *58*, 1252–1260.

42. Valette, C.; Devismes, M.-F.; Voytovych, R.; Eustathopoulos, N. Interfacial reactions in alumina/CuAgTi braze/CuNi system. *Scr. Mater.* **2005**, *52*, 1–6.

43. Landry, K.; Eustathopolos, N. Dynamics of wetting in reactive metal/ceramic systems: Linear spreading. *Acta Mater.* **1996**, *44*, 3923–3932.

44. Mortensen, A.; Drevet, B.; Eustathopoulos, N. Kinetics of diffusion-limited spreading of sessile drops in reactive wetting. *Scr. Mater.* **1997**, *36*, 645–651.

45. Koltsov, A.; Hodaj, F.; Eustathopoulos, N.; Dezellus, A.; Plaindoux, P. Wetting and interfacial reactivity in Ag-Zr/sintered AlN system. *Scr. Mater.* **2003**, *48*, 351–357.

4

Experimental Investigations on the Influence of Adhesive Oxides on the Metal-Ceramic Bond

Susanne Enghardt [1], Gert Richter [1], Edgar Richter [2], Bernd Reitemeier [1] and Michael H. Walter [1,*]

[1] Department of Prosthetic Dentistry, Faculty of Medicine, Technische Universität Dresden, Fetscherstr, 74 D-01307 Dresden, Germany; E-Mails: s.enghardt@t-online.de (S.E.); gert.richter@uniklinikum-dresden.de (G.R.); bernd.reitemeier@uniklinikum-dresden.de (B.R.)

[2] Forschungszentrum Dresden-Rossendorf e.V., Institute of Ion Beam Physics and Materials Research, Bautzner Landstr, 400 D-01328 Dresden, Germany; E-Mail: edgar.richterdd@gmx.net

* Author to whom correspondence should be addressed; E-Mail: michael.walter@uniklinikum-dresden.de

Academic Editor: Marta Ziemnicka-Sylwester

Abstract: The objective of this study was to test the influence of selected base metals, which act as oxide formers, on the metal-ceramic bond of dental veneer systems. Using ion implantation techniques, ions of Al, In and Cu were introduced into near-surface layers of a noble metal alloy containing no base metals. A noble metal alloy with base metals added for oxide formation was used as a reference. Both alloys were coated with a low-temperature fusing dental ceramic. Specimens without ion implantation or with Al_2O_3 air abrasion were used as controls. The test procedures comprised the Schwickerath shear bond strength test (ISO 9693-1), profile height (surface roughness) measurements (ISO 4287; ISO 4288; ISO 25178), scanning electron microscopy (SEM) imaging, auger electron spectroscopy (AES) and energy dispersive X-ray analysis (EDX). Ion implantation resulted in no increase in bond strength. The highest shear bond strengths were achieved after oxidation in air and air abrasion with Al_2O_3 (41.5 MPa and 47.8 MPa respectively). There was a positive correlation between shear bond strength and profile height. After air abrasion, a pronounced structuring of the surface occurred compared to ion implantation. The established concentration shifts in alloy and ceramic could be reproduced. However, their positive effects on shear bond strength were not confirmed. The mechanical bond appears to be of greater importance for metal-ceramic bonding.

Keywords: alloy; ion implantation; shear bond strength; metal-ceramic bond; mechanical bond; chemical bond

---

1. Introduction

Dental ceramics are bioinert. Their biocompatibility in the oral environment is widely accepted. The metal-ceramic bond is a typical material bond used in practice for more than 40 years for dental crowns and bridges. There are numerous publications on *in vitro* and *in vivo* studies [1–5]. Metal-ceramic restorations combine the positive properties of ceramics with the high mechanical stability of the metallic framework. The advantages of cured ceramic masses are above all their good durability in the mouth with respect to food, their satisfactory aesthetics in the visible parts of dental prostheses, low thermal conductivity, neutrality toward the gingiva and mucous membranes and resistance to mechanical and other stresses. It was previously believed that the chemical bond was of primary importance in adhesion. In this respect, base metals, such as In, Sn and Fe, were particularly important for the formation of adhesive oxides. The clinical applicability of these alloys for dental prostheses is well documented [1,2]. Mechanical and adhesive bonding factors are considered to play only a secondary role in the metal-ceramic bond [6]. Base metals appear to be important for good bonding stability. However, they can also be cytotoxic and, in sensitive persons, can act as allergens [7–10]. In recent publications, there are evidences that the surface roughness of the metal substrate is important in the metal-ceramic bond. There are also reports that during the firing of the ceramic, enough oxides are formed on the surface of the alloy, so that those formed during oxidation in air should be abraded or stripped before applying the ceramic [11].

Various treatments/conditions have been tested to evaluate the adhesion of ceramics to metal with focus on the effect of base metals. To avoid negatively influencing the volumetric properties, such as the thermal expansion coefficient, the ion implantation method was chosen [12]. Mihoc and coworkers [13] reported positive results for the combination of Ti and Si ions.

Therefore, the goals of the present study were to evaluate statistically the bond strength of a noble metal alloy without base metals and with base metal ions introduced by ion implantation in order to explain the role of base metal implantation. Moreover, the effect of various surface treatments on the composition of the alloy surface was characterized in this study.

2. Materials and Methods

2.1. Materials

Two commercial dental alloys were used. One was a noble metal alloy without base metals (Primallor 3), the other one noble metal alloy with base metals added to form adhesive oxides (Degutan), which served as the reference alloy. Both alloys were coated with the same low-temperature fusing dental ceramic type, Symbio ceram (Ducera, Rosbach, Germany, Table 1). This is a hydrothermal ceramic designed for low firing temperatures [14].

Table 1. Materials used.

Material	Trade Name	Manufacturer	Composition as Percent Mass
Alloy	Primallor 3	DeguDent Hanau, Germany	Au: 70.00; Pd 15.00; Pt: 7.50; Ag: 7.68; Rh: 3.22; Ir: 0.43
	Degutan	DeguDent Hanau, Germany	Au: 80.20; Pd: 13.50; Pt: 4.00; Ir: 0.20; Sn: 2.10
Dental ceramic	Symbio ceram	Ducera Rosbach, Germany	SiO_2: 60–70; Al_2O_3: 10–15; K_2O: 5–10; Na_2O: 10–15; CaO: 0–0.2; SnO_2: 0–0.2; F: 0–0.2; B_2O_3: 0–1; CeO_2: 0–1;

Table 2. Summary of the various specimen treatments. The abbreviation "St" in Series 1 and 9 stands for standard treatment of the alloy surface, *i.e.*, in accordance with the manufacturer's specifications, before application of the ceramic. at% = atomic percent.

Series No.	Series	Alloy	Treatment
1	P St	Primallor 3	—Air abrasion (110 μm Al_2O_3, 2 bar) —Oxidation in air (980 °C, 10 min) —Air abrasion (110 μm Al_2O_3, 2 bar)
2	P	Primallor 3	—Preparation with 1200 grit abrasive paper
3	P Al I	Primallor 3	—Preparation with 1200 grit abrasive paper —Implantation of Al ions (~ 5 at%)
4	P Al II	Primallor 3	—Preparation with 1200 grit abrasive paper —Implantation of Al ions (~ 15 at%)
5	P Cu I	Primallor 3	—Preparation with 1200 grit abrasive paper —Implantation of Cu ions (~ 5 at%)
6	P Cu II	Primallor 3	—Preparation with 1200 grit abrasive paper —Implantation of Cu ions (~ 5 at%) —Oxidation in air (980 °C, 10 min)
7	P In	Primallor 3	—Preparation with 1200 grit abrasive paper —Implantation of In ions (~ 5 at%)
8	Dg	Degutan	—Preparation with 1200 grit abrasive paper
9	Dg St	Degutan	—Air abrasion (110 μm Al_2O_3, 2 bar) —Oxidation in air (980 °C, 10 min) —Air abrasion (110 μm Al_2O_3, 2 bar)

The models for the test specimens were made from the synthetic material, Erkodur (Erkodent, Pfalzgrafenweiler, Germany). This material leaves no traces after firing. The dimensions of synthetic material dies met the specifications of ISO 9693 and ISO 9693-1 [15,16]. Embedding was done with the embedding mass, Deguvest CF (DeguDent, Hanau, Germany). Pre-heating (burnout) of the muffles was done in a 5636 muffle furnace (KaVo-EWL, Leutkirch, Germany). Melting of the alloys was done with an open flame (single-orifice propane oxygen torch). The noble alloy specimens were prepared with a centrifugal casting machine (Multicast Compact; DeguDent, Hanau, Germany). After deflasking, the sprues were cut off and the specimens were cleaned with a spray of silica (50 μm particle size/diameter) at a pressure of 2 bar. Afterward, the specimens were subjected to various treatments (Table 2). Each group encompassed 8 specimens, which were prepared following the procedure recommended by the manufacturer:

- Air abrasion with Al_2O_3 (particle size: 110 µm) at 2 bar pressure;
- Oxidation in air at 980 °C for 10 min;
- Air abrasion again (as above).

All specimens were subjected to wet abrasion in a Roto-Pol-22 (Struers, Rødovre, Denmark) using 1200 grit abrasive paper, analogous to the conditioning for corrosion testing according to ISO 1562 [17]. The elements Al, In and Cu were implanted into the specimens using a Danfys 1090 implanter (Danfysik, Jyllinge, Denmark) at the Institute for Ion Beam Physics and Materials Research at the Rossendorf Research Center. The implantation conditions are summarized in Table 3. Ceramic coating was performed using a ceramic furnace of model type Austromat 3001 (Dekema, Freilassing, Germany).

Table 3. Implantation conditions for the elements Al, Cu and In.

Series	Implanted Ion	Ion Energy (keV)	Application Rate (ions/cm^{-2})
P Al I	Al	200	4×10^{16}
P Al II	Al	200	2×10^{17}
P Cu I	Cu	200	4×10^{16}
P Cu II	Cu	200	4×10^{16}
P In	In	200	4×10^{16}

2.2. Testing

The following test procedures were used:

- Mechanical testing of the metal-ceramic bond was conducted using the Schwickerath method. The specimen preparation for this was in accordance with ISO 9693 and ISO 9693-1 [15,16]. Three-point bending tests to determine the shear bond strength were performed with a universal strength testing machine, TIRA test 2720 (Industriegerätewerk, Rauenstein, Germany). The testing speed was 1.5 mm/min. The evaluation of the results for the shear bond strength was done in accordance with ISO 9693 and ISO 9693-1 (Young's modulus of Primallor 3: 97 kN/mm²; Degutan: 83 kN/mm²) [15,16].
- Surface roughness measurement of the test specimens was performed with a profilometer of the type Hommel-Tester T6000 (Hommelwerke, Schwennigen, Germany) in roughness mode (TKE 100/17 probe). The direction of measurement was at a right angle to the direction of abrasion. The error of measurement was determined prior to each series with a surface roughness reference calibration standard (No. 230747/3511, Hommelwerke, Schwennigen, Germany). The mean error was 3.5%, which is within the allowed tolerances. Measurements were in accordance with ISO 4287, 4288 and 25178 [18–20].
- Raster electron microscopy (REM) and SEM (XL 30 ESEM, Phillips, Eindhoven, The Netherlands) were used in order to analyze the surface microstructure combined with the chemical composition of specimens. One specimen surface was evaluated for each treatment

condition. The effects of ion implantations were studied by comparing implanted and non-implanted surfaces.

- The element distribution was analyzed using AES, applying low-energy electrons [21]. The depth profile of the implanted elements was produced with a microlab 310 F (Fissons Instruments, Uckfield, UK). The depth profiles of individual elements were plotted directly after implantation, after implantation and oxidation in air, as well as after application and firing of the dental ceramic. Using Profile-Codes™, a computer program (Implant Sciences Inc., Wilmington, WA, USA), the concentration profiles for each of the ions to implant was calculated beforehand, in order to derive the target implantation depth in relation to the initial implantation conditions mentioned.

- EDX was used to analyze the composition of the alloy surfaces. One specimen coated with ceramic was studied for each series. The specimen was embedded in a synthetic resin of the type Speci Fix 20 (Struers, Rødovre, Denmark). After hardening, the specimen resin block was sectioned using an Accutom 50 (Struers, Rødovre, Denmark) and prepared for EDX analysis. An Edwards Sputter Coater S 158 B (Edwards High Vacuum International, Crawby, West Sussex, UK) was used for carbon sputtering. The EDX analyses were performed with the aforementioned scanning electron microscope. In both the alloy and the ceramics, chemical composition was analyzed in 6 points situated on a line with various distance from the alloy-ceramic interface, 0.5, 1.0, 1.5, 2.0, 5.0 and 10 (mm) micrometers, respectively. An elemental analysis was derived from each measurement.

2.3. Statistics

For statistical testing of the results for shear bond strength and roughness, the Mann–Whitney U-test ($\alpha = 0.05$) with an adjustment according to Bonferroni–Holm, as well as a correlation analysis (Spearman's rank-order correlation, level of significance $\alpha = 0.01$) were used. The method according to Bonferroni–Holm is an adapted fixation of the significance level [22,23].

3. Results

3.1. Shear Bond Strength

Tables 4 and 5 summarize the results of all test series and their statistical relationships. Ion implantation of the base metals used resulted in no increase in bond strength. The highest values were achieved for both Degutan and Primallor 3 with the "standard" treatment conditions. Both of these series also had the best repeatability, expressed by the lowest standard deviations.

Table 4. Shear bond strength and maximum roughness profile height (Rz). All test series $n = 8$.

Series		Primallor 3							Degutan	
		P St	P	P Al I	P Al II	P Cu I	P Cu II	P In	Dg St	Dg
Shear Bond	Mean	41.5	29.0	31.4	19.2	30.3	29.1	29.0	47.8	36.6
Strength (MPa)	Standard Deviation	2.6	5.0	4.6	3.2	4.6	3.2	5.2	1.7	3.8
Maximum Roughness	Mean	9.5	1.3	1.4	1.4	1.6	2.6	0.5	12.7	1.2
Profile Height (Rz) (μm)	Standard Deviation	0.9	0.2	0.1	0.1	0.3	1.1	0.1	0.6	0.2

Table 5. Statistical tests of the shear bond strengths of all implanted and non-implanted test series; p-values (s = significant, ns = not significant).

Series	P Al I	P Al II	P Cu I	P Cu II	P In
P St	s. (0.002)	s. (0.001)	s. (0.001)	s. (0.001)	s. (0.001)
P	n.s. (0.172)	s. (0.002)	n.s. (0.600)	n.s. (1.000)	n.s. (0.834)
Dg St	s. (0.001)	s. (0.001)	s. (0.001)	s. (0.001)	s. (0.001)
Dg	n.s. (0.003)	s. (0.001)	n.s. (0.003)	n.s. (0.004)	n.s. (0.004)

3.2. Roughness

The results of the profile depth measurements included the parameter currently found to be most significant, R_z (maximum roughness profile height). Tables 4 and 6 summarize the data for the profile depth of the alloy surfaces of all test series and their statistical relationships. As expected, ion implantation results in minimal profile depth. With respect to this finding, the test series with standard conditioning differ in a statistically significant manner.

Table 6. Statistical tests of the R_z values of all implanted and non-implanted test series (s = significant, ns = not significant); p-values.

Series	P Al I	P Al II	P Cu I	P Cu II	P In
P St	s. (0.002)	s. (0.002)	s. (0.002)	s. (0.002)	s. (0.001)
P	n.s. (0.010)	n.s. (0.010)	n.s. (0.008)	s. (0.004)	s. (0.001)
Dg St	s. (0.001)	s. (0.002)	s. (0.002)	s. (0.002)	s. (0.001)
Dg	n.s. (0.007)	n.s. (0.005)	n.s. (0.012)	s. (0.0019)	s. (0.001)

3.3. Correlation Analysis Shear Bond Strength: Roughness

A correlation analysis was performed to test the relationship between roughness of the alloy surface and the shear bond strength values found. The Spearman's rank-order correlation coefficient was +0.5614 (the level of significance was $\alpha = 0.01$). This positive result means that shear bond strength increases with the roughness of the alloy surface (Figure 1).

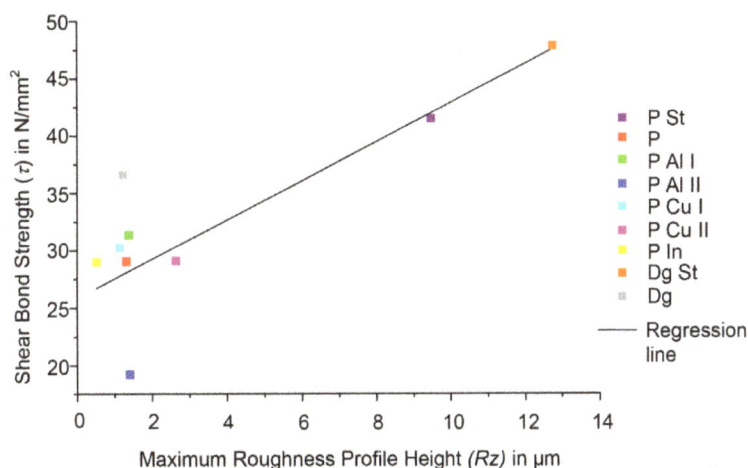

Figure 1. Correlation plot of the maximum roughness profile height *vs.* shear bond strength.

3.4. Concentration Analyses

In the concentration analyses with AES and EDX, the concentration shifts in alloy and dental ceramic could be observed. Figures 2 and 3 were included as examples. The vertical line, which represents the alloy ceramic interface, was defined through the clear differences in brightness between the two materials.

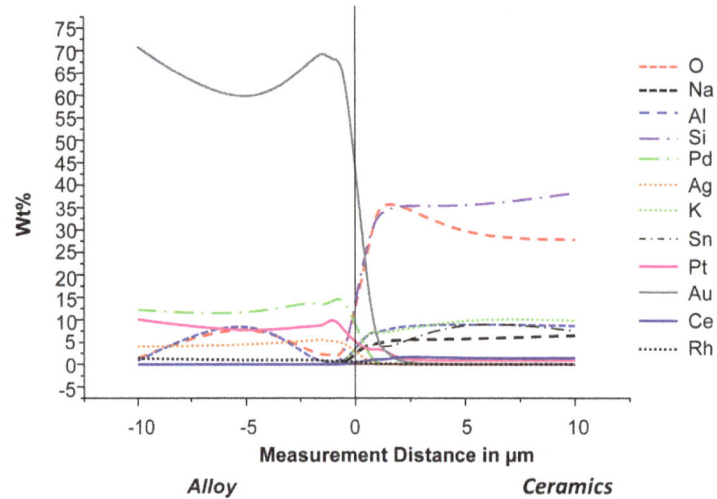

Figure 2. Concentration profiles of all elements near the metal-ceramic interface for the series P St, measured using EDX.

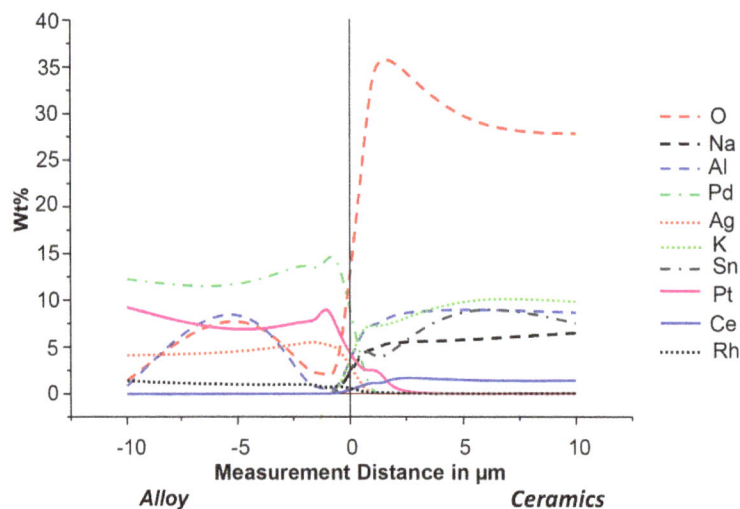

Figure 3. Concentration profiles near the metal-ceramic interface for the series P St, measured using EDX. For better representation of the chemical elements present at a lower concentration, the high concentration elements in Figure 2, Au and Si, were excluded.

4. Discussion

4.1. Shear Bond Strength and Roughness

The shear bond strength test by the Schwickerath method has been used for other low-fusing dental ceramics in previous studies [24]. In all series investigated in our study, except for one having specimens implanted with a high dose of Al ions, the values of shear bond strength are above 25 MPa, which is the minimum required in ISO standard [15,16]. The values measured for the reference alloy, Degutan, are similar to other metal-ceramic combinations with low fusing point dental ceramics. In addition to Sn and Fe, In was selected as an alloy component, because it ensures chemical bonding of the ceramic as a result of adhesive oxide formation in the firing process [25].

Shear bond strengths for the implanted series were found to be higher in this study than those described in a previous investigation [26]. In different metal-ceramic systems, Derfert found the highest shear bond strengths for noble metal alloys and low-temperature fusing dental ceramics with median values up to 51 MPa [27]. The Au content of the noble metal alloys had no significant influence in this material combination. Within the base metal alloys, higher values up to 42 MPa were found for Co-Cr alloys compared to Ni-Cr alloys [27].

The series blasted with Al_2O_3, which were characterized by the highest roughness, possessed superior shear bond strength values. This is consistent with the results from other studies [28,29]. They all came to the conclusion that the best adhesion could be achieved by oxidation and Al_2O_3 air abrasion.

The analyses resulted in a positive correlation between shear bond strength and roughness. Similar relationships were also described in the literature [30–32].

The investigations demonstrated that Al_2O_3 air abrasion of the alloy surface improves the shear bond strength for both alloys by approximately 30%. This means that air abrasion with Al_2O_3 is an important aspect in surface preparation prior to ceramic application, affecting the shear bond strength.

4.2. Chemical Bonding: Mechanical Bonding

The notion that chemical bonding, which originates from adhesive oxide formation and diffusion, contributes the major part of the metal-ceramic bond strength could not be supported. The standard surface treatment with air abrasion (110 μm Al_2O_3, 2 bar), oxidation in air (980 °C, 10 min) and repeated air abrasion (110 μm Al_2O_3, 2 bar) led to significantly higher shear bond strength values than with the ion implanted specimens. Therefore, ion implantation revealed not to be an effective method to improve adhesion and bond strength. The results presented indicate that the concentration of non-noble components in dental alloys can be minimized without deterioration of the shear bond strength. However, the expected resulting improvement of the biocompatibility of the respective alloys has to be verified experimentally and clinically. Long-term investigations and controlled clinical studies are necessary [33–35]. Additionally, it has to be noted that the incorporation of secondary base metal elements in the compositions of noble alloys for metal-ceramic applications also provides strengthening by the mechanisms of solid solution hardening and secondary phase formation (precipitates), in addition to their roles as potential oxide-formers.

Because a correlation between surface roughness and the degree of adhesion between alloy and ceramic could be determined, it is important to promote mechanical bonding by conditioning the surface with Al_2O_3 air abrasion. Based on experimental results, it can be assumed that an effective range of surface roughness resulting in high shear bond strength values was around Rz values from 9 to 13 μm. Since this was the highest value of roughness investigated, the effect of a further increase is unknown, so it is impossible to give an optimum roughness.

In reviewing published experimental results, also contradictory conclusions related to bonding mechanisms can be found. This suggests significant differences in the way the experiments were carried out, which were not described in the publications appraised [36]. It is to be supposed that these differences are secondary in nature and reflect influences, such as environmental conditions (temperature, humidity, *etc.*), temporal factors in the experiments, the manual skills of the experimenters and similar matters. Thus, it appears to be of significant importance that investigations of bonding strength are carried out according to simple, unified procedures that can be easily reproduced at any time and in each laboratory.

This study exhibits the typical limitations of an *in vitro* trial. Furthermore, no defects in the metal-ceramic interface were examined. However, these defects are frequently caused by flaws in the processing of dental restorations [37]. We tried to avoid respective problems by exclusively using one dental ceramic that was processed by only one experienced and specially-trained dental technician. Despite these limitations, our results are considered valid and may create the basis for further studies on metal-ceramic bonding of noble metal alloys without base metals for oxide formation. Additionally, prospective *in vivo* studies in the oral environment would be valuable [38].

5. Conclusions

The highest shear bond strength for a noble metal alloy without base metals was found for surfaces treated according to a standard protocol with air abrasion and oxidation in air. The implantation of ions and oxidation in air create a minimal increase in roughness. Implantation of base metal alloys leads to significantly lower bond strengths and does not positively influence and actually deteriorates bonding. The positive correlation between surface roughness and bond strength shows the necessity of mechanical retention through air abrasion with Al_2O_3.

We cannot support the hypothesis that the major part of the metal-ceramic bond originates from the chemical bond through adhesive base metal oxides. Instead, the results of the study suggest that the concentration of base metal components in dental alloys can be minimized without negatively affecting the material bond.

Acknowledgments

We would like to thank DeguDent GmbH for the company's support of the investigations. The authors thank Ursula Range (Institute for Medical Information Technology and Biometrics, Medical Faculty, Technische Universität Dresden) for the cooperation.

Author Contributions

Susanne Enghardt: substantial contribution to the production and analysis of the results, preparation of the manuscript; Gert Richter: substantial contribution to the production and analysis of the results, preparation of the manuscript; Edgar Richter: substantial contribution to the production and analysis of the results (especially ion implantation), editing the manuscript; Bernd Reitemeier: substantial contribution to the design of the study, planning of the experiments, roughness measurements, organization of the interdisciplinary study, preparation of the manuscript; Michael Walter: substantial contribution to the design of the study, planning of the experiments, preparation of the manuscript.

Conflicts of Interest

The authors declare no conflict of interest.

References

1. Coornaert, J.; Adriaens, P.; De Boever, J. Long-term clinical study of porcelain-fused-to-gold restorations. *J. Prosthet. Dent.* **1984**, *51*, 338–342.

2. Walter, M.; Reppel, P.D.; Böning, K.; Freesmeyer, W.B. Six-year follow-up of titanium and high-gold porcelain-fused-to-metal fixed partial dentures. *J. Oral. Rehabil.* **1999**, *26*, 91–96.

3. Walton, T.R. An up to 15-year longitudinal study of 515 metal-ceramics FDPs: Part 2. Modes of failure and influence of various clinical characteristics. *Int. J. Prosthodont.* **2003**, *16*, 177–182.

4. Holm, C.; Tidehag, P.; Tillberg, A.; Molin, M. Longevity and quality of FDPs: A retrospectivestudy of restorations 30, 20, and 10 years after insertion. *Int. J. Prosthodont.* **2003**, *16*, 283–289.

5. Napankangas, R.; Raustia, A. An 18-year retrospective analysis of treatment outcomes with metal-ceramic fixed partial dentures. *Int. J. Prosthodont.* **2011**, *24*, 314–319.

6. Eichner, K. Gegenwärtiger Stand der werkstoffkundlichen und klinischen Metallkeramik. *ZWR* **1997**, *97*, 477–485.

7. Bumgardner, J.D.; Lucas, L.C. Corrosion and cell culture evaluations of nickel-chromium dental casting alloys. *J. Appl. Biomater.* **1994**, *5*, 203–213.

8. Wataha, J.C.; Hanks, C.T. Biological effects of palladium and risk of using palladium in dental casting alloys. *J. Oral. Rehabil.* **1996**, *23*, 309–320.

9. Wataha, J.C. Biocompatibility of dental casting alloys: A review. *J. Prosthet. Dent.* **2000**, *83*, 223–234.

10. Al-Hiyasat, A.S.; Bashabsheh, O.M.; Darmani, H. An investigation of the cytotoxic effects of dental casting alloys. *Int. J. Prosthodont.* **2003**, *16*, 8–12.

11. Wataha, J.C. Alloys for prosthodontics restorations. *J. Prosthet. Dent.* **2002**, *87*, 351–363.

12. Möller, W.; Richter, E. Praktische Anwendungen der Ionenimplantation. *Galvanotechnik* **1989**, *89*, 3–11.

13. Mihoc, R.; Hobkirk, J.A.; Armitage, D.; Jones, F.H. Influence of ion implantation on physicochemical processes at titanium surfaces. *Int. J. Prosthodont.* **2006**, *19*, 24–28.

14. Hohmann, W. Dentalkeramik Auf Der Basis Hydrothermaler Gläser. Quintessenz: Berlin, Germany, 1993.

15. *Metal-Ceramic Dental Restorative Systems*; Beuth: Berlin, Germany, 1999; ISO 9693.

16. *Dentistry—Compatibility Testing, Metal-Ceramic Systems*; Beuth: Berlin, Germany, 2012; ISO 9693-1.

17. *Dentistry—Casting Gold Alloys*; Beuth: Berlin, Germany, 2004; ISO 1562.

18. *Geometrical Product Specifications (GPS)—Surface Texture : Profile Method—Terms, Definitions and Surface Texture Parameters*; Beuth: Berlin, Germany, 2010; ISO 4287.

19. *Geometrical Product Specifications (GPS)—Surface Texture: Profile Method—Rules and procedures for the assessment of surface texture*; Beuth: Berlin, Germany, 1998; ISO 4288.

20. *Geometrical Product Specifications (GPS)—Surface Texture: Areal—Part 2: Terms, Definitions and Surface Texture Parameters; Surface Texture: Areal—Part 3: Specification Operators; Surface Texture: Areal—Part 6: Classification of Methods For Measuring Surface Texture*; Beuth: Berlin, Germany, 2012; ISO 25178.

21. Briggs, D.; Seah, M.P. Practical Surface Analysis: Vol. 1. In *Auger and X-ray Photoelectron Spectroscopy*; Wiley: Chichester, UK, 1990.

22. Holm, S. A simple sequentially rejective multiple test procedure. *Scand. J. Stat.* **1979**, *6*, 65–70.

23. Armitage, P.; Colton, T. *Encyclopaedia of Biostatistics*, 2nd ed.; Wiley: Chichester, UK, 2005; pp. 7872–7878.

24. Gilbert, J.L.; Covey, D.A.; Lautenschläger, E.P. Bond characteristics of porcelain fused to milled titanium. *Dent. Mater.* **1994**, *10*, 134–140.

25. Hautaniemi, J.A. The effect of indium on porcelain bonding between porcelain and Au-Pd-In alloy. *J. Sci. Mat. Med.* **1995**, *66*, 46–50.

26. Petridis, H.; Garefis, P.; Hirayama, H.; Kafantaris, N.M; Koidis, P.T. Bonding indirect resin composites to metal: Part I. Comparison of shear bond strengths between different metal-resin bonding systems and a metal-ceramic system. *Int. J. Prosthodont.* **2003**, *16*, 635–639.

27. Derfert, B. Vergleichende Untersuchungen zur Verbundfestigkeit von verschiedenen Metall-Keramik-Kombinationen auf der Basis von EM- und NEM-Legierungen in Verbindung mit herkömmlicher und niedrigschmelzender Keramik. Ph.D. Thesis, Freie Universität Berlin, Berlin, Germany, 12 December 2003.

28. Mc Lean, J.W.; Sced, I.R. Bonding of Dental Porcelain to metal I. The Gold Alloy Porcelain Bond. *Trans. J. Brit. Ceram. Soc.* **1973**, *72*, 229–233.

29. Charnay, R.; Guiraldenq, P.; Enriore, J.M.; Brugirard, J.; Heberard, X. Mechanical properties of the ceramic-metal bonding of a gold-platinum dental alloy and chemical characterization of the interface. *Mem. Etud. Rev. Met.* **1992**, *12*, 797–803.

30. Mc Lean, J.W. *The Science and Art of Dental Ceramics*; Quintessence: Chicago, IL, USA, 1980; pp. 65–90.

31. Anusavice, K.J. *Phillip's Science of Dental Materials*, 12th ed.; Saunders: Philadelphia, PA, USA, 2012.

32. Schwickerath, H. Das Festigkeitsverhalten von aufbrennfähigen Keramiken. *Dtsch. Zahnärztl. Z* **1985**, *40*, 996–1003.

33. Karlsson, S. Failures and length of service in fixed Prosthodontics after Long-term function. A longitudinal clinical study. *Swed. Dent. J.* **1989**, *13*, 185–192.

34. Jokstad, A.; Esposito, M.; Coulthard, P.; Worthington, H.V. The reporting of randomised controlled trials in prosthodontics. *Int. J. Prosthodont.* **2002**, *15*, 230–242.

35. Glantz, P.O.; Nilner, K.; Jendresen, M.D.; Sundberg, H. Quality of fixed prosthodontics after twenty two years. *Acta. Odontol. Scand.* **2002**, *60*, 213–218.

36. Bullard, J.T.; Dill, R.E.; Marker, V.A.; Payne, E.V. Effects of sputtered metal oxide films on the ceramic-to-metal-bond. *J. Prosthet. Dent.* **1985**, *54*, 776–778.

37. Walter, M. Zur Porenbildung in der keramischen Verblendung von Palladium-Silber-Legierungen. *Dtsch. Zahnärztl. Z* **1988**, *43*, 145–149.

38. Reitemeier, B.; Hänsel, K.; Kastner, C.; Weber, A.; Walter, M.H. A prospective 10-year study of metal ceramic single crowns and fixed dental prosthesis retainers in private practice settings. *J. Prosthet. Dent.* **2013**, *109*, 149–155.

Effects of Different Heat Treatment on Microstructure, Mechanical and Conductive Properties of Continuous Rheo-Extruded Al-0.9Si-0.6Mg (wt%) Alloy

Di Tie [1], Ren-guo Guan [1,*], Ning Guo [2], Zhouyang Zhao [1], Ning Su [1], Jing Li [3] and Yang Zhang [1]

[1] School of Metal and Metallurgy, Northeastern University, Wenhua Rd. No.3, Shenyang 110003, China; E-Mails: tie-di@hotmail.com (D.T.); zyoungneu@sina.com (Z.Z.); salex-neu@outlook.com (N.S.); zyukineu@sina.com (Y.Z.)

[2] Postgraduate Academy, Shenyang Ligong University, Nanpingzhong Rd. No.6, Shenyang 110000, China; E-Mail: guoning-slu@hotmail.com

[3] CNPC Institute, Taiyanggongnan St. NO.23, Beijing 100010, China; E-Mail: ljane_cnpc@sina.com

* Author to whom correspondence should be addressed; E-Mail: guanrg@smm.neu.edu.cn

Academic Editor: Anders E. W. Jarfors

Abstract: Al-0.9Si-0.6Mg (wt%) alloy conductive wires were designed and produced by continuous rheo-extrusion process. The effects of different heat treatment on microstructure, mechanical and conductive properties of the wires were studied. Results show that, after T6 heat treatment, conductive property of the alloy increased while elongation decreased with the higher aging temperature and longer aging time. After T8 and T9 heat treatment, acicular strengthening phase β''-Mg_2Si homogeneously precipitated, which effectively improved mechanical and conductive property of the alloy. The tensile strength, elongation and resistivity of T8 heat treated alloy reached 336 MPa, 13.7% and 29.3 nΩm respectively. After T9 heat treatment, the alloy's tensile strength, elongation and resistivity was 338 MPa, 6.0% and 30.2 nΩ·m respectively.

Keywords: Al-0.9Si-0.6Mg (wt%) alloy; continuous rheo-extrusion; heat treatment; microstructure; mechanical properties

1. Introduction

In the past few decades, the All-Aluminum Alloy Conductor (AAAC) was more widely used in electrical engineering area. Compared with traditional All Aluminum Conductor (AAC), AAAC can afford better sag characteristics due to its high strength/weight rate [1]. From corrosion property's point of view, AAAC also exhibits higher corrosion-resistance than AAC and another widely-used conductor material Aluminum Conductors Steel Reinforced Conductors (ACSR) [2]. With promising mechanical and conductive properties, Al-Si-Mg alloys have been applied for manufacturing AAAC for transferring electrical energy over long distances since last century. Furthermore, Al-Si-Mg alloys' mechanical and conductive properties could still be improved by optimum of processing and heat treatment.

As a low-cost semisolid metal processing technique, continuous rheo-extrusion process has gained increasing research interest in recent years [3–5]. Heat treatment after rheo-extrusion process has been proved effective method to adjust aluminum alloys' mechanical property under high temperature, and therefore could be used to improve Al-Si-Mg alloy's mechanical and conductive property [6]. Since heat treatment conditions have great influences on the material's microstructure, mechanical and conductive properties, the conditions of heat treatment should be carefully investigated [7,8]. In this paper, by using a self-designed test machine, an originally designed Al-0.9Si-0.6Mg(wt%) alloy wire was achieved by continuous rheo-extrusion process, and two different solution treatments as well as three different aging treatments (T6, T8 and T9) were performed and compared. After investing the treated alloy's mechanical and conductive properties, the effects of heat treatment were concluded and the optimized technical parameters of heat treatment were gained.

2. Experimental Section

2.1. Continuous Rheo-Extrusion

A self-designed D-350 CSEP machine was used in the experiment. The principle of the machine is shown in Figure 1. Melt alloy flowed out of the tundish and entered the roll–shoe cavity, and it solidified continuously under the cooling by the water-cooled work roll and the shoe. Due to the actions of shearing and cooling by the roll, semisolid slurry consisting of non-dendritic solid phase and liquid phase was prepared. When the slurry met the block, its flow direction turned 90°, so the extending mold could be continuously filled with the slurry. The extending cavity was firstly filled by the slurry, and then the slurry was forced to fill the welding cavity through the splitflow orifices. Finally, the slurry was forced to flow out of the channel formed by the mold orifice and the wires were obtained.

Al-0.9Si-0.6Mg (wt%) alloy designed by our own which was melted by commercial pure aluminum ingots, pure magnesium ingots and Al-Si master alloy. The chemical composition of the alloy was analyzed using an inductively coupled plasma-optical emission spectrometer (ICP-OES; VARIAN, Palo Alto, CA, USA). In order to assure the alloy wires' high mechanical performance, alloying elements and impurity contents were strictly controlled during melting process. The solidus and liquidus temperatures of the alloy were *ca.* 607 °C and 654 °C, respectively. After being melted, the melt was refined by hexachloroethane for 3–5 min to remove the oxides, impurities and hydrogen.

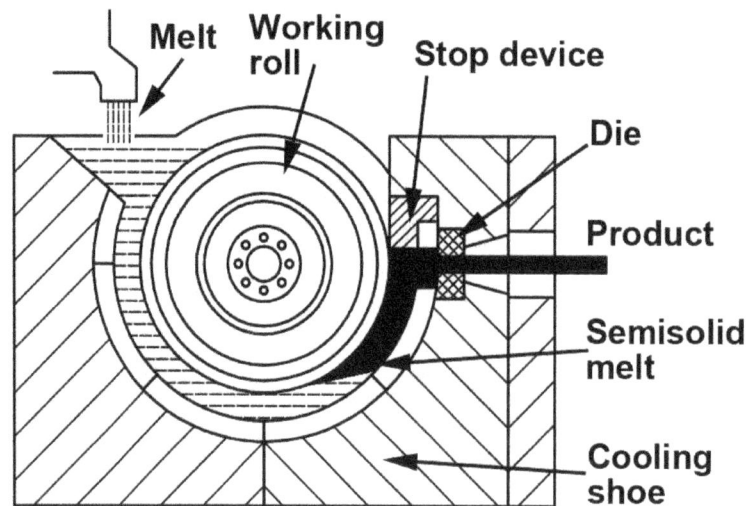

Figure 1. Schematic diagram of the continuous rheo-extrusion machine.

2.2. Heat Treatment Process

After extrusion, continuous solution treatment was applied for the alloy at 520–540 °C for 6 h. Aging treatment was then performed to obtain the optimized mechanical and conductive properties. Three kinds of aging treatment were carried out and compared: T6, T8 and T9, and the detailed technical parameters could be found in Table 1.

Table 1. Conditions of T6, T8 and T9 aging treatment.

Heat treatment	Solution temperature (°C)	Aging temperature (°C)	Aging time (h)
T6	520	150–190	2–14
T8	520	150–170	3–7
T9	540	150–170	3–7

2.3. Microstructure Analysis

Materials after different treatment were machined for next metallurgical and mechanical tests. Samples were first polished and etched, then analyzed by scanning electron microscopy (SEM; SSX-550, Shimadzu, Japan) for determining the distribution of phases. The corrosive solution used for corroding the alloys was: 15 mL HCl + 56 mL C_2H_5OH + 47 mL H_2O. Samples were processed into Φ 3 mm × 0.5 mm discs using a linear cutting machine; the discs were then ground to a thickness of 80 μm and thinned using an ion milling machine. The microstructure of the precipitate phase in the alloy was then analyzed by high-resolution transmission electron microscopy (HRTEM; G2 F20, Tecnai, Delft, The Netherland).

2.4. Mechanical and Conductive Properties

The mechanical performance of these aged alloys formed using continuous rheo-rolling was investigated using an electronic tensile testing machine according to the manufacture's protocol

(Huaxing Experimental Equipments Co. Ltd., Jinan, China). Conductive properties were determined according to standard JB /T8640-1997.

3. Results and Discussion

3.1. Optimum of Technical Parameters of Continuous Rheo-Extrusion

The experimentally measured composition of the alloy was: Si 0.93 ± 0.01 wt.%, Mg 0.58 ± 0.02 wt.% and balanced Al. Impurities mainly composed by Ni (<0.003 wt.%), Fe (<0.002 wt.%) and Cu (<0.002 wt.%). Optimized technical parameters of continuous rheo-extrusion were obtained after experiment, which including: cooling water flow rate was 10–15 L/Min; pouring temperature was 680–720 °C and rolling speed was 0.18 m/s. By using these technical parameters, the extruded wires were manufactured continuously without surface defect and break. The outlook of the wire is shown in Figure 2.

Figure 2. Outlook of Al-0.9Si-0.6Mg (wt%) alloy conductive wire produced by continuous rheo-extrusion.

3.2. Influence of T6 Aging Treatment on Microstructure and Properties

Distributions of Mg_2Si precipitate in alloy's matrix after T6 treatment by different aging time were observed by SEM and TEM microscope and present in Figure 3. At the beginning of aging treatment at 2 h under 150 °C, second phases precipitated homogeneously in spot form as shown in Figure 3a. With increasing aging time, spots form phases turned to spin form phases, and this trend could be clearly observed after 14 h aging (Figure 3d).

Materials' tensile strength, elongation ratio and electrical resistivity were summarized in Figure 4. It can be found that the wire's ultimate tensile strength (UTS) value turned bigger and then smaller with increasing aging time. This is due to the bigger size of the secondary phases brought by aging. According to Orowan's theory, growth of secondary precipitates promotes the impedance of dislocations, and therefore increases the UTS [9]. When aging time was beyond the peak time, the density of secondary phases became smaller due to the fast growth of phases, movement of dislocations happens between phases, and then made the tensile strength of the alloy decreased.

The elongation as well as the electrical resistivity always decreases with longer aging time. This trend was resulted by faster atom movement with higher aging temperature. As the size of secondary phases grew bigger, the difference deforming properties between secondary phases and matrix decrease the elongation [10]. Another effect brought by faster atom movement is of dislocations' movement was also accelerated. As a result, the density of dislocations and grain boundary decreased, so the conductive properties were accordingly improved (electrical resistivity is smaller, Figure 4) [11].

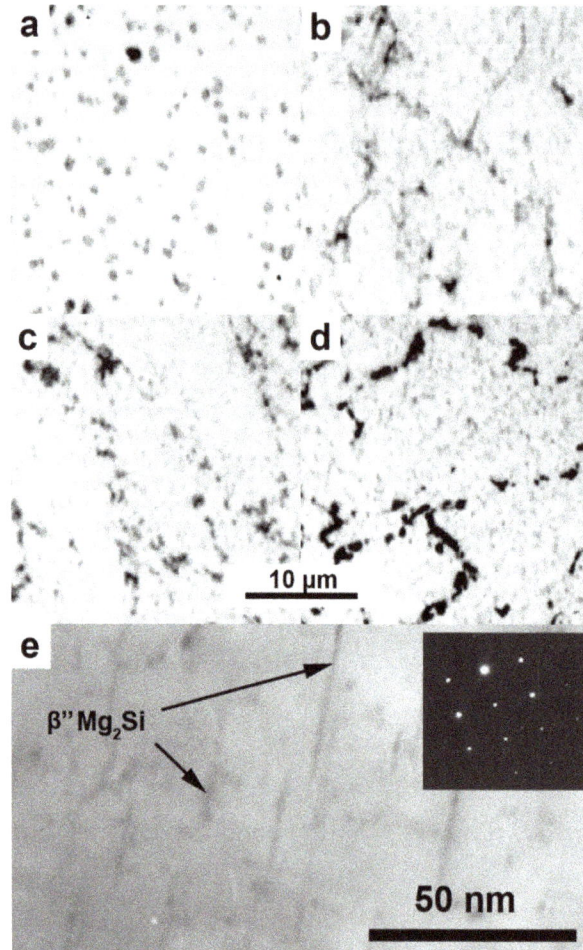

Figure 3. SEM and TEM images of precipitated phases in matrix after artificially aging treatment at 150 °C for different time: (**a**) 2 h; (**b**) 6 h; (**c**) 10 h; (**d**) 14 h; (**e**) 14 h TEM image with according diffraction pattern.

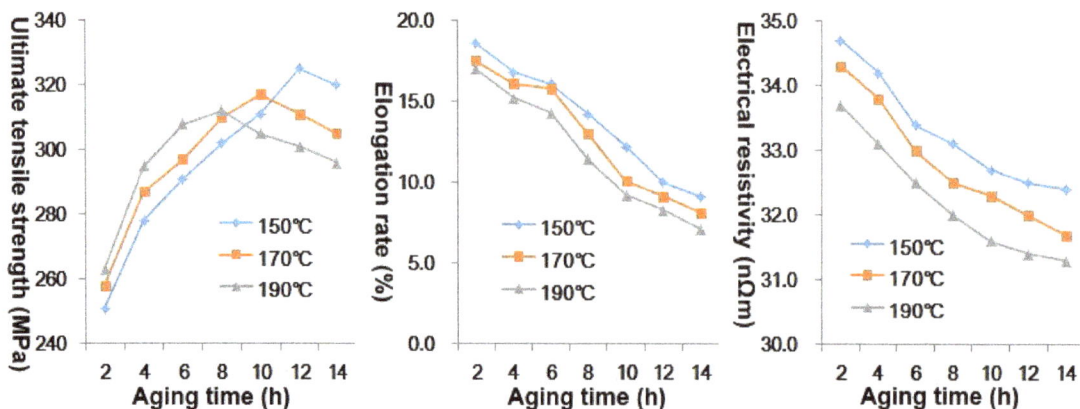

Figure 4. Relationship between tensile strength, elongation ratio and electrical resistivity of alloy conductors and T6 aging time.

3.3. Influence of T8 and T9 Aging Treatment on Microstructure and Properties

The effects of T8 treatment time on ultimate tensile strength, elongation and electrical resistivity of the alloy are summarized in Figure 5 (upper line). After complete T8 aging treatment, UTS of the alloy reached 336 MPa while elongation and electrical resistivity was 13.7% and 29.3.5 nΩm. In Figure 5 (lower line), the effects of T9 treatment time on ultimate tensile strength, elongation and electrical resistivity of the alloy are concluded. After complete T9 aging treatment, UTS of the alloy increased to 338 MPa while elongation and electrical resistivity was 6.0% and 30.2 nΩm respectively. Compare with the standard JB/T8640-1997 (UTS = 295 MPa; Electrical resistivity = 32.2 nΩm), the treated alloy's UTS increased by 13.2% and the conductive properties increased by 5.9%.

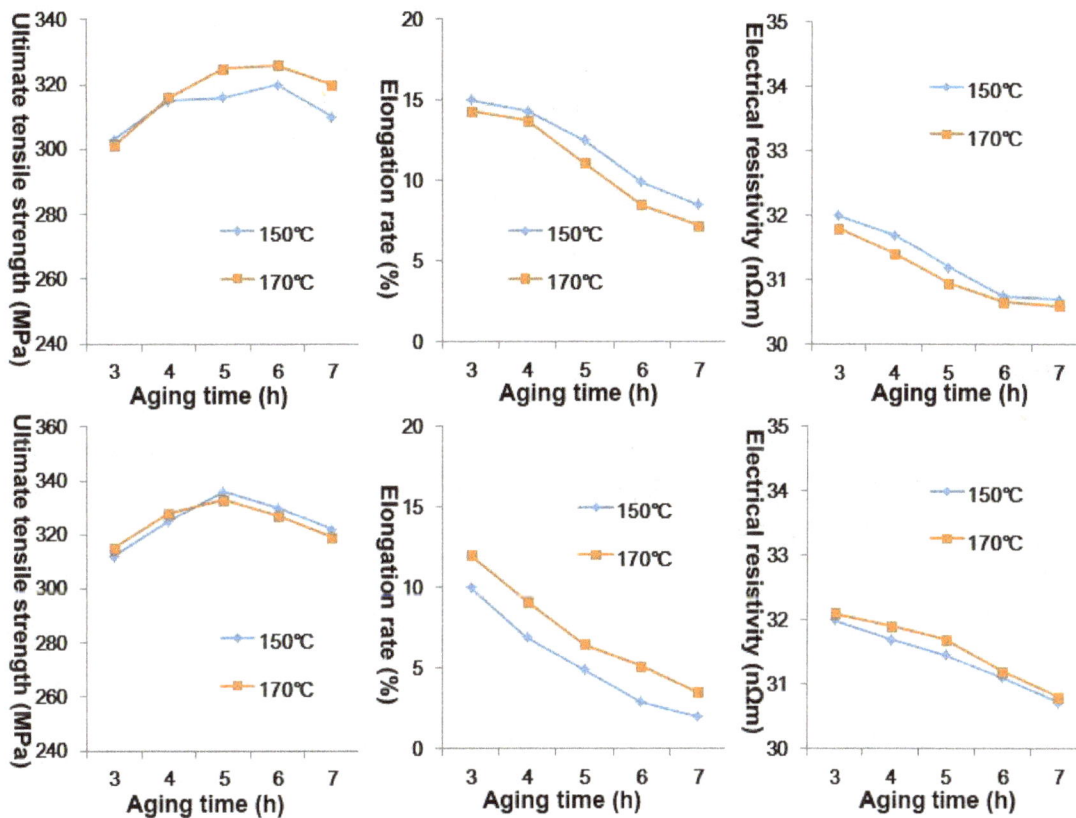

Figure 5. Relationship between tensile strength, elongation ratio and electrical resistivity of alloy conductors and T8 (upper)/T9 (lower) aging time.

The comprehensive mechanical properties and conductivity of the alloy wire can be effectively improved by the low temperature heat treatment, such as T8 process and T9 process. UTS of the T9 treated wire is the highest while elongation is the lowest. Conversely, UTS of the T8 treated wire is lower while elongation is higher. This result is related to a typical deformation structure: with the fragmentation of grain by extrusion and high density dislocation brought by T8 process or T9 process, dislocation grid was formed. Strength of the alloy wire increased due to work hardening by extrusion and precipitation strengthening by aging treatment. This process benefits from both aging strengthening and work hardening, so the tensile strength of T9 treated alloy climbed to the highest value. However, its elongation was lower due to serious lattice distortion. When aging temperature is

150 °C, less influence of strain hardening could be observe, so UTS of the T8 treated alloy is lower than that of T9 treated one. The smaller precipitated phases by lower aging temperature brought the electron scattering effect [12], leading to the decrease of electrical conductivity (Figure 4). Therefore, it is crucial to obtain balanced mechanical and conductive properties by adjusting the aging temperature. When aging temperature remained constant, aging time was the most important parameter which affected conductivity of the alloy as well as mechanical properties.

4. Conclusions

Optimized technical parameters of continuous rheo-extrusion include: cooling water flow rate of 10–15 L/Min, pouring temperature of 680–720 °C and rolling speed of 0.18 m/s. During T6 treatment to rheo-extruded alloy, conductivity increased with higher aging temperature and aging time, whilst the elongation decreased. After T8 and T9 heat treatment, the conductive and mechanical properties of the alloy both increased due to more homogeneous precipitation and distribution of secondary Mg_2Si β phase. After complete T8 aging treatment, UTS of the alloy reached 336 MPa while elongation and electrical resistivity was 13.7% and 29.3 nΩm respectively. After complete T9 aging treatment, UTS of the alloy increased to 338 MPa while elongation and electrical resistivity was 6.0% and 30.2 nΩm respectively.

Acknowledgments

The authors thank for the supports of National Natural Science Foundation of China under Grant Nos.51222405 and 51474063, and the Fundamental Research Funds for the Central Universities, Northeastern University No. N130302007.

Author Contributions

Di Tie and Renguo Guan conceived and designed the experiments; Xiang Wang, Yang Zhang and Zhouyang Zhao performed the experiments; Ning Su and Jing Li analyzed the data; Di Tie drafted the manuscript.

Conflicts of Interest

The authors declare no conflict of interest.

References

1. Yuan, W.; Liang, Z. Effect of zr addition on properties of Al–Mg–Si aluminum alloy used for all aluminum alloy conductor. *Mater. Des.* **2011**, *32*, 4195–4200.
2. Azevedo, C.R.F.; Cescon, T. Failure analysis of aluminum cable steel reinforced (ACSR) conductor of the transmission line crossing the paraná river. *Eng. Failure Anal.* **2002**, *9*, 645–664.
3. Guan, R.-G.; Wen, J.-L.; Wang, S.-C.; Liu, X.-H. Microstructure behavior and metal flow during continuously extending-extrusion forming of semisolid A2017 alloy. *Trans. Nonferrous Metals Soc. China* **2006**, *16*, 382–386.

4. Kerr, A.; Watson, L.M.; Szasz, A.; Muller, H.; Kirchmayr, H. On the electronic stability of the Al-Mg-Si age-hardened alloys. *J. Phys. Chem. Solids* **1996**, *57*, 1285–1292.

5. Flemings, M. Behavior of metal alloys in the semisolid state. *MTB* **1991**, *22*, 269–293.

6. Xu, C.; Schroeder, S.; Berbon, P.B.; Langdon, T.G. Principles of ECAP–Conform as a continuous process for achieving grain refinement: Application to an aluminum alloy. *Acta Mater.* **2010**, *58*, 1379–1386.

7. Mrówka-Nowotnik, G.; Sieniawski, J. Influence of heat treatment on the microstructure and mechanical properties of 6005 and 6082 aluminium alloys. *J. Mater. Process. Technol.* **2005**, *162–163*, 367–372.

8. Guyot, P.; Cottignies, L. Precipitation kinetics, mechanical strength and electrical conductivity of AlZnMgCu alloys. *Acta Mater.* **1996**, *44*, 4161–4167.

9. Jiang, D.; Wang, C. Influence of microstructure on deformation behavior and fracture mode of Al–Mg–Si alloys. *Mater. Sci. Eng. A* **2003**, *352*, 29–33.

10. Tie, D.; Guan, R.G.; Cui, T.; Ling, C.; Wang, X.; Guan, X.H. Optimisation of composition and cast temperature for continuous semisolid extruded Al–Sc–Zr electrical conductor. *Mater. Res. Innov.* **2014**, *18*, S4-926–S4-928.

11. Suzuki, S.; Shibutani, N.; Mimura, K.; Isshiki, M.; Waseda, Y. Improvement in strength and electrical conductivity of Cu–Ni–Si alloys by aging and cold rolling. *J. Alloys Compd.* **2006**, *417*, 116–120.

12. Zhou, W.W.; Cai, B.; Li, W.J.; Liu, Z.X.; Yang, S. Heat-resistant Al–0.2Sc–0.04Zr electrical conductor. *Mater. Sci. Eng. A* **2012**, *552*, 353–358.

6

Experimental Investigation of the Equal Channel Forward Extrusion Process

Mahmoud Ebrahimi [1,*]**, Faramarz Djavanroodi** [1,2]**, Sobhan Alah Nazari Tiji** [1]**, Hamed Gholipour** [1] **and Ceren Gode** [3]

[1] Department of Mechanical Engineering, Iran University of Science and Technology, Tehran 16846-13114, Iran; E-Mails: fdjavanroodi@pmu.edu.sa (F.D.); s.nazari86@yahoo.com (N.T.); h_golipour@yahoo.com (H.G.)

[2] Department of Mechanical Engineering, Prince Mohammad Bin Fahd University, Al Khobar 31952, Saudi Arabia

[3] School of Denizli Vocational Technology, Program of Machine, Pamukkale University, Denizli 20100, Turkey; E-Mail: cgode@pau.edu.tr

* Author to whom correspondence should be addressed; E-Mail: mebrahimi@iust.ac.ir

Academic Editor: Heinz Werner Höppel

Abstract: Among all recognized severe plastic deformation techniques, a new method, called the equal channel forward extrusion process, has been experimentally studied. It has been shown that this method has similar characteristics to other severe plastic deformation methods, and the potential of this new method was examined on the mechanical properties of commercial pure aluminum. The results indicate that approximate 121%, 56%, and 84% enhancements, at the yield strength, ultimate tensile strength, and Vickers micro-hardness measurement are, respectively, achieved after the fourth pass, in comparison with the annealed condition. The results of drop weight impact test showed that the increment of 26% at the impact force, and also decreases of 32%, 15%, and 4% at the deflection, impulse, and absorbed energy, are respectively attained for the fourth pass when compared to the annealed condition. Furthermore, the electron backscatter diffraction examination revealed that the average grain size of the final pass is about 480 nm.

Keywords: SPD; ECFE; mechanical properties; impact behavior; grain size

1. Introduction

In the last decade, production and application of ultra-fine grain (UFG) and nano-structure (NS) metals and alloys have been deeply studied by researchers and scientists in the material science field [1,2]. These materials possess improved mechanical properties at the room temperature and enhanced superplastic behavior at higher temperatures [3,4]. In general, there are two main processing categories to fabricate UFG and NS materials, called bottom-up and top-down manners. In the bottom-up method, the UFG or NS materials are synthesized, atom-by-atom, and, also, in layer-by-layer arrangement, and that these samples possess small dimensions with porous structures, which are rarely appropriate for industrial applications. In the top-down approach, micro-structure (MS) materials at the industry scale have been altered to UFG, and even NS ones, using severe plastic deformation (SPD) techniques [1,5].

Up to now, numerous SPD methods have been proposed, experimented, and investigated in detail. Based on the geometry of the work-piece, SPD techniques can be divided into the three groups nominating bulk, sheet, and tube classifications. For group 1: equal channel angular pressing (ECAP) [1], high pressure torsion (HPT) [2], twist extrusion (TE) [6], accumulative back extrusion (ABE) [7]; for group 2: equal channel angular rolling (ECAR) [8], accumulative roll bonding (ARB) [9], constrained groove pressing (CGP) [10]; and for group 3: high pressure tube twisting (HPTT) [11], accumulative spin bonding (ASB) [12], tubular channel angular pressing (TCAP) [13] are the major examples. It should be pointed out that the principal rule of all SPD methods consists of imposing shear stress to the sample, increasing the dislocation density in deformed material, formation of dense dislocation walls and then low angle grain boundaries (LAGBs), and, finally, transformation of LAGBs into high angle grain boundaries (HAGBs) [1,14–17].

Recently, a novel SPD method called the equal channel forward extrusion (ECFE) process has been proposed by authors to fabricate UFG and NS bulk materials [18]. This research focuses on the capabilities of this new method. Hence, mechanical and microstructural investigations have been carried out to observe and compare the characteristics of this new method with the other SPD processes. For this aim, commercial pure aluminum billets have been ECFEed up to four passes and then the potential of this approach has been investigated by means of a tensile test, hardness measurements, an impact test, and microstructural observations.

2. Principle of the ECFE Process

The equal channel forward extrusion process is schematically represented in Figure 1a. As can be observed, the ECFE die consists of three major parts: inlet or entry channel, main deformation zone (MDZ), and outlet or exit channel. A sample with a rectangular cross-section is placed in the entry channel and then pushed by punch to pass through the entrance channel and enter the MDZ, where the material is subjected to intense shear stress. As the material passes through the MDZ, the cross-section of the sample incrementally expands in the width direction, and reduces in the length direction, simultaneously. Figure 1b represents the alteration of both the length and width of billets' cross-section to each other at the MDZ section, where the area of the rectangular sample remains constant during the process. It needs to be emphasized that there is no sample rotation during the process.

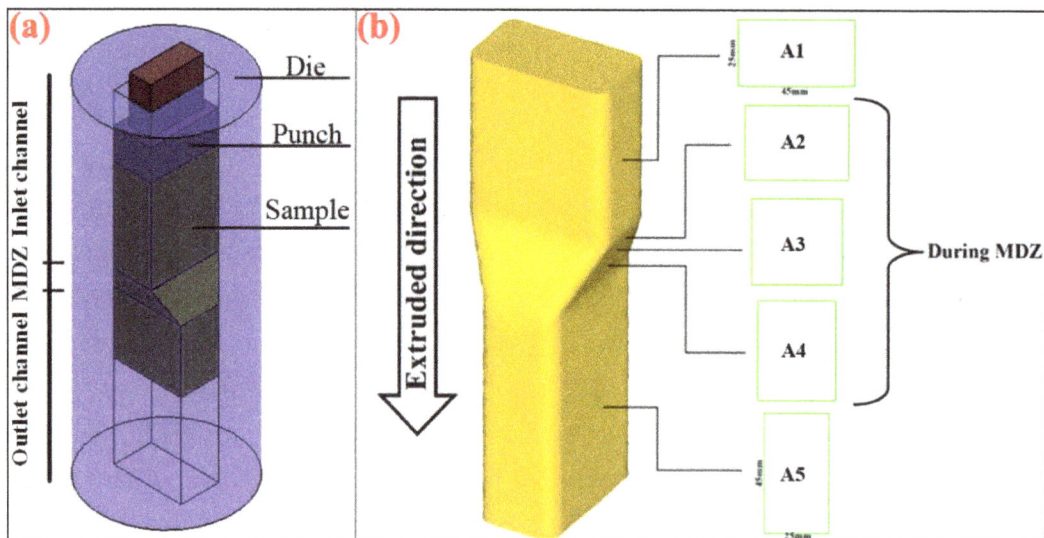

Figure 1. (a) The schematic representation of equal channel forward extrusion process and **(b)** the length to width alteration at the billet's cross-section in the MDZ section during the ECFE process where A1 = A2 = A3 = A4 = A5.

It is noted that the length to width ratio of the sample, and, also, the magnitude of MDZ's height are the two major parameters that influence the strain behavior, mechanical properties, and microstructural characteristics of the deformed materials.

3. Experimental Procedure

3.1. Materials

Commercial pure (CP) aluminum (Al1070) billets with the chemical composition of (in wt.%) 0.185% Fe, 0.09% Si, 0.012% Mg, 0.011% Zn, 0.008% Ti, 0.008% V, 0.004% Ni, 0.004% Cu, 0.003% Mn, and Al as the balance, were prepared with the dimensions of 25 mm × 45 mm × 140 mm. Before extrusion, all the samples were annealed at 380 °C for 1.5 h and then cooled slowly in a furnace to room temperature [19,20]. This leads to a homogeneous and uniform structure with good ductility in all specimens before operation.

3.2. ECFE Die

The equal channel forward extrusion set-up, which includes die and punch parts, was designed and manufactured to the following specifications: (1) The die was made of 1.2510 tool steel (hardened up to 50 HRC) with the channel's cross-section being 25 mm × 45 mm in the three separate parts, namely entrance channel, MDZ, and exit channel; and (2) The punch was constructed with 1.2344 tool steel (hardened up to 55 HRC). Figure 2 displays the hydraulic press, ECFE die, and CP aluminum samples after extrusion up to the four passes. As can be seen, there is no considerable change in the dimensions of the billet samples after the process. The ECFE process was performed at ambient temperature. Molybdenum disulfide (MoS_2) was applied as a lubricant to reduce the frictional influence between the die and billet, and, also, the punch speed was equal to 2 mm/s during the operation.

Figure 2. The hydraulic press, ECFE die parts (entrance channel, MDZ and exit channel), and CP aluminum billets after the ECFE process up to the four passes.

3.3. Microstructural Testing

Optical microscopy (OM) was applied by use of Clemex Vision PE software (Clemex, Denizli, Turkey) in accordance with the ASTM E112 to measure the average grain size of the annealed condition. Additionally, the classic Williamson-Hall technique of the X-ray diffraction (XRD, Philips, Tehran, Iran) patterns was employed to theoretically calculate the cell/sub-grain size of the deformed billets. The XRD analysis was conducted on the polished sections of the deformed samples in a Philips X-ray diffractometer, equipped with a graphite monochromator using CuKα radiation (1.541 Å) with an initial angle of 4°, step size of 0.05°, and step time of 3 s. Cell/sub-grain size and lattice distortion are two prominent crystalline imperfections of the materials. These imperfections lead to peak broadening in X-ray diffraction (XRD) patterns [21]. Cell/sub-grain size and lattice micro-strain can be calculated by measuring the deviation of the line profile from the perfect crystal diffraction. In the classic Williamson-Hall technique, cell/sub-grain size of material can be estimated using Equation (1) [22]:

$$\beta\cos\theta = \frac{k\lambda}{d} + f(\varepsilon)\sin\theta \tag{1}$$

$$\beta_{exp}^2 \cong \beta_{ins}^2 + \beta^2 \tag{2}$$

where, β is the integral breadth of the XRD profile, k is the shape factor (0.9), d is the cell/sub-grain size, λ is the wavelength, ε is the lattice strain, θ is the Bragg angle, and $f(\varepsilon)$ is a defined function. Furthermore, by considering that the experimental profile (β_{exp}) is the convolution of the instrumental profile (β_{ins}) and the intrinsic profiles (β), the intrinsic profile can be attained by unfolding the experimental profile via the Gaussian assumption, as is represented by Equation (2) [22,23]. Moreover, using Equation (1), a line can be fitted by plotting βCosθ against Sinθ and the intercept gives the cell size. This line is known as Williamson-Hall graph. Thus, the above equation can be rewritten as:

$$\begin{cases} Y = \beta\cos\theta \\ X = \sin\theta \\ a = f(\varepsilon) \\ b = \dfrac{0.9\lambda}{d} \end{cases} \Rightarrow \beta\cos\theta = \frac{0.9\lambda}{d} + f(\varepsilon)\sin\theta \Rightarrow Y = b + aX \tag{3}$$

Then, both cell/sub-grain size and strain can be estimated from the y-intercept and the slope of the line, respectively. Figure 3a displays the XRD patterns of the Al1070 after being subjected to one and four passes of the ECFE process, respectively. This figure also shows the normalized XRD patterns of the highest intensity peak, indicating the peak shift, which is related to long-range stress, as well as cell/sub-grain boundaries produced by the ECFE process. Furthermore, the grain size of the final pass, at the central portion of billet's cross-section, was measured by electron backscatter diffraction (EBSD) via scanning electron microscopy (SEM) with a field emission (FE) gun, which was operated at an accelerating voltage of 15 kV, beam current of 10 nA, and step size of 50 nm, using TSL OIM analysis version 5.2 (EDAX, Denizli, Turkey). Prior to the EBSD test, the surface was ground with SiC paper up to a grit size of 4000, and then electro-polished in a solution of 100 mL $HClO_4$ + 900 mL CH_3OH at −20 °C at a voltage of 50 V.

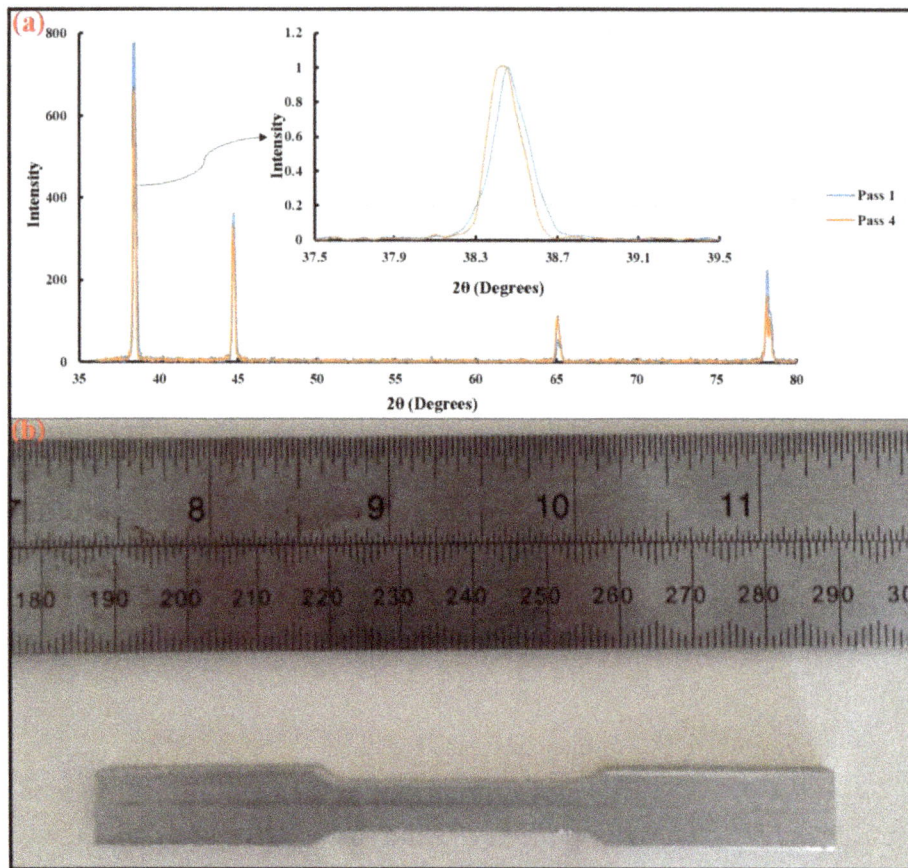

Figure 3. (**a**) The XRD pattern of the CP aluminum after the first and fourth passes of the ECFE process and (**b**) the prepared tensile testing sample of a CP aluminum billet.

3.4. Mechanical Testing

After the extrusion of Al1070, up to the four passes, various mechanical and microstructural tests were performed to determine the behavior of CP aluminum material before and after the process. A tensile test was carried out, according to the ASTM B557M, to obtain yield strength (YS), ultimate tensile strength (UTS), and elongation to failure (El%). As can be seen in Figure 3b, the gage length, gage width and sample thickness were 25 mm, 6 mm, 2 mm, respectively. In addition, Vickers micro-hardness (HV) examination was done according to ASTM E92 at the cross-section of the specimens, before and after the ECFE process, to evaluate the hardness properties. The magnitudes of imposed load and dwell time were 100 gf and 15 s, respectively. In order to prepare hardness specimens,

SiC paper up to the 1000 grit and afterwards, the 3 μm diamond paste were used. HV measurements were performed ten times for each pass sample, and the average magnitude was reported. The wire-cut type of electro-discharge machining (EDM) was used to prepare the tensile test samples.

Moreover, a low velocity drop weight impact test using Instron's Dynatup 9250 HV machine (Instron, Denizli, Turkey) was employed to investigate the influence of the ECFE process on the impact behavior of CP aluminum. During this test, a ball-end dart or tup is raised to a specific height and dropped suddenly on to the test sample, as can be schematically seen in Figure 4. The magnitude of imposed impact energy was constant and equal to 40 J for all aluminum samples. The shape of the impact tup was a hemisphere with a diameter of 12.5 mm, and, in additiona, a sample with a thickness of 5 mm was circumferentially clamped using a pneumatic clamp. It was considered to be a fixed-fixed support. The time histories of the impact load and impact velocity were respectively measured and recorded via a load cell and a pair of photoelectric diodes, and the magnitudes of absorbed energy, tup velocity, and deflection at the center were derived and achieved by the use of motion equations [24,25]. These tests were performed to investigate the material grain size refinement, mechanical properties, and impact behavior of CP aluminum material during the ECFE process.

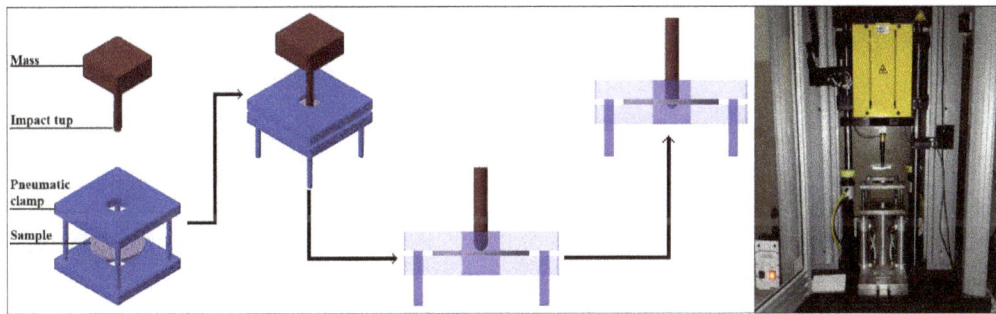

Figure 4. The schematic representation of the drop weight impact test, and, in addition, the used device.

4. Results and Discussion

4.1. Mechanical Properties

As mentioned above, the ECFE process has been successfully performed on Al1070 billets up to four passes. The engineering stress–strain curves of the ECFE aluminum, up to four passes, are presented in Figure 5, and the results including the tensile strengths, elongations to failure, and the magnitudes of hardness measurements, are listed in Table 1 and also shown in Figure 6. By considering Table 1, it can be said that enhancement of the strength values (yield strength (YS) and ultimate tensile strength (UTS)), improvement of the HV magnitudes, and reduction of the elongation percentage were achieved by utilizing this new SPD process.

Table 1. The mechanical properties and grain size of CP aluminum billets before and after the ECFE process up to four passes.

Pass Number	YS (MPa)[SD]	UTS (MPa)[SD]	El (%)[SD]	HV [SD]	Cell/sub-grain size (nm)
0	47 [1.63]	78 [0.78]	43 [0.28]	25 [1.05]	~2000
1	84 [1.34]	103 [0.78]	13.7 [0.14]	38 [1.82]	~460
2	97 [1.70]	109 [1.13]	10 [0.14]	42 [1.37]	-
3	101 [1.27]	116 [0.56]	8 [0.21]	45 [1.56]	-
4	104 [1.98]	122 [0.92]	14.7 [0.35]	46 [1.63]	~350

[SD] indicates the magnitude of standard deviation.

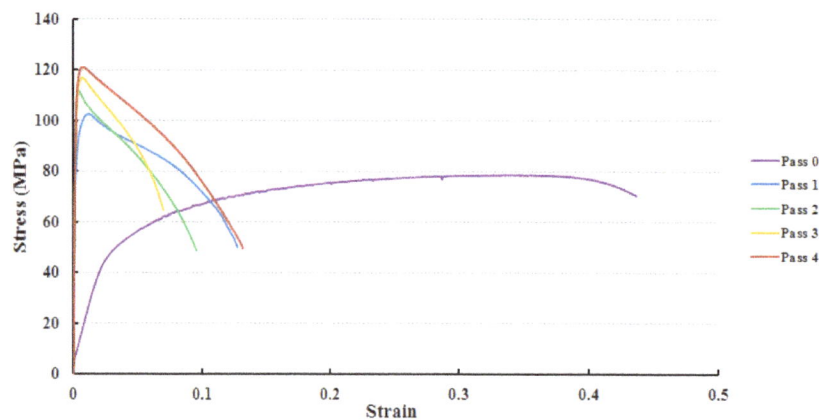

Figure 5. The engineering stress–strain curves for CP aluminum billets after the ECFE process, up to four passes.

It is obvious that, after the 1st and 4th passes of the ECFE process, 79% and 121% enhancements of the YS, 32% and 56% improvements of the UTS, and 68% and 66% reductions of the El, have been attained as compared to the annealed condition, respectively. In addition, about 52% and 84% growths of the HV magnitude were obtained after the first and fourth passes of the process, in comparison with the condition of the as-received materials, respectively. It should be pointed out that the major YS, UTS, and HV increments are achieved after the first pass, and further passes lead to improvements of these properties at a slower rate (see Figure 6). Additionally, Figure 5 shows that the improvement of the yield strength is more profound than the ultimate tensile strength. This means that the uniform plastic area, which is located between the YS and UTS points, has been limited and causes restrictions for various metal forming processes. On the other hand, an intense decrease at the elongation to failure can be seen by imposing the ECFE process. It can be noted that the reduction of the elongation to failure is high after the first pass and then it diminishes at a slower rate. The low magnitude for the elongation to failure property results in a sizeable loss of ductility after the ECFE process. Although the elongation to failure magnitude is increased slightly at the fourth pass, its magnitude is 66% lower than the aluminum under annealed conditions. The slight increase in the elongation to failure for the fourth-pass ECFE aluminum, as compared to the first pass (7%) can be related to the occurrence of grain boundary recovery (conversion of LAGBs to HAGBs) [26–28]. This can be due to a decreasing of the internal strains and accumulation of internal energy, therefore, increasing the possibility for crack formation, or it can be related to the increase of the strain rate sensitivity of the material, which causes resistance to neck formation. This improvement of the mechanical properties of material behavior using the ECFE process has also been reported for Al1050, Al6061, Al7075, and nickel during the ECAP process [26,27,29,30] and Al-3%Mg-0.2%Sc during the HPT process [31]. In addition, the ECAP process on the same material, reported by Tolaminejad and Dehghani [32], indicated that about 64% and 108% enhancement of the hardness value and approximately 202% and 267% improvement at the yield strength magnitude have been achieved after the first and fourth passes when compared to the annealed condition. The corresponding percent for the ECFE process is 52% and 84% for the HV, and 79% and 121% for the YS, respectively.

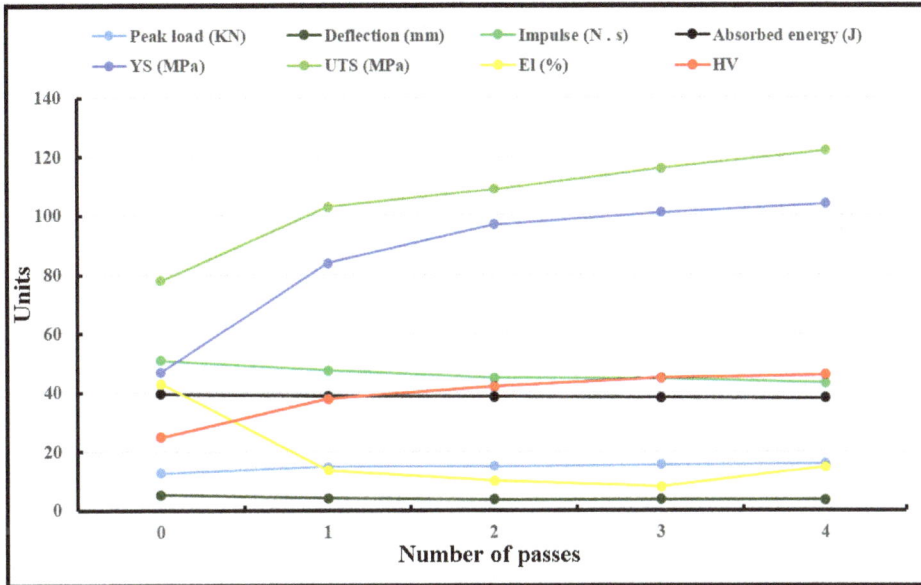

Figure 6. The variation of tensile properties, hardness behaviors, and impact characteristics of the CP aluminum, before and after the ECFE process, up to four passes.

Low velocity drop weight impact tests have been performed on the CP aluminum samples before and after the ECFE process to evaluate their impact behavior. Figure 7a,b represent the curves of load *versus* time and deflection. As seen in this figure, each curve has two regions, including loading and unloading regions. In the first part, the force slowly increases up to the maximum peak load and then it decreases sharply in the second region. The ascending region occurs due to the sample resistance to the impact force, and the descending region is related to the rebound of the tup from the sample. In other words, the required load for penetration is increased when the tup makes contact with the sample's surface. Table 2 lists the magnitudes of peak load, deflection, impulse, and absorbed energy for each pass of the ECFE sample and the trend of impact behavior *versus* pass number is also shown in Figure 6. The results indicate that the impact load increases and the deflection decreases by adding an ECFE pass number, and that the effects of the first pass ECFE process on the impact behavior is more significant than with other passes. About a 26% increment of the impact force and also 32%, 15%, and 4% decreases at the deflection, impulse, and absorbed energy, have been obtained, respectively, after the four passes of the ECFE process in comparison with the as-received condition. Thus, it can be said that the highest impact load, and also the lowest deflection, impulse, and absorbed energy can be achieved for the fourth-pass ECFE sample, which means that an aluminum sample with an enhanced strength and brittle behavior has been achieved. These results confirm the obtained tensile properties.

Interestingly, the smooth slope of the curves of the annealed and first pass samples in Figure 7b indicates that the deflection is continued by the same load, which is observed in the ductile materials.

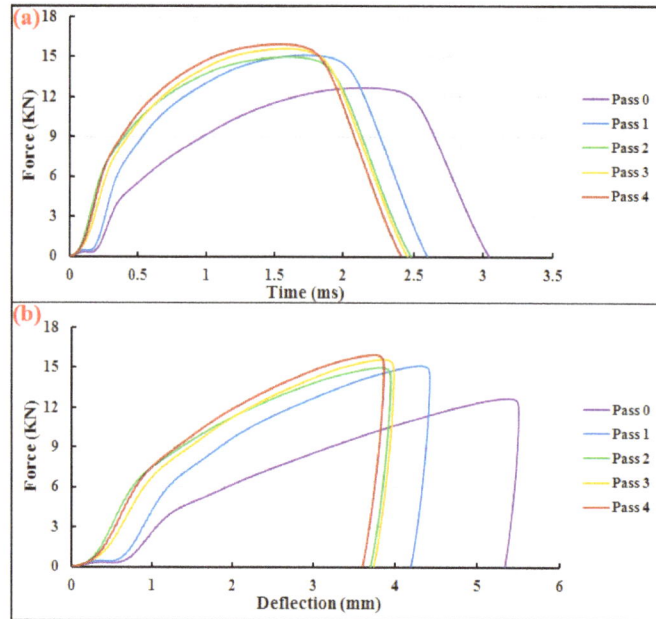

Figure 7. The (**a**) force–time and (**b**) force–deflection curves of CP aluminum billets, before and after the ECFE process, up to four passes, achieved by a drop weight impact test.

Table 2. The impact characteristics of CP aluminum billets before and after the ECFE process up to the four passes.

Pass Number	Peak load (KN)[SD]	Deflection (mm)[SD]	Impulse (N·s)[SD]	Absorbed energy (J)[SD]
0	12.72 [0.03]	5.33 [0.06]	50.99 [0.03]	39.63 [0.02]
1	15.03 [0.04]	4.19 [0.06]	47.64 [0.06]	38.92 [0.01]
2	15.17 [0.06]	3.71 [0.01]	45.03 [0.02]	38.56 [0.02]
3	15.64 [0.06]	3.75 [0.01]	44.88 [0.06]	38.31 [0.03]
4	15.98 [0.01]	3.61 [0.03]	43.30 [0.06]	38.16 [0.03]

[SD] indicates the magnitude of standard deviation.

4.2. Microstructural Characteristics

The proposed strength mechanism for the SPD materials (modified Hall–Petch relationship) is combined with the contribution of the incidental dislocation boundaries due to the statistical trapping of dislocations (LAGBs) and the geometrically necessary boundaries, because of the difference in the slip system operating with the neighboring slip systems, or local strain differences within each grain (HAGBs) [33]. This superior strengthening behavior is accompanied with a dramatic grain size reduction. Williamson-Hall analyses on the XRD patterns have been used to calculate the crystalline size of the ECFE billets, as can be seen in Figure 3a. Only three high intensity peaks of the XRD patterns of the first and final pass samples have been considered. In addition, optical microscopy (OM) has been employed to measure the average grain size of the as-received aluminum billet. Table 1 lists the average grain size of Al1070 before and after the ECFE process for the first and final passes. The results show that the ECFE process results in 77% and 82% reductions at the grain size for the 1st and 4th passes of the deformed CP Al as compared to the non-ECFE condition. Furthermore, the EBSD orientation color map reveals that the magnitude of the average grain size is about 480 nm after four passes of the ECFE process, as is represented in Figure 8. Although the calculated XRD method gives the cell or sub-grain size, and the EBSD analysis expresses the average grain size [34], it seems that the results of the EBSD image are more accurate than the theoretical approach because the XRD procedure could be affected by the strain, stress, and energy density and also other relevant parameters influencing the measurement. It is clearly observed that the fraction number of LAGBs is minor and the HAGBs are the major fraction, *i.e.*, the large angle grain boundaries occupy about 87% of the microstructure. Similar results have been reported for the ECAP process of Al1070, rapid increment of the boundary misorientation angle, and the addition of HAGB fractions up to the four passes [32].

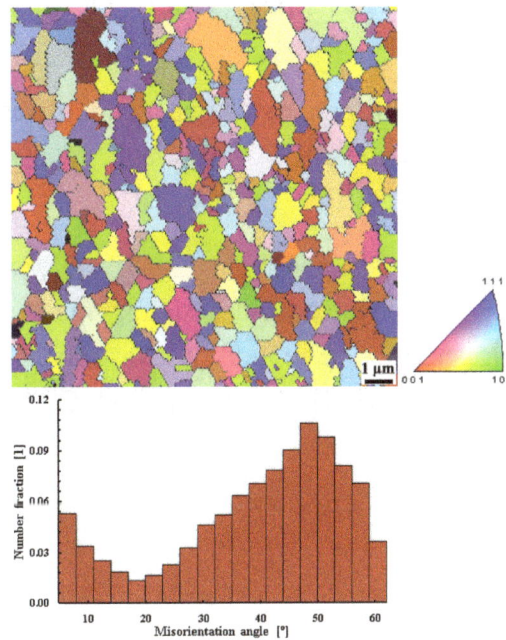

Figure 8. The orientation color map of electron backscatter diffraction for the CP aluminum billet after the four passes of the ECFE process.

5. Conclusions

In this research, the equal channel forward extrusion (ECFE) process had been proposed and introduced as a novel technique of SPD methods to fabricate UFG materials. After designing and manufacturing the die set-up, the capability of this new method has been investigated via tensile test, hardness examination, drop weight impact test, and grain size measurement on commercial pure aluminum billets, which were extruded up to four passes at room temperature. The main conclusions of this research are as follows:

- The magnitudes of yield strength, ultimate tensile strength, and Vickers micro-hardness have increased from 47 MPa, 78 MPa, and 25 HV for the annealed condition to 104 MPa, 122 MPa, and 46 HV for the fourth pass of the ECFE process, which indicate improvements of about

121%, 56%, and 84%, respectively. In addition, there is about a 66% reduction at the elongation to failure in this way. Additionally, significant enhancements in the strengthening of CP aluminum billets was achieved after the first pass of the ECFE process, which is in agreement with the hardness measurements.

- During the drop weight impact test, the magnitudes of peak load, deflection, impulse, and absorbed energy have increased from 12.72 KN, 5.33 mm, 50.99 N·s, and 39.63 J, to 15.98 KN, 3.61 mm, 43.3 N·s, and 38.16 J after four passes of the ECFE process, which means that material with the enhanced strength and brittle behavior has been attained.

- This superior improvement in the mechanical properties of the ECFE CP aluminum billet is accompanied with grain size reduction. The use of the classic Williamson–Hall method on the XRD patterns indicates that about 77% and 82% reductions have been obtained in the cell/sub-grain size of the first and fourth passes of the aluminum billets, in comparison with the annealed condition. Additionally, the EBSD scan of the final pass indicates an average grain size of about 480 nm.

The above outcomes denote that the ECFE process can be a suitable candidate as one of the SPD methods for grain refinement and the production of UFG materials.

Author Contributions

M. Ebrahimi and F. Djavanroodi were conceived and designed this study. M. Ebrahimi wrote this manuscript and contributed in all activities. The edition was done by F. Djavanroodi. Also, experiments were performed by S.A. Nazari, H. Gholipour and C. Gode. All authors read and approved the manuscript.

Conflicts of Interest

The authors declare no conflict of interest.

References

1. Valiev, R.Z.; Langdon, T.G. Principles of equal-channel angular pressing as a processing tool for grain refinement. *Prog. Mater. Sci.* **2006**, *51*, 881–981.
2. Zhilyaev, A.P.; Langdon, T.G. Using high-pressure torsion for metal processing: Fundamentals and applications. *Prog. Mater. Sci.* **2008**, *53*, 893–979.
3. Kim, W.J.; An, C.W.; Kim, Y.S.; Hong, S.I. Mechanical properties and microstructures of an AZ61 Mg Alloy produced by equal channel angular pressing. *Scr. Mater.* **2002**, *47*, 39–44.
4. Sergueeva, A.V.; Stolyarova, V.V.; Valiev, R.Z.; Mukherjee, A.K. Enhanced superplasticity in a Ti-6Al-4V alloy processed by severe plastic deformation. *Scr. Mater.* **2000**, *43*, 819–824.
5. Rafizadeh, E.; Mani, A.; Kazeminezhad, M. The effects of intermediate and post-annealing phenomena on the mechanical properties and microstructure of constrained groove pressed copper sheet. *Mater. Sci. Eng. A* **2009**, *515*, 162–168.
6. Beygelzimer, Y.; Varyukhin, V.; Synkov, S.; Orlov, D. Useful properties of twist extrusion. *Mater. Sci. Eng. A* **2009**, *503*, 14–17.

7.   Kwan, C.C.F.; Wang, Z. Cyclic deformation of ultra-fine grained commercial purity aluminum processed by accumulative roll-bonding. *Materials* **2013**, *6*, 3469–3481.

8.   Lee, J.C.; Seok, H.K.; Suh, J.Y. Microstructural evolutions of the Al strip prepared by cold rolling and continuous equal channel angular pressing. *Acta Mater.* **2002**, *50*, 4005–4019.

9.   Saito, Y.; Tsuji, N.; Utsunomiya, H.; Sakai, T.; Hong, R.G. Ultra-fine grained bulk aluminum produced by accumulative roll-bonding (ARB) process. *Scr. Mater.* **1998**, *39*, 1221–1227.

10. Shin, D.H.; Park, J.J.; Kim, Y.S.; Park, K.T. Constrained groove pressing and its application to grain refinement of aluminum. *Mater. Sci. Eng. A* **2002**, *328*, 98–103.

11. Tóth, L.S.; Arzaghi, M.; Fundenberger, J.J.; Beausir, B.; Bouaziz, O.; Arruffat-Massion, R. Severe plastic deformation of metals by high-pressure tube twisting. *Scr. Mater.* **2009**, *60*, 175–177.

12. Mohebbi, M.S.; Akbarzadeh, A. Accumulative spin-bonding (ASB) as a novel SPD process for fabrication of nanostructured tubes. *Mater. Sci. Eng. A* **2010**, *528*, 180–188.

13. Faraji, G.; Mashhadi, M.M.; Kim, H.S. Tubular channel angular pressing (TCAP) as a novel severe plastic deformation method for cylindrical tubes. *Mater. Lett.* **2011**, *65*, 3009–3012.

14. Xue, Q.; Beyerlein, I.J.; Alexander, D.J.; Gray, G.T. Mechanisms for initial grain refinement in OFHC copper during equal channel angular pressing. *Acta Mater.* **2007**, *55*, 655–668.

15. Zhao, G.; Xu, S.; Luan, Y.; Guan, Y.; Lun, N.; Ren, X. Grain refinement mechanism analysis and experimental investigation of equal channel angular pressing for producing pure aluminum ultra-fine grained materials. *Mater. Sci. Eng. A* **2006**, *437*, 281–292.

16. Su, C.W.; Lu, L.; Lai, M.O. A model for the grain refinement mechanism in equal channel angular pressing of Mg alloy from microstructural studies. *Mater. Sci. Eng. A* **2006**, *434*, 227–236.

17. Shin, D.II.; Kim, I.; Kim, J.; Park, K.T. Grain refinement mechanism during equal-channel angular pressing of a low-carbon steel. *Acta Mater.* **2001**, *49*, 1285–1292.

18. Ebrahimi, M.; Djavanroodi, F. Experimental and numerical analyses of pure copper during ECFE process as a novel severe plastic deformation method. *Prog. Nat. Sci. Mater. Inter.* **2014**, *24*, 68–74.

19. Wang, J.W.; Duan, Q.Q.; Huang, C.X.; Wu, S.D.; Zhang, Z.F. Tensile and compressive deformation behaviors of commercially pure Al processed by equal-channel angular pressing with different dies. *Mater. Sci. Eng. A* **2008**, *496*, 409–416.

20. Reihanian, M.; Ebrahimi, R.; Moshksar, M.M.; Terada, D.; Tsuji, N. Microstructure quantification and correlation with flow stress of ultrafine grained commercially pure Al fabricated by equal channel angular pressing (ECAP). *Mater. Charact.* **2008**, *59*, 1312–1323.

21. Mukherjee, P.; Sarkar, A.; Barat, P.; Bandyopadhyay, S.K.; Sen, P.; Chattopadhyay, S.K.; Chatterjee, P.; Chatterjee, S.K.; Mitra, M.K. Deformation characteristics of rolled zirconium alloys: A study by X-ray diffraction line profile analysis. *Acta Mater.* **2004**, *52*, 5687–5696.

22. Zhang, Z.; Zhou, F.; Lavernia, E.J. On the analysis of grain size in bulk nanocrystalline materials via X-ray diffraction. *Metall. Mater. Trans. A* **2003**, *34*, 1349–1355.

23. Hosseini, E.; Kazeminezhad, M. Nanostructure and mechanical properties of 0–7 strained aluminum by CGP: XRD, TEM and tensile test. *Mater. Sci. Eng. A* **2009**, *526*, 219–224.

24. Múgica, J.I.; Aretxabaleta, L.; Ulacia, I.; Aurrekoetxea, J. Impact characterization of thermoformable fibre metal laminates of 2024-T3 aluminum and AZ31B-H24 magnesium based on self-reinforced polypropylene. *Compos. Part A* **2014**, *61*, 67–75.

25. Liu, B.; Villavicencio, R.; Soares, C.G. On the failure criterion of aluminum and steel plates subjected to low-velocity impact by a spherical indenter. *Int. J. Mech. Sci.* **2014**, *80*, 1–15.

26. Shokuhfar, A.; Nejadseyfi, O. A comparison of the effects of severe plastic deformation and heat treatment on the tensile properties and impact toughness of aluminum alloy 6061. *Mater. Sci. Eng. A* **2014**, *594*, 140–148.

27. Krasilnikov, N.; Lojkowski, W.; Pakiela, Z.; Valiev, R. Tensile strength and ductility of ultra-fine-grained nickel processed by severe plastic deformation. *Mater. Sci. Eng. A* **2005**, *397*, 330–337.

28. Bystrzycki, J.; Fraczkiewicz, A.; Lyszkowski, R.; Mondon, M.; Pakiela, Z. Microstructure and tensile behavior of Fe–16Al-based alloy after severe plastic deformation. *Intermetallics* **2010**, *18*, 1338–1343.

29. Puertas, I.; Luis Pérez, C.J.; Salcedo, D.; León, J.; Fuertes, J.P.; Luri, R. Design and mechanical property analysis of AA1050 turbine blades manufactured by equal channel angular extrusion and isothermal forging. *Mater. Des.* **2013**, *52*, 774–784.

30. Shaeri, M.H.; Salehi, M.T.; Seyyedein, S.H.; Abutalebi, M.R.; Park, J.K. Microstructure and mechanical properties of Al-7075 alloy processed by equal channel angular pressing combined with aging treatment. *Mater. Des.* **2014**, *57*, 250–257.

31. Harai, Y.; Edalati, K.; Horita, Z.; Langdon, T.G. Using ring samples to evaluate the processing characteristics in high-pressure torsion. *Acta Mater.* **2009**, *57*, 1147–1153.

32. Tolaminejad, B.; Dehghani, K. Microstructural characterization and mechanical properties of nanostructured AA1070 aluminum after equal channel angular extrusion. *Mater. Des.* **2012**, *34*, 285–292.

33. Luo, P.; McDonald, D.T.; Xu, W.; Palanisamy, S.; Dargusch, M.S.; Xia, K. A modified Hall–Petch relationship in ultrafine-grained titanium recycled from chips by equal channel angular pressing. *Scr. Mater.* **2012**, *66*, 785–788.

34. Azimi, A.; Tutunchilar, S.; Faraji, G.; Besharati Givi, M.K. Mechanical properties and microstructural evolution during multi-pass ECAR of Al1100–O alloy. *Mater. Des.* **2012**, *42*, 388–394.

7

A Comparative Characterization of the Microstructures and Tensile Properties of As-Cast and Thixoforged *in situ* AM60B-10 vol% Mg₂Sip Composite and Thixoforged AM60B

Suqing Zhang, Tijun Chen *, Faliang Cheng and Pubo Li

Key Laboratory of Advanced Processing and Recycling of Nonferrous Metals,
Lanzhou University of Technology, Lanzhou 730050, China;
E-Mails: zhangsuqing1985@163.com (S.Z.); chengfaliang@yahoo.com (F.C.);
lipubogs@163.com (P.L.)

* Author to whom corresponding should be addressed; E-Mail: chentj@lut.cn

Academic Editor: Anders E. W. Jarfors

Abstract: The microstructure and tensile properties of the thixoforged *in situ* Mg₂Sip/AM60B composite were characterized in comparison with the as-cast composite and thixoforged AM60B. The results indicate that the morphology of α-Mg phases, the distribution and amount of β phases and the distribution and morphology of Mg₂Si particles in thixoforged composite are completely different from those in as-cast composite. The Mg₂Si particles block heat transfer and prevent the α-Mg particles from rotation or migration during reheating. Both the thixoforged composite and thixoforged AM60B alloy exhibit virtually no porosity in the microstructure. The thixoforged composite has the highest comprehensive tensile properties (ultimate tensile strength (UTS)) of 209 MPa and an elongation of 10.2%. The strengthening mechanism of the Mg₂Si particle is the additive or synergetic effect of combining the load transfer mechanism, the Orowan looping mechanism and the dislocation strengthening mechanism. Among them, the load transfer mechanism is the main mechanism, and the latter two are minor. The particle splitting and interfacial debonding are the main damage patterns of the composite.

Keywords: thixoforged; *in situ* Mg₂Sip/AM60B composite; tensile properties

1. Introduction

Magnesium alloys are the lightest commercially-used metals, offer excellent cast-ability, machinability, low density, high specific strength and stiffness, electromagnetic shielding characteristics and are, thus, attractive for applications in the transportation industry, electronic products, portable tools, sporting goods and aerospace vehicles [1–3]. Unfortunately, with the rapid expansion of magnesium applications, magnesium alloys suffer from the challenge of meeting the requirements of strength, ductility, fatigue and creep properties under high temperature. Metal-matrix composites possess many advantages over monolithic materials, such as high-temperature mechanical strength, good wear resistance and dimensional stability, and they have been widely used in aircraft, space, defense and automotive industries [4,5]. Thus, the fabrication of magnesium-based composite is a maneuverable and reliable way to overcome these shortcomings by taking full advantage of magnesium alloys [6]. There are several methods to fabricate magnesium-based composite, such as self-propagating high temperature synthesis (SHS), directed reactive synthesis (DRS), mechanical alloying (MA) and reaction spontaneous infiltration (RSI), blend press sintering, disintegrated melt deposition and *in situ* synthesis [7–15]. Alternatively, the *in situ* synthesis process produces the desired reinforcements via reaction synthesis in melting alloy by adding grain refiner during traditional casting and is a relatively good potential technique with good maneuverability and reliability for industrial manufacturing, because it does not need special treating procedures and equipment. In the authors' previous investigations, uniform distribution and dispersion of fine-grained, *in situ* Mg_2Si_p/AM60B composite have been achieved via traditional gravity casting by the addition of 0.5 wt% Sr and 0.2 wt% SiC_p [16]. Modified Mg_2Si particles with a grain size of 20~40 μm were uniformly distribute in the matrix.

As a novel metal process manufacturing, thixoforging involves the advantages of both casting and forging technologies and significantly decreases or even eliminates porosities [2,17]. Consequently, superior tensile properties of the thixoforged components resulting from the pore-free fine microstructure can be achieved. Moreover, the amount and size of the β phase ($Mg_{17}Al_{12}$), which is harmful for the tensile properties of the alloy, can also be reduce. Therefore, it can be supposed that fabrication of thixoforged *in situ* Mg_2Si_p/AM60B composite is a promising way to further expand the applications of AM60B alloy. However, limited information on thixoforged *in situ* Mg_2Si_p/AM60B composite is available in the open literature.

This article presents the progress of an ongoing research work on thixoforged *in situ* Mg_2Si_p/AM60B composite. The microstructure, tensile properties and fracture behavior of *in situ* Mg_2Si_p/AM60B composite are studied. The informative results are compared with identical as-cast composites and thixoforged AM60B alloy, in order to elucidate the strengthening mechanisms of the thixoforging technique and Mg_2Si particles.

2. Experimental Section

2.1. As-Cast Preparation

The *in situ* Mg_2Si_p/AM60B composites used in this work were prepared by the traditional gravity casting route using commercial AM60B magnesium alloy, pure Mg (99.9 wt%) and Al-30Si (all of them were provided by Changfeng factory in Lanzhou, China). Homemade Mg-30Sr master alloys and

Mg-25SiC$_p$ press cake (mixture powders) were used as a modifier and grain refiner for the Mg$_2$Si phase and α-Mg phase, respectively. The AM60B alloy was prepared by this method, also, using commercial AM60B magnesium alloy by the addition of 0.2 wt% SiC$_p$. The chemical compositions of those materials are listed in Table 1.

A quantity of AM60B alloy, pure Mg and Al-30Si master alloy was melted in an electric resistance furnace at 790 °C and then modified by 0.5% Sr (using Mg-30Sr master alloy). The melt was then isothermally held for 20 min, and 0.2% SiC$_p$ (using the pressed cake of Mg-25SiC$_p$ mixture powders) was introduced. Finally, the resulting melt was degassed using C$_2$Cl$_6$ and poured into a steel mold with a cavity of φ 50 mm × 500 mm after it had been held for 10 min. Thus, the as-cast composite with 10 vol% Mg$_2$Si$_p$ was obtained (as-cast *in situ* AM60-10 vol% Mg$_2$Si$_p$ composite). The melting process of AM60B alloy is similar to the composite, melted at 790 °C and refined by 0.2% SiC$_p$. Owing to the Mg$_2$Si phase being absent, the Mg-30Sr is not necessary to add to this melting alloy. Then, after the melt was degassed by C$_2$Cl$_6$, it was poured into the same mold. A covering agent of RJ-2 (Hongguang Company, Shanghai, China) designed for magnesium alloys was used for protecting the melt from oxidation during both of the melting processes.

Table 1. Chemical composition (in wt%) of the materials studied.

Materials	Al	Si	Mn	Zn	Cu	Ni	Fe	Mg
As-cast composite	5.5–6.5	2.00	0.24–0.6	≤0.22	≤0.010	≤0.002	≤0.005	Balance
Thixoforged composite	5.5–6.5	2.00	0.24–0.6	≤0.22	≤0.010	≤0.002	≤0.005	Balance
Thixoforged AM60B	5.5–6.5	≤0.05	0.24–0.6	≤0.22	≤0.010	≤0.002	≤0.005	Balance

2.2. Thixoforging Process

For the thixoforging, some ingots with dimensions of φ 42 mm × 30 mm were cut from the as-cast rods and then reheated in a resistant furnace under argon gas protection at a semisolid temperature of 600 °C for 60 min. The obtained semisolid feedstocks were quickly handed into a die with a cavity of φ 50 mm × 20 mm and then thixoforged using a hydraulic press. The preheating temperature of the die was 300 °C; the applied punch velocity and pressure were 60 mm/s and 192 MPa, respectively. The holding time was 20 s. Repeating the above experimental procedures, thixoforging composite (10 vol% Mg$_2$Si$_p$ containing) and AM60B alloy were obtained.

2.3. Microstructural Analysis

The metallographic specimens were cut from the center region of each product and the cross-section polished by standard metallographic techniques. Subsequently, they were chemically etched using 4% nitric acid ethanol solution and observed on an optical microscope (OM, Nikon Instruments, Shanghai, China) and a scanning electron microscope (SEM, NEC Electronics Corporation, Tokyo, Japan). The compositions of the primary α-Mg phase in the microstructures were examined by energy dispersive spectroscopy (EDS, NEC Electronics Corporation, Tokyo, Japan) using spot scan mode in the SEM. The average of at least five primary α-Mg phases was taken as the composition of each specimen. Porosity was evaluated via measuring the optical micrograph of the un-etched metallographic specimens. The related images were analyzed by Image-Pro Plus 5.0 software (Media Cybernetics Company,

Silver Spring, MD, USA), and the percentage of the porosity to the whole was quantitatively examined and the results based on the average of three images.

2.4. Tensile Testing

The mechanical properties of the materials were evaluated by tensile testing, which was performed at ambient temperature on a universal material testing machine with a loading velocity of 1 mm/s. Samples for tensile testing with a cross-section of 1.2 mm × 2.5 mm and a gauge length of 10 mm were machined by a Computer Numerical Control (CNC) wire-cut machine (Taizhou Dengfeng CNC Machine Company, Taizhou, China) from the center of each product. The tensile properties of each product, including ultimate tensile strength (UTS) and elongation to failure (Ef), were obtained based on the average of at least five tests. Some typical fracture surfaces and side views of fracture surface were observed on the SEM and OM, respectively, to ascertain the nature of the fracture mechanisms.

3. Results and Discussion

3.1. Microstructural Analysis

Figure 1 presents the microstructures of the as-cast composite, the thixoforged composite and AM60B revealed by OM and SEM, respectively. The microstructure of the as-cast composite mainly consists of primary α-Mg dendrites, Mg_2Si particles and eutectic phases (Figure 1a). The size of the primary α-Mg dendrites is around 70~90 μm, which is relatively large compared with the reported value in the literature [18,19]. This is primarily due to the somewhat slow solidification rate taking place in the current work. The diameter of the mold in this work is 50 mm, which is only 16 mm in the literature. The primary Mg_2Si particles with a size of 15~30 μm were located at the primary α-Mg dendrite boundaries. The SEM result (Figure 1b) displays that the β-$Mg_{17}Al_{12}$ eutectic phase (bright contrast) belongs to divorced eutectics and tend to form a network surrounding the α-Mg phase (dark contrast). According to the Mg-Al binary phase diagram [20], it is known that AM60B alloy is a hypoeutectic alloy, since its Al content and composition is far away from the eutectic point. Under this circumstance, the residual liquid amount should be very low when the eutectic reaction occurs and exists in thin layers between the primary α-Mg dendrites and dendrite arms. Then, the eutectic α phase preferentially directly grows on the primary α-Mg phase without renucleation, and only the eutectic β and eutectic Mg_2Si phases are left in the interdendritic regions. The previously-formed primary Mg_2Si particles are pushed to the interdendritic regions by the growing interface.

As shown in Figure 1c, the microstructure of the thixoforged composite is composed of primary α-Mg particles, secondarily solidified structures and Mg_2Si particles. The morphology of the primary α-Mg particles and the distribution of both the eutectic β phases and Mg_2Si particles are completely different from those present in the as-cast composite. The primary α-Mg particles coarsen and connect to each other, the size being approximately 90~120 μm, which is significantly larger than the primary α-Mg dendrites in as-cast microstructure. The Mg_2Si particles coarsen, as well, and their size is about 30~40 μm. The sharp edges and corners become blunt. However, the Mg_2Si particles in the thixoforged coupon not only surround the primary α-Mg particles, but also some of them locate inside the primary α-Mg particles. The size and amount of the β phase clearly decrease in the thixoforged specimen, which is located at the boundaries and inside the primary α-Mg particles, as well (shown in Figure 1d). Figure 1e presents the microstructure of the thixoforged AM60B alloy. It indicates that the primary α-Mg particles slightly coarsen, and the outline becomes indistinct (comparing Figure 1f with 1d). The secondarily solidified structures almost disappear and only can be found in some triple point. As shown in Figure 1f, the β phase size and amount of both the thixoforged composite and AM60B alloy are at a comparable level.

The morphological change of the α-Mg phase occurs during the reheating process. The thixoforged composites are subjected to a partial remelting treatment. During this technical process, the primary α-Mg dendrites transform into spheroidal primary α-Mg particles uniformly suspended in the liquid phase. During the subsequent thixoforging, the liquid solidifies to form secondarily-solidified structures. Coarsening of the primary α-Mg grains in the thixoforged specimen should be attributed to the following two aspects. One is Ostwald ripening and the coalescence of the nearby primary α-Mg grains during reheating, driven by minimizing the interfacial energy [21]. Grain growth by coalescence by grain boundary migration is dominant for short times after the liquid is formed, and Ostwald ripening is dominant for longer times [22]. This is a common phenomenon in thixoforged materials [23–25]. The Mg_2Si particles and the liquid might be engulfed by the merged primary α-Mg grains, so Mg_2Si particles or the eutectic β phase would distribute inside the primary α-Mg grains in the thixoforged specimens. The other is the solidification behavior of the thixoforged materials. Table 2 gives the compositions of α-Mg phase under these three methods. It reveals that the Al concentration in these two thixoforged materials is significantly higher than that in the as-cast coupon. This is because the eutectic phase is dissolving towards the primary α-Mg grains during reheating, which results in the Al concentration increase in the primary α-Mg grains and a decrease in the liquid. In this case, the formed secondarily primary α-Mg phase (to differentiate the primary α-Mg grains, the primary α-Mg phase solidified form of the liquid is named the secondarily primary α-Mg phase) and the eutectic α-Mg phase should increase, accompanied by the decrease of the eutectic β phase (the Al element is a necessary constituent for forming the eutectic β phase). The secondarily primary α-Mg phase should preferentially directly grow on the surfaces of the primary α-Mg grains, and then, the eutectic α-Mg phase also preferentially attaches on the surfaces of the secondarily primary α-Mg phase, which leads to the primary α-Mg grains coarsening and connecting with each other.

Although the reheating temperature of 600 °C is lower than the eutectic point of Mg-Mg_2Si (637.6 °C) [20], the primary Mg_2Si particles and eutectic Mg_2Si phases should partially melt due to the penetration of the liquid and diffusion of Mg and Si atoms between Mg_2Si and liquids, especially at the sharp edges and corners. During the thixoforging, the melted Mg_2Si (including primary and eutectic Mg_2Si phases) grows as halos surrounding the nearly spherical Mg_2Si particles. Therefore, the Mg_2Si particles are somewhat coarse and spherical in the thixoforged specimen compared to those in the as-cast specimen.

With regard to the Mg_2Si particles, two primary factors affect the semisolid microstructural evolution, as described in this section. Firstly, the Mg_2Si particle acting as a ceramic phase with the low thermal conductivity, which uniformly distributes in the matrix, would block the heat transfer from the edge to the center of the semisolid ingot [26]. This would suggest that the heating rate is delayed, so that the phase transformation rate is reduced. It is known that the microstructure evolution closely depends on the phase transformation. Therefore, coarsening from Ostwald ripening is suppressed. This would also affect the composition of the primary α-Mg grains simultaneously. On the other hand, the pin effect of the Mg$_2$Si particles in the primary α-Mg boundary prevent the primary α-Mg boundary from rotation or migration [27]. Thus, the coalescence of the contacted primary α-Mg grains through mergence would be suppressed. As a result, the size and Al solubility of the primary α-Mg grains in thixoforged composite are slightly lesser than in the thixoforged AM60B alloy. Based on these standpoints, it is no difficult to understand the resultant microstructures under those different processing technologies.

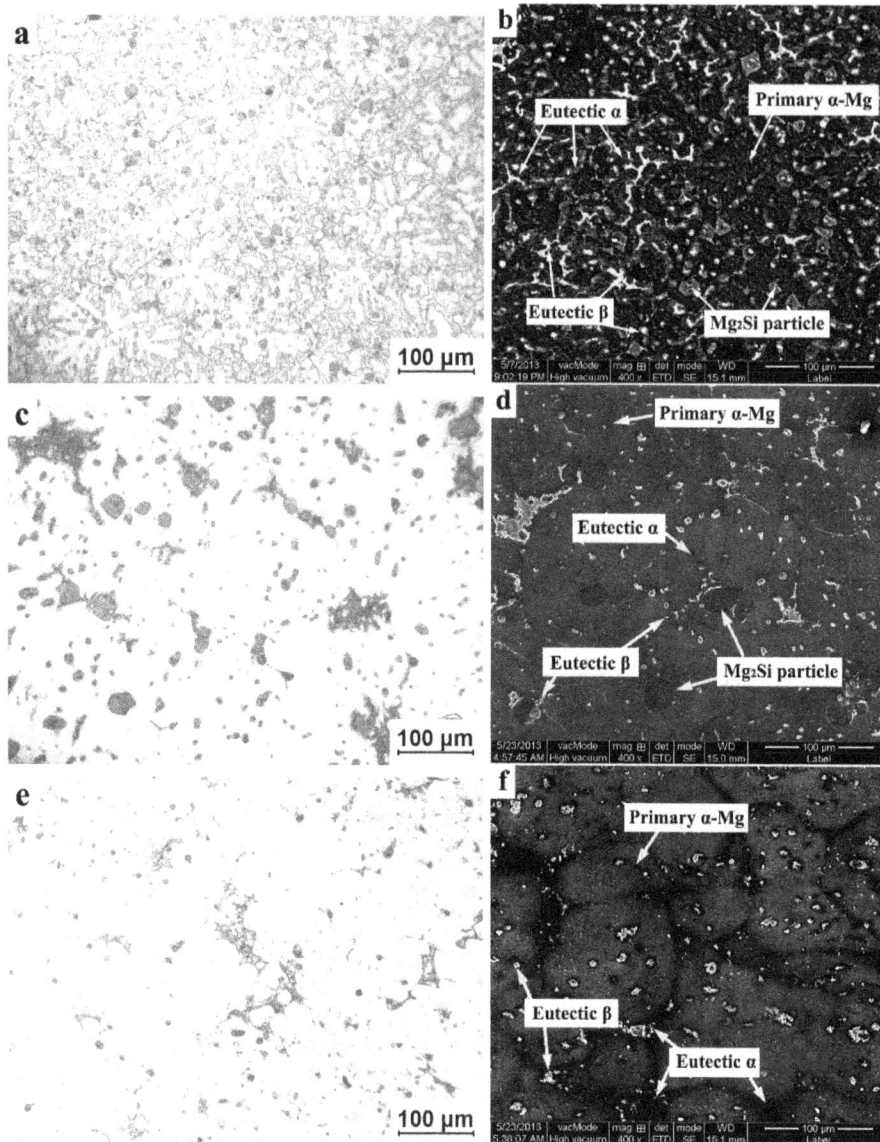

Figure 1. OM micrographs and SEM images of: (**a,b**) as-cast composite; (**c,d**) thixoforged composite; and (**e,f**) thixoforged AM60B.

Table 2. Compositions of primary α-Mg dendrites or the grains of these materials.

Materials	Compositions/wt%	
	Al	Mg
As-cast composite	4.74	95.26
Thixoforged composite	6.87	93.13
Thixoforged AM60B	7.20	92.80

3.2. Porosity Evaluation

Figure 2 reveals the porosity distribution in the polished specimens. Representative pores can be easily spotted in the as-cast composite, as indicated in Figure 2a. However, it is evidently shown in Figure 2b,c that the thixoforged coupons are virtually free of gas and shrinkage porosities. In comparison with that (4.00%) of the as-cast composite, the porosity percentage of the thixoforged composite is 0.15%, and the thixoforged AM60B is 0.12%.

The porosity elimination of thixoforged ingots should result from the following reasons. The first is the high applied pressure during solidification and low filling velocity during mold filling. The high applied pressure reduces the shrinkage porosity by squeezing the liquid metal into the last region of the casting to solidify. Thus, the feeding ability to solidification shrinkage is enhanced. The purpose of the low filling velocity is to effectively avoid air entrapment. The second is the inherent characteristic of the semisolid forming [2,5,17]. The spherical morphology of the primary α-Mg grains would be more favorable to liquid penetration for feeding [28]. The proper liquid fraction of the thixoforging, which is lower than that of the traditional casting, should reduce the probability of the solidification shrinkage and effectively avoid entrapped gas, as well.

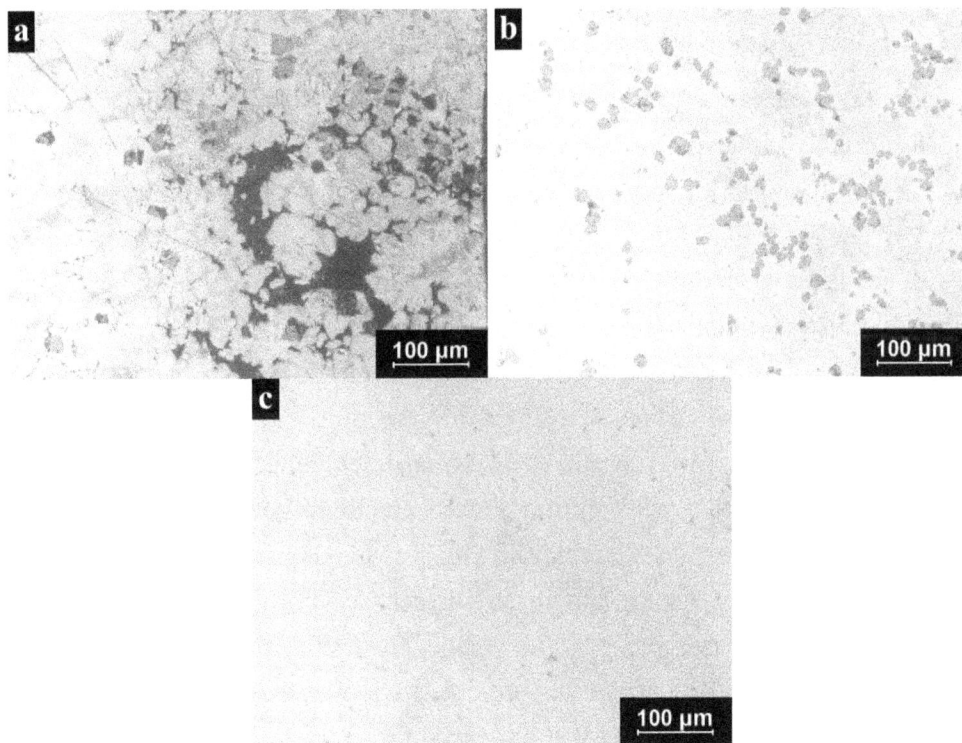

Figure 2. Optical micrographs showing porosity in: (**a**) as-cast composite; (**b**) thixoforged composite; and (**c**) thixoforged AM60B.

3.3. Tensile Properties

Figure 3 gives the tensile properties of those three materials. It can be evidently seen from Figure 3 that the UTS of the thixoforged composite is 209 MPa, which is significantly higher than that of the as-cast conditions (108 MPa) and the thixoforged AM60B alloy (146 MPa). However, the elongation of the thixoforged AM60B is 13.3%, which is the maximum value among these three materials.

The thixoforged composite has the highest comprehensive tensile properties, UTS of 209 MPa and elongation of 10.2%.

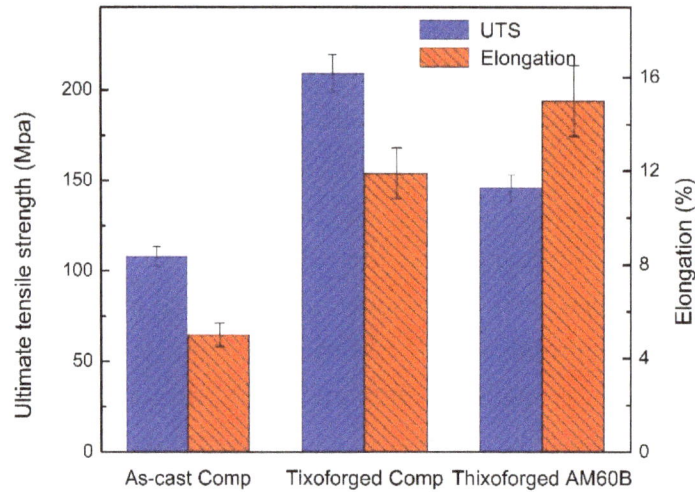

Figure 3. Tensile properties of the materials at room temperature. UTS, ultimate tensile strength.

Figure 4 shows the typical fractographs of these three materials. It indicates that the fracture surface of the as-cast composite is characterized by porosity features (Figure 4a), which are somewhat brittle in nature. As mentioned above, the porosities can be obviously observed from the micrograph of as-cast composite, which should generate in the last solidified zones, *i.e.*, the eutectic structures between the primary α-Mg dendrites. The porosities serve as the initiation point of cracks in the as-cast specimens. Then, the cracks grow and propagate along the eutectic structures during tensile testing, which is depicted in Figure 5a. The failure is mainly attributed to the intergranular fracture and is partially caused by the segregation of the brittle eutectic β phase at dendrite boundaries. As described in the previous section, the thixoforged composite is virtually free of gas and shrinkage porosities. Correspondingly, the porosity characteristics on the fracture surface of the thixoforged composite disappear and are substituted by small dimples (Figure 4b). Plenty of fractured Mg_2Si particles could be observed on the fracture surface. The crack propagation path turns from along the eutectic structures into across the primary α-Mg grains (Figure 5b, marked by arrows). The failure transfers to the transgranular fracture mode. The resultant fractographs and side-view of the fractured surfaces of both the as-cast and thixoforged composite are in good agreement with the data of tensile properties shown in Figure 3. It is well established that the tensile properties of the material are determined by their microstructures. Therefore, it is concluded that the elimination of porosity, the decrease of the eutectic β phase and enhancement of solution strengthening are responsible for the superior tensile properties of the thixoforged composite.

Figure 4c illustrates the fracture surface of the thixoforged AM60B alloy, which is characterized by small dimples and flat facets. The dimples result from localized microvoid coalescence, due to the dislocation motion or grain boundary sliding. The microvoids grow and connect, eventually leading to the creation of cracks. The flat facets are caused by cracks moving through the primary α-Mg grains. Cracks propagate between the primary α-Mg grains (marked by A) and occasionally across the primary α-Mg grains (marked by B) in some local zones (Figure 5c), which is consistent with the fracture surface. The fracture of the thixoforged AM60B belongs to a mixture of transgranular and intergranular modes. The inferior UTS of the thixoforged AM60B should be ascribed to there being no reinforcing phase to strengthen the α-Mg phase. A large amount of eutectic β phase is a harmful effect for UTS. However, the amount of the β phase reduced to a given value may also decrease the UTS, owing to the strengthening role of the β phase being reduced. This is also the same reason that results in the superior elongation of this alloy. However, the adopted technologic parameters (reheating time and temperature, mold preheating temperature, *etc.*) in this work are optimum for the thixoforged composite. There may not be optimized parameters for the thixoforged AM60B alloy. It can be reasonably concluded that the tensile properties of the thixoforged AM60B alloy will be further improved through adjusting the technologic parameters. This will be discussed in further work.

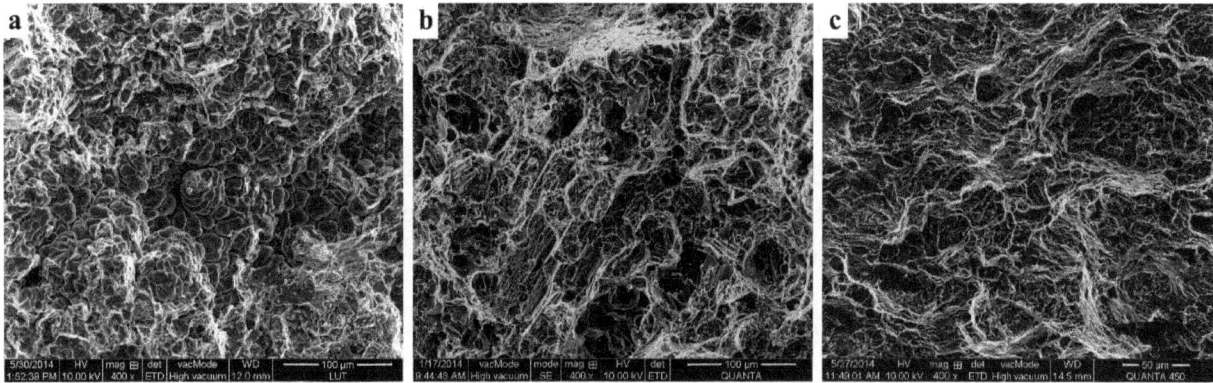

Figure 4. SEM fractographs of: (**a**) as-cast composite; (**b**) thixoforged composite and (**c**) thixoforged AM60B.

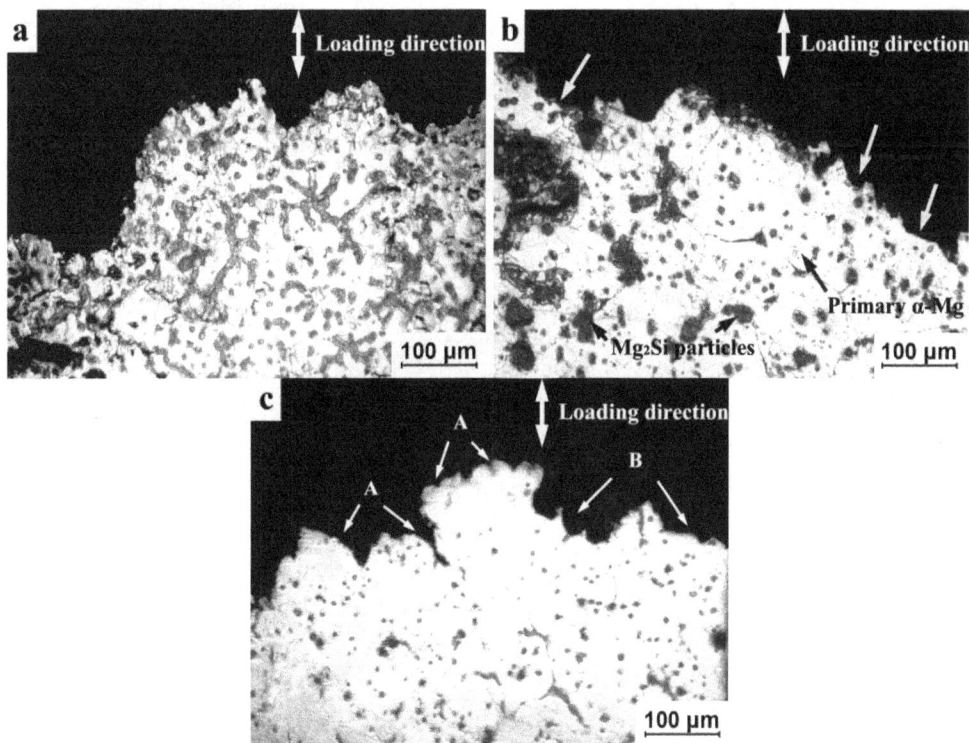

Figure 5. OM images reveal the side-view of the fracture surface: (**a**) as-cast composite; (**b**) thixoforged composite; and (**c**) thixoforged AM60B.

3.4. Strengthening Mechanisms of the Mg₂Si Grains

For the purpose of verifying the strengthening mechanism of Mg_2Si particles for the α-Mg matrix, some typical fracture surfaces are carefully observed with high magnification (Figure 6). Pay attention to the composite in the as-cast condition: the Mg_2Si particles can only be found in some local zone on the fracture surface. The Mg_2Si particles are surrounded by the deformed α-Mg matrix and keep their original morphology (Figure 6a). During tensile testing, the pin effect of the Mg_2Si particles in the grain boundaries keeps them from sliding. The interfaces of the Mg_2Si_p/matrix belong to an incoherent interface, due to the differences of the crystal structures and the lattice constants [29]. Therefore, local stress concentration should be preferentially generated at the sharp edges and corners of the Mg_2Si particles. Therefore, the interfaces of the Mg_2Si_p/matrix are easily debonded under this local stress (marked by arrows in Figure 6a). Subsequently, the debonding areas extend and connect with the cracks, which initiate from the eutectic structure or porosity, eventually leading to the final fracture. Owing to the non-compact microstructure of the as-cast composite, the reinforcement of Mg_2Si particles of the matrix has not fully taken part in the contribution to the improved the tensile properties.

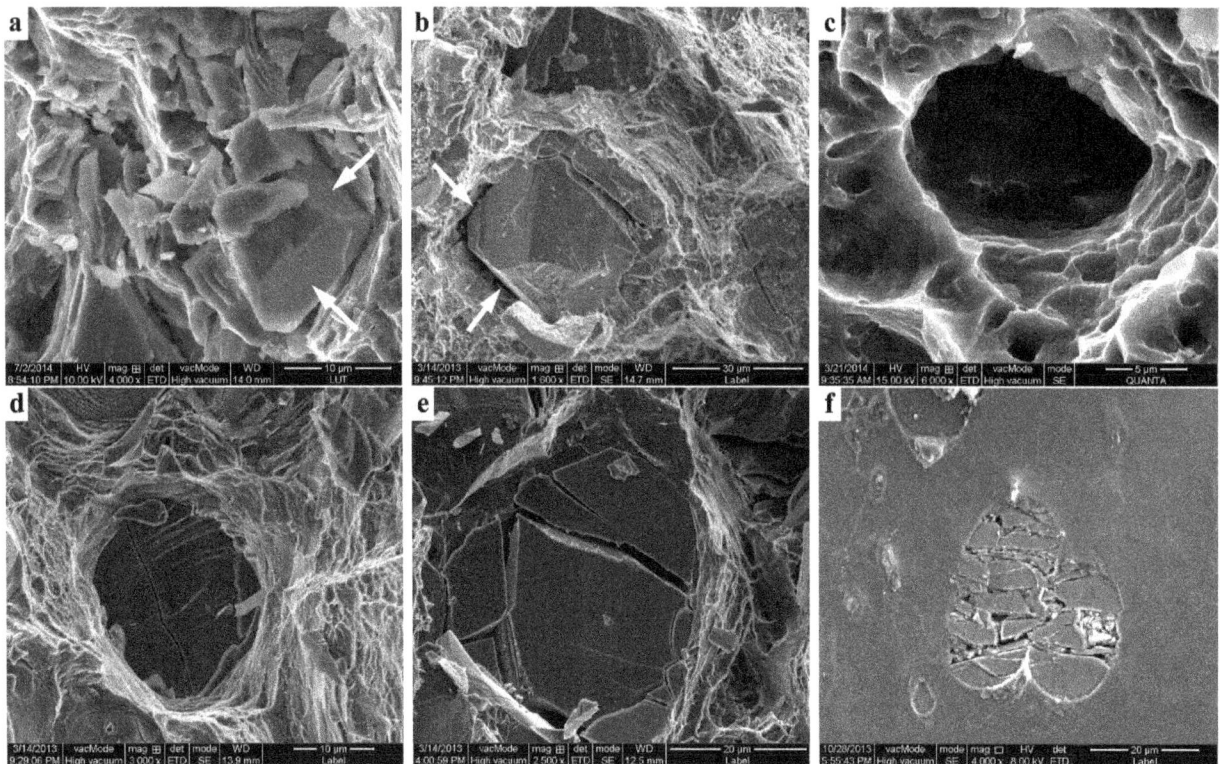

Figure 6. Typical SEM images showing Mg_2Si particles in: (**a**) as-cast composite; (**b**–**e**) the thixoforged composite fracture surface; (**f**) the thixoforged composite tensile bar.

There are two kinds of failure behaviors related to the Mg_2Si particles, which can be found on the thixoforged composite fracture surface. One is the interfacial debonding of Mg_2Si_p/matrix (Figure 6b). As mentioned above, the sharp edges and corners become blunt. Therefore, the local stresses are uniformly distributed along the Mg_2Si_p/matrix interfaces and increase as the tensile stress increases. Microvoids should generate in the surrounding matrix under this local stress. Then, the localized microvoids' coalescence results in the debonding of the interface (Figure 6c). The other is the fragmentation of Mg_2Si particles. There are two ways for the Mg_2Si particle to fracture: either fractured parallel to the fracture surface of the whole specimen (Figure 6d) or broken into several parts (Figure 6e). The composite mainly contains two phases with very different mechanical behaviors: the α-Mg phase and Mg_2Si particles. While the soft magnesium alloy deforms plastically during tensile testing, the Mg_2Si particles are rigid and deform only elastically. The reinforcing Mg_2Si particles prevent the straining of the surrounding matrix. Thus, the composite is plastically non-homogeneous, because the plastic deformation gradient is imposed on the Mg_2Si_p/matrix interfaces. Therefore, a high stress concentration generates at the interface and gives rise to the increase of tensile strain. The existing investigation supposed that the stress concentration was two- to four-times higher than that of the α-Mg matrix [30]. When the stress concentration exceeds some value, this should even result in the fragmentation of Mg_2Si particles (Figure 6d,e). Figure 6f reveals that the Mg_2Si particles split into several parts and that there are no visible cracks in the surrounding matrix. The interfacial debonding and particle fragmentation are the indications of the absorption of energy and the relaxation of local stress concentration. This is just due to the formation of local stress concentration at the Mg_2Si/matrix interface and the subsequent relaxation; the growth rate and nucleation of the crack in the surrounding matrix are delayed. Thus, the strengthening of the Mg_2Si particles of the matrix should primarily be the result of the load transfer mechanism. The particle splitting and interfacial debonding are the main damage patterns of the composite.

The contribution of Mg_2Si particles to improving the mechanical properties should also be attributed to other mechanisms. One is the Orowan looping mechanism, which is described as the interaction between dislocations and fine particles: the resistance of the reinforcing particles to the passage of dislocations from a balance between the force acting on the dislocation and the force coming from the line tension acting on both sides of the reinforcing particle [31]. However, this mechanism is only effective in action when the reinforcing particles are located within the grains. In the composite employed in this work, the Mg_2Si particles are mainly located at the boundaries of primary α-Mg grains and occasionally located inside of them. Therefore, the strengthening effect due to the Orowan mechanism will be minor. The other is the dislocation strengthening mechanism, which results from the coefficient of thermal expansion (CTE) mismatch between Mg_2Si particles and the matrix: dislocations are created, due to the relaxation of thermal expansion mismatch between reinforcing particles and matrix, and may cause an increase in the dislocation density [32]. This effect can impede the dislocation movement, also playing a very important role in strengthening the matrix.

Although the contribution of each strengthening mechanism to improving the mechanical properties has not been calculated separately, it should also be believed that an additive or synergetic effect probably occurs by combining several mechanisms. Among them, the load transfer mechanism should primarily take part in strengthening the matrix, and the Orowan looping mechanism and dislocation strengthening mechanism should be minor.

4. Conclusions

(1). In comparison with the as-cast composite, the morphology of α-Mg phases, the distribution and amount of β phases and the distribution and morphology of Mg_2Si particles in thixoforged composite are completely different from those in as-cast composite.

(2). The α-Mg dendrites evolve into spheroidal α-Mg grains uniformly suspended in the liquid phase during reheating. The liquid solidifies to form a secondarily-solidified structure. The eutectic structure dissolving towards the α-Mg grains results in the β phases' decrease, and the Al concentration in primary α-Mg grains increases. The β phases and Mg_2Si particles are entrapped within the merged α-Mg grains. The coarsening of α-Mg grains results from coalescence, Ostwald ripening and subsequent solidification behavior.

(3). The Mg_2Si particles block heat transfer, so as to delay the Ostwald ripening. The pin effect of the Mg_2Si particles prevents the α-Mg grains from rotation or migration, so as to reduce the probability of α-Mg grain mergence. The resulting α-Mg grains in the thixoforged composite are slightly finer than that in the thixoforged AM60B alloy.

(4). The porosity elimination in the thixoforged component is attributed to the low filling velocity during mold filling, the applied high press during solidification, the enhanced feeding ability of spherical primary α-Mg grains and the low liquid fraction of the semisolid slurry.

(5). The UTS of the thixoforged composite is 209 MPa, which is significantly higher than that of the as-cast conditions (108 MPa) and the thixoforged AM60B alloy (146 MPa). The thixoforged AM60B has the maximum value of elongation (13.3%). The thixoforged composite has the highest comprehensive tensile properties, UTS of 209 MPa and elongation of 10.2%.

(6). The strengthening mechanism of the Mg_2Si particles is the additive or synergetic effect combining the load transfer mechanism, the Orowan looping mechanism and the dislocation strengthening mechanism. Among them, the load transfer mechanism is the main mechanism, and the latter two are minor.

Acknowledgments

The authors wish to express thanks for the financial support from the National Basic Research Program of China (Grant No. G2010CB635106), the Program for New Century Excellent Talents in University of China (Grant No. NCET-10-0023) and the Program for Hongliu Outstanding Talents of Lanzhou University of Technology.

Author Contributions

Tijun Chen and Suqing Zhang conceived and designed the experiments; Suqing Zhang and Faliang Cheng performed the experiments; Faliang Cheng and Pubo Li analyzed the data; Suqing Zhang wrote the paper.

Conflicts of Interest

The authors declare no conflict of interest.

References

1. Hao, Y.; Chen, T.J.; Ma, Y.; Li, Y.D.; Yan, F.Y.; Huang, X.F. Some key issues and accesses to the application of magnesium alloys. *Int. J. Modern Phys. B* **2010**, *24*, 2237–2242.

2. Fan, Z. Development of the rheo-diecasting process for magnesium alloys. *Mater. Sci. Eng. A* **2005**, *413*, 72–78.

3. Eliezer, D.; Aghion, E.; Froes, F. The science, technology, and applications of magnesium. *JOM-US* **1998**, *50*, 30–34.

4. Ji, S.; Qian, M.; Fan, Z. Semisolid processing characteristics of AM series Mg alloys by rheo-diecasting. *Metal. Mater. Trans. A* **2006**, *37A*, 779–787.

5. Flemings, M.C. Solidification processing. *Metal. Mater. Trans. B* **1974**, *5*, 2121–2134.

6. Gadow, R. Lightweight engineering with advanced composite materials-ceramic and metal matrix composites. *Adv. Sci. Technol.* **2006**, *50*, 163–173.

7. Song, M.S.; Zhang, M.X.; Zhang, S.G.; Huang, B.; Li, J.G. In situ fabrication of TiC particulates locally reinforced aluminum matrix composites by self-propagating reaction during casting. *Mater. Sci.Eng. A* **2008**, *473*, 166–171.

8. Contreras, A.; Leon, C.A.; Drew, R.A.L.; Bedolla, E. Wettability and spreading kinetics of Al and Mg on TiC. *Scr. Mater.* **2003**, *48*, 1625–1630.

9. Nie, K.B.; Wang, X.J.; Wu, K.; Xu, L.; Zheng, M.Y.; Hu, X.S. Fabrication of SiC particles-reinforced magnesium matrix composite by ultrasonic vibration. *J. Mater. Sci.* **2012**, *47*, 138–144.

10. Fahami, A.; Nasiri-Tabrizi, B. Characterization of mechanothermal-synthesized hydroxyapatite-magnesium titanate composite nanopowders. *J. Adv. Ceram.* **2013**, *2*, 63–70.

11. Vallauri, D.; Deorsola, F.A. Synthesis of TiC-TiB2-Ni cermets by thermal explosion under pressure. *Mater. Res. Bull.* **2009**, *44*, 1528–1533.

12. Wu, B.; Wang, Z.; Gong, Q.M.; Song, H.H.; Liang, J. Fabrication and mechanical properties of *in situ* prepared mesocarbon microbead/carbon nanotube composites. *Mater. Sci. Eng. A* **2008**, *487*, 271–277.

13. Wang, H.M.; Li, G.R.; Zhao, Y.T.; Chen, G. In situ fabrication and microstructure of Al₂O₃ particles reinforced aluminum matrix composites. *Mater. Sci. Eng. A* **2010**, *527*, 2881–2885.

14. Hassan, S.F.; Tun, K.S.; Gupta, M. Effect of sintering techniques on the microstructure and tensile properties of nano-yttria particulates reinforced magnesium nanocomposites. *J. Alloy. Compd.* **2011**, *509*, 4341–4347.

15. Tham, L.M.; Gupta, M.; Cheng, L. Influence of processing parameters during disintegrated melt deposition processing on near net shape synthesis of aluminium based metal matrix composites. *Mater. Sci. Technol.* **1999**, *15*, 1139–1146.

16. Chen, T.; Zhang, S.; Chen, Y.; Li, Y.; Ma, Y.; Hao, Y. Effects of reheating duration on the microstructures and tensile properties of thixoforged *in situ* Mg₂Siₚ/AM60B composites. *Acta Metal. Sin.* **2014**, *27*, 957–967.

17. Fan, Z.; Fang, X.; Ji, S. Microstructure and mechanical properties of rheo-diecast (RDC) aluminium alloys. *Mater. Sci. Eng.* **2005**, *412*, 298–306.

18. Chen, T.; Ma, Y.; Wang, R.; Li, Y.; Hao, Y. Microstructural evolution during partial remelting of AM60B magnesium alloy refined by MgCO₃. *Trans. Nonferr. Met. Soc. China* **2010**, *20*, 1615–1621.

19. Chen, T.J.; Lu, W.B.; Ma, Y.; Huang, H.J.; Hao, Y. Semisolid microstructure of AM60B magnesium alloy refined by SiC particles. *Int. J. Mater. Res.* **2011**, *102*, 1459–1467.

20. Bennett, L.H.; Massalski, T.B.; Murray, J.L.; Baker, H. *Binary Alloy Phase Diagrams*; American Society For Metals: Russell, OH, USA, 1986; Volume 2.

21. Evangelos, T.; Antonios, Z. Evolution of near-equiaxed microstructure in the semisolid state. *Mater. Sci. Eng.* **2000**, *289*, 228–240.

22. Evangelos, T.; Antonios, Z. Mechanical behavior of alloys with equiaxed microstructure in the semisolid state at high solid content. *Acta. Mater.* **1999**, *47*, 517–528.

23. Chen, T.J.; Hunag, L.K.; Huang, X.F.; Ma, Y.; Hao, Y. Effects of reheating temperature and time on microstructure and tensile properties of thixoforged AZ63 magnesium alloy. *Mater. Sci. Technol.* **2014**, *30*, 96–108.

24. Pillai, R.M.; Pai, B.C.; Satyanarayana, K.G. Smisolid processing of aluminium and composites. *Met. Mater. Proc.* **2001**, *13*, 279–290.

25. Chen, Q.; Yuan, B.G.; Zhao, G.Z.; Shu, D.Y.; Hu, C.K.; Zhao, Z.D.; Zhao, Z.X. Microstructural evolution during reheating and tensile mechanical properties of thixoforged AZ91D-Re magnesium alloy prepared by squeeze casting-solid extrusion. *Mater. Sci. Eng. A* **2012**, *537*, 25–38.

26. Chen, T.J.; Hao, Y.; Sun, J.; Li, Q.L. Microstructural evolution of SiC_p/ZA27 composite modified by Zr during partial remelting. *Acta Mater. Compo. Sin.* **2003**, *5*, 1–7.

27. Hong, T.W.; Kim, S.K.; Ha, H.S.; Kim, M.G.; Lee, D.B.; Kim, Y.J. Microstructural evolution and semisolid forming of SiC particulate reinforced AZ91HP magnesium composites. *Mater. Sci. Technol.* **2000**, *16*, 887–892.

28. Browne, D.J.; Hussey, M.J.; Carr, A.J.; Brabazon, D. Direct thermal method: New process for development of globular alloy microstructure. *Int. J. Cast. Metal. Res.* **2003**, *16*, 418–426.

29. Mabuchi, M.; Higashi, K. Strengthening mechanisms of Mg-Si alloys. *Acta. Mater.* **1996**, *44*, 4611–4618.

30. Arsenault, R.J. Interfaces in metal matrix composites. *Scr. Metall.* **1984**, *18*, 1131–1134.

31. Arsenault, R.J. Strengthening and deformation mechanisms of discontinuous metal matrix composites. *Key Eng. Mater.* **1993**, *79–80*, 265–278.

32. Arsenault, R.J.; Shi, N. Dislocation generation due to differences between the coefficients of thermal expansion. *Mater. Sci. Eng.* **1986**, *81*, 175–187.

8

Formulation of the Effect of Different Alloying Elements on the Tensile Strength of the *in situ* Al-Mg₂Si Composites

Halil Ibrahim Kurt * and Murat Oduncuoglu

Technical Sciences, University of Gaziantep, 27310 Gaziantep, Turkey;
E-Mail: oduncuoglu@gmail.com

* Author to whom correspondence should be addressed; E-Mail: hiakurt@gmail.com

Academic Editor: Hugo F. Lopez

Abstract: In this paper, the effect of different alloying elements on the ultimate tensile strength of Al-Mg₂Si composites is theoretically studied. The feed forward back propagation neural network with sigmoid function is used. The extensive experimental results taken from literature are modeled and mathematical formula is presented in explicit form. In addition, it is observed that magnesium and copper have a stronger effect on the ultimate tensile strength of Al-Mg₂Si composites comparison to other alloying elements. The proposed model shows good agreement with test results and can be used to find the ultimate tensile strength of Al-Mg₂Si composites.

Keywords: Al-Mg₂Si; composites; metal matrix composites; modeling

1. Introduction

In the production of composite materials, aluminum (Al), magnesium (Mg), titanium (Ti) and nickel (Ni) alloys are commonly used as metal matrix. Among the materials, Al and its alloys are the most commonly used matrix material in the production of metal matrix composites (MMCs). The composites are manufactured with the diffusion bonding, power metallurgy and casting (also known as liquid metal infiltration) processes [1]. MMCs are widely used in various industries, especially in the automotive, energy and aerospace applications, as they have excellent mechanical properties. The need to reduce emissions while enhancing performance has driven manufacturers to use more Al in industry.

This effort has been accompanied by the development of new Al alloys specifically tailored for these applications.

Al matrix composites (AMCs) reinforced by ceramic particles are prepared by *in situ* and *ex situ* methods. In the former method, the reinforce phase is synthesized internally in the matrix during the composite fabrication. In the latter method, the reinforce phase is synthesized externally and then added into the matrix during composite fabrication. The composites fabricated by these methods have important advantages such as good corrosion, high wear resistance, low cost, greater strength, compared to unreinforced materials [2,3]. Composites the produced by *in situ* technique exhibit the better particle wetting, even distribution of the reinforcing phase and thermodynamically stable system [4,5].

Al-Mg$_2$Si composites constitute a new category of superlight materials attracting significant interest for potential applications. The Mg$_2$Si intermetallic compound exhibits high melting point, low density, high hardness, low thermal expansion coefficient and reasonably high elastic modulus. The presence of Mg and silicon (Si) particles in the composite matrix with different alloying elements is considered to obtain the appropriate strength values and mechanical properties [6–8]. Additionally, the mechanical behaviors of the composites reinforced with particles were found to be a function of the matrix structure, addition alloying element, the volume fraction, particle size and shape of reinforcement [9].

The use of numerical modeling technique represents a new methodology in many different applications including materials science. One of the most used models is the artificial neural network (ANN), a form of artificial intelligence that has the ability to auto-analyze the relationship between multi-variable inputs without any hypothesis [10]. In many studies, the researchers reported that the ANN can be used as an efficient tool in predicting the properties of composite under given conditions and prescribed materials and the comparison of the designed NN and experimental results shows good agreement [11–15]. Expensive and time consuming tests are required for the determination of tensile properties of Al-based composites containing different additive element. The type and percent weight of alloying elements in the composition affect the ultimate tensile strength of the composite materials. Therefore, it is very important to select and add an element in different composition to obtain the maximum strength. The aim of this detailed theoretical study is to investigate the effect of different alloying elements on the tensile properties of *in situ* Al-Mg$_2$Si composites.

2. Experimental Section

The high cost-time for production and test is one of the most important barriers in the production new types of materials and composites. ANNs can accommodate multiple input variables to predict multiple output variables and are able to learn key information patterns within a multi information domain.

ANN is a mathematical model that performs a computational simulation of the behavior of neurons as in human brain and in nervous system. ANNs are capable of learning patterns by training with a number of known patterns. The learning ability of NNs procures an advantage in solving complex problems that are too difficult to solve with the analytical or numerical methods [16]. NN consists of the three components, namely: an activation function, weights and bias. Each neuron receives inputs, attached with a weight w_i, which shows the connection strength for that input for each connection.

Each input is multiplied by the corresponding weight of the neuron connection. Next, a bias (b_i) value is added to the summation of inputs and corresponding weights (u) according to following equation:

$$u_i = \sum_{j=1}^{H} w_{ij} x_j + b_i \tag{1}$$

The summation u_i is converted as the output with an activation (transfer) function, $f(u_i)$ yielding a value called the unit's "activation", as following formula:

$$O = f(u_i)O \tag{2}$$

Dataset and Processing

As the preprocessing of the data for ANN training, testing and validation of the models, each input and output variables are scaled to the range of 0 to 1 by the following the formula:

$$x_N = \frac{x - x_{min}}{x_{max} - x_{min}} \tag{3}$$

where x_N is the normalized value of variable x, x_{max} and x_{min} are the maximum and minimum values of the variables, respectively. Output values resulted from ANN also in the range [0,1] and transformed to its equivalent values based on reverse method of normalization technique [3]. The unnormalized method is as:

$$x = x_N(x_{max} - x_{min}) + x_{min} \tag{4}$$

The ANN is trained and implemented using fully developed feed forward back propagation with sigmoid function. Neural network toolbox in Matlab is used in training of the ANN. The back-propagation is an effective, supervised and the most popular learning method that consists of an input layer, one or more hidden layers and an output layer [17–19]. Sigmoid function is an activation function joins curvilinear, linear and constant behavior depending on the values of the input in ANN system [20,21]. In the ANN, mean square error (MSE) and mean absolute error (MAE) are used as error evaluation criteria in order to facilitate the comparisons between predicted values and desired values. MSE and MAE were calculated by the program.

3. Results and Explicit Formulation of NN Model

The aim of this study is nominal strength prediction of Al-Mg2Si composite materials containing different alloying elements. Therefore, an extensive literature survey has been performed for available experimental results [22–31]. The experimental datasets are divided into three sets as training, validation and test dataset to avoid the over fitting problems. The datasets for training, validation and test are randomly selected from among experimental results where 40 sets are training set as shown in Table 1, 9 sets are validation and test sets as shown in Table 2.

Table 1. Training input data (wt.%).

Data Number	Mg	Si	Cu	Mn	Cr	P	Be	B	Li	Y	Na	Al
1	16.2	9.1	0.01	0.03	0.02	-	-	-	-	-	-	Rem.
2	16.2	9.1	0.01	0.03	0.02	0.1	-	-	-	-	-	Rem.
3	16.2	9.1	0.01	0.03	0.02	0.5	-	-	-	-	-	Rem.
4	16.2	9.1	0.01	0.03	0.02	3	-	-	-	-	-	Rem.
5	9.5	5.5	0.01	0.01	0.02	-	0.1	-	-	-	-	Rem.
6	9.5	5.5	0.01	0.01	0.02	-	0.5	-	-	-	-	Rem.
7	9.68	5.7	0.02	0.02	0.01	-	-	-	-	-	-	Rem.
8	9.68	5.7	0.01	0.01	0.52	-	-	-	-	-	-	Rem.
9	9.68	5.7	0.01	0.01	1.02	-	-	-	-	-	-	Rem.
10	9.68	5.7	0.01	0.01	2.02	-	-	-	-	-	-	Rem.
11	9.5	5.5	0.01	0.01	0.02	-	-	0.1	-	-	-	Rem.
12	9.5	5.5	0.01	0.01	0.02	-	-	0.5	-	-	-	Rem.
13	9.7	5.5	0.01	0.01	0.02	-	-	-	-	-	-	Rem.
14	9.7	5.5	0.01	0.51	0.02	-	-	-	-	-	-	Rem.
15	9.7	5.5	0.01	1.01	0.02	-	-	-	-	-	-	Rem.
16	9.7	5.5	0.01	2.01	0.02	-	-	-	-	-	-	Rem.
17	9.7	5.5	0.01	3.01	0.02	-	-	-	-	-	-	Rem.
18	9.7	5.5	0.01	5.01	0.02	-	-	-	-	-	-	Rem.
19	9.7	5.5	0.01	0.01	0.02	-	-	-	-	-	-	Rem.
20	9.7	5.5	0.01	0.01	0.02	-	-	-	-	-	-	Rem.
21	9.7	5.5	0.01	0.01	0.02	-	-	-	-	-	-	Rem.
22	9.7	5.5	0.01	0.01	0.02	-	-	-	0	-	-	Rem.
23	9.7	5.5	0.01	0.02	0.01	-	-	-	5.5	-	-	Rem.
24	9.74	6	0.01	0.01	0.03	-	-	-	6.12	-	-	Rem.
25	9.47	7	0.01	0.01	0.02	-	-	-	7.11	-	-	Rem.
26	9.54	11	0.01	0.02	0.03	-	-	-	1.37	-	-	Rem.
27	9.7	5.5	0.01	0.02	0.01	-	-	-	-	-	-	Rem.
28	9.7	5.5	0.01	0.01	0.02	-	-	-	-	0.1	-	Rem.
29	9.7	5.5	0.01	0.01	0.02	-	-	-	-	0.5	-	Rem.
30	9.7	5.5	0.01	0.01	0.02	-	-	-	-	1	-	Rem.
31	9.7	5.5	0.01	0.01	0.02	-	-	-	-	-	-	Rem.
32	9.7	5.5	0.01	0.01	0.02	-	-	-	-	-	0.01	Rem.
33	9.7	5.5	0.01	0.01	0.02	-	-	-	-	-	0.05	Rem.
34	9.7	5.5	0.01	0.01	0.02	-	-	-	-	-	0.08	Rem.
35	9.7	5.5	0.01	0.01	0.02	-	-	-	-	-	0.15	Rem.
36	9.82	5.7	0.11	0.01	0.01	-	-	-	-	-	-	Rem.
37	9.82	5.7	0.31	0.01	0.02	-	-	-	-	-	-	Rem.
38	9.82	5.7	0.51	0.01	0.02	-	-	-	-	-	-	Rem.
39	9.82	5.7	1.01	0.01	0.02	-	-	-	-	-	-	Rem.
40	9.82	5.7	5.01	0.01	0.02	-	-	-	-	-	-	Rem.

The input (independent) variables are Al, Mg, Si, copper (Cu), manganese (Mn), chromium (Cr), phosphorus (P), beryllium (Be), boron (B), lithium (Li), yttrium (Y) and sodium (Na) wt.%. The output (dependent) variable is the ultimate tensile strength (UTS) in unit of MPa. Levenberg–Marquardt (Trainlm) algorithm with back propagation is used in the training of NN.

Table 2. Validation and test input data (wt.%).

Part	Data Number	Mg	Si	Cu	Mn	Cr	P	Be	B	Li	Y	Na	Al
	1	16.2	9.1	0.01	0.03	0.02	1	-	-	-	-	-	Rem.
	2	9.5	5.5	0.01	0.01	0.02	-	-	-	-	-	-	Rem.
	3	9.68	5.7	0.01	0.01	5.02	-	-	-	-	-	-	Rem.
	4	9.7	5.5	0.01	0.01	0.02	-	-	-	-	-	-	Rem.
Validation	5	9.7	5.5	0.01	0.01	0.02	-	-	-	-	-	-	Rem.
	6	9.66	6.5	0.01	0.01	0.02	-	-	-	6.45	-	-	Rem.
	7	9.52	13	0.02	0.02	0.03	-	-	-	12.7	-	-	Rem.
	8	9.7	5.5	0.01	0.01	0.02	-	-	-	-	0.3	-	Rem.
	9	9.7	5.5	0.01	0.01	0.02	-	-	-	-	-	-	Rem.
	1	16.2	9.1	0.01	0.03	0.02	0.1	-	-	-	-	-	Rem.
	2	9.5	5.5	0.01	0.01	0.02	-	0.3	-	-	-	-	Rem.
	3	9.68	5.7	0.01	0.01	3.02	-	-	-	-	-	-	Rem.
	4	9.5	5.5	0.01	0.01	0.02	-	-	-	-	-	-	Rem.
Test	5	9.5	5.5	0.01	0.01	0.02	-	-	0.3	-	-	-	Rem.
	6	9.5	5.5	0.01	0.01	0.02	-	-	1	-	-	-	Rem.
	7	9.62	7.5	0.01	0.01	0.02	-	-	-	7.65	-	-	Rem.
	8	9.7	5.5	0.01	0.01	0.02	-	-	-	-	-	0.2	Rem.
	9	9.82	5.7	3.01	0.01	0.02	-	-	-	-	-	-	Rem.

The performance of an NN is affected by the network architecture, initial weight and learning rate. One of the most difficult tasks in NN works is the determination of the number of hidden layers and the number of neurons in per layer. There is no well-defined procedure to find the optimal settings of parameters and network architecture. The trial and error approach is used to determine the number of neurons in the hidden layer. Three different neuron numbers in one hidden layer (12, 13 and 14) are used in this study. The training data set (70%) is used to determine the weights and learning ability of the network. It is known that increasing the data used in training process of NN enhances the learning ability of NN. After the network is trained, the validation dataset is used to measure the performance and generalization ability of the network. The test dataset is used to verify the effectiveness of the network and to estimate the expected performance in the future. It is observed that the optimal NN architecture is found to be 12–12–1 NN architecture with logistic sigmoid transfer function.

The experimental results are compared with the predicted results for the performance of NN model. Figure 1 shows the correlation of NN and experimental results for training set.

In the all stages of NN work, the effects of the percent weight of alloying elements on the strength of Al-Mg2Si composites are quantified. The prediction accuracy of NN for training set is quite satisfactory. Severe deviations between the experimental and theoretical results are observed in training of NN. These can be attributed to the sizes, volume fraction and morphology of Mg2Si phase and other phases formed in the matrix and variation in experimental conditions. It is known that the various geometric shapes of Mg₂Si crystals and the formation and morphology of intermetallic phases have an important effect on the strength of Al-Mg₂Si composites [32].

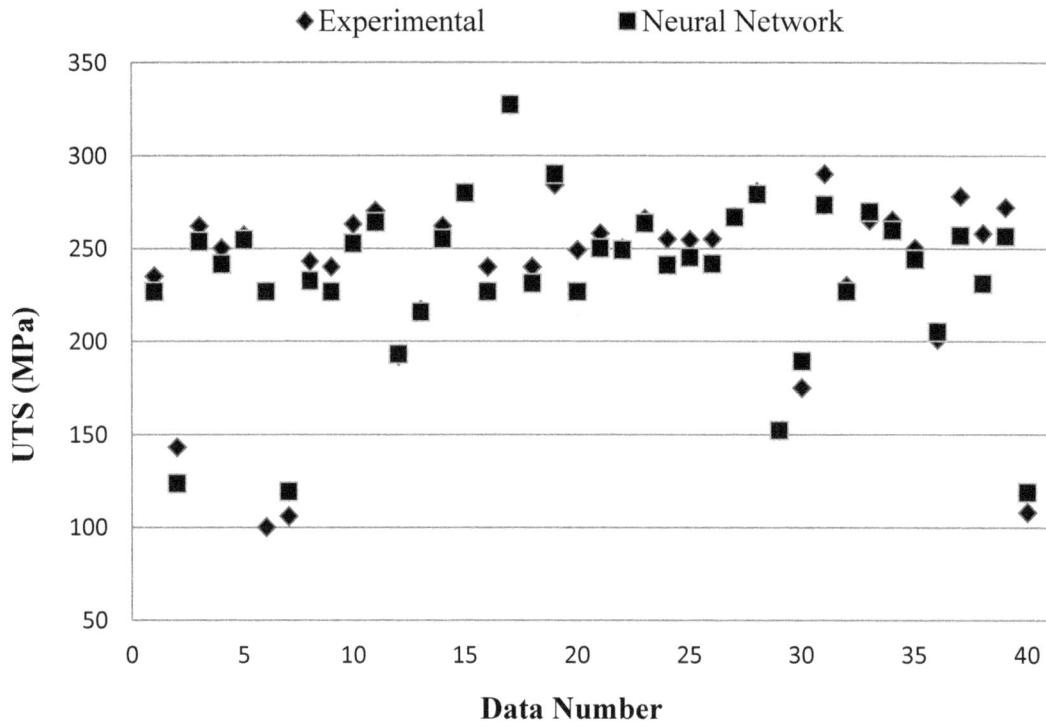

Figure 1. Correlation of NN and experimental results for training set.

The correlation of NN and experimental results for validation and test sets are shown in Figure 2a,b, respectively. After the network is trained, the validation data (Figure 2a) are used to check that the model behaves correctly when presented with previously unseen data. The validation (15%) and test (15%) data sets are randomly selected by program from among 58 experimental data. It is not interfered to the program in the selection stage of the data sets. It is clearly seen from Figure 2 that the experimental and predicted values developed from ANN for UTS are very close to each other and it can also be seen a few minor deviations. The predictability ratio of proposed NN increases from training to test set.

The statistical parameters of training, validation and test datasets of the NN model is given in Table 3. In the train set, the observed correlation coefficient (R) is 0.899, which means that the performance of trained network model is acceptable. The model is verified against the cases in the test dataset, which are independent of the cases in the train dataset. The predicted results are plotted versus the experimental results. As shown in Table 3, the correlation coefficients of validation and test sets are 0.932 and 0.951, indicating that the network can predict the UTS of Al-Mg₂Si composite materials with high accuracy and reliability. It can be said that the using of all the alloying elements with the different weight ratio in the training stage of NN is contributed to the increasing of R values of the validation and test sets.

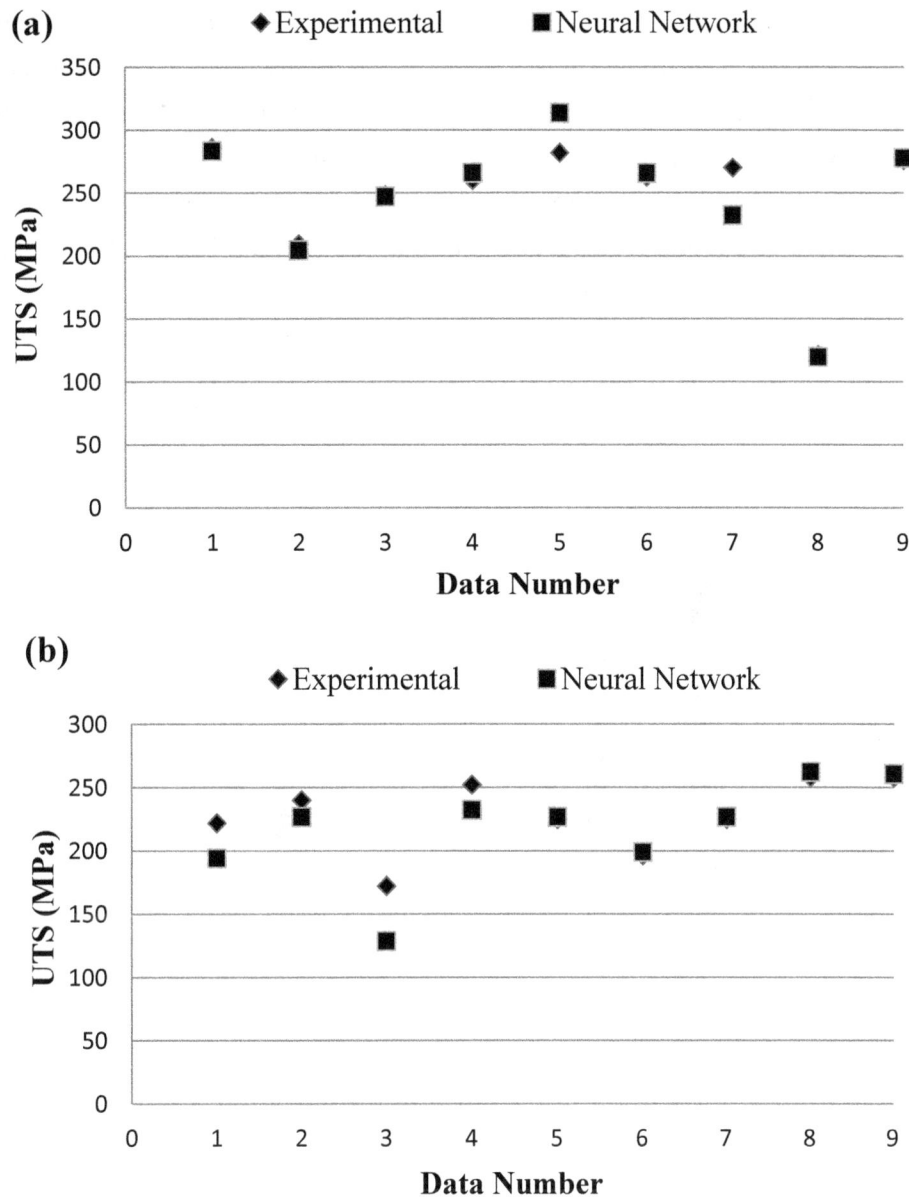

Figure 2. Correlation of neural network (NN) and experimental results for (**a**) validation and (**b**) test sets.

Table 3. Statistical parameters of train, validation and test sets.

Part	R	MSE	MAE
Train set	0.899	1.005	5.123
Validation set	0.932	0.706	5.765
Test set	0.951	0.537	4.385

As mentioned before, the effects of the percent weight of alloying elements on the strength of Al-Mg$_2$Si are considered. The change in UTS values are probably due to the sizes, volume fraction and morphology of the phases and it needs extensive studies. Nevertheless, R values of training, validation and test sets indicate that the learning ability of NN is well enough and the proposed NN model has high accuracy.

MSE and MAE are used to fix the performance of the proposed NN in prediction technique. MSE is 1.005% for training set, 0.706% for validation set and 0.537% for test set. If the MSE reaches zero, the performance of model is regarded as the excellent [13]. MAE is 5.123% for training set, 5.765% for validation set and 4.385% for test set. It is clear that the level of error decreases from training set to test set. These error criteria show that the main source of prediction error is the "noise" in the experimental data. These levels of error can be accepted and considered to be satisfactory. It can be said that the UTS of Al-Mg$_2$Si composites can be predicted by ANN model with 95.1% accuracy and the presented NN model is in good agreement with the experimental data and all errors are within acceptable ranges. The sensitivity of input vectors on UTS of Al-Mg$_2$Si composites is given in Figure 3.

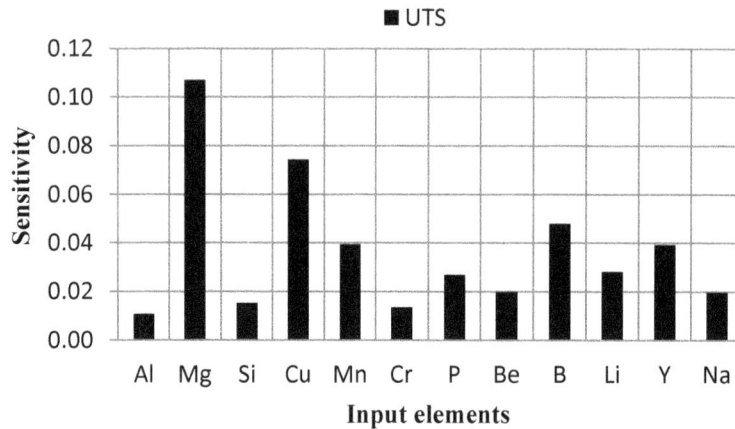

Figure 3. The sensitivity of input vectors.

The mechanical property of the *in situ* composites has a great relationship with the size and morphology of the Mg$_2$Si phases [33]. It is seen that Mg has more impact on UTS of Al-Mg$_2$Si composites than the other alloying elements. Any change in Mg and Cu levels will have significant effect the UTS.

The main focus is to obtain the explicit formulation of UTS for Al-Mg$_2$Si composites as a function of addition alloying elements. The proposed equation below is obtained using a developed macro in Matlab program. It should be noted that the proposed explicit formulation is valid for the ranges of training set.

$$Y = 227 \times \left(\frac{1}{1+e^{-w}} \right) + 100 \tag{5}$$

where Y is the ultimate tensile strength and w is

$w = (-2.92) * \left(\frac{1}{1+e^{-u1}} \right) + (0.19) * \left(\frac{1}{1+e^{-u2}} \right) + (-2.66) * \left(\frac{1}{1+e^{-u3}} \right) + (1.18) * \left(\frac{1}{1+e^{-u4}} \right) + (0.66) * \left(\frac{1}{1+e^{-u5}} \right) +$

$(-0.47) * \left(\frac{1}{1+e^{-u6}} \right) + (1.30) * \left(\frac{1}{1+e^{-u7}} \right) + (0.28) * \left(\frac{1}{1+e^{-u8}} \right) + (2.29) * \left(\frac{1}{1+e^{-u9}} \right) + (-4.70) * \left(\frac{1}{1+e^{-u10}} \right) + (0.88) *$

$\left(\frac{1}{1+e^{-u11}} \right) + (-1.01) * \left(\frac{1}{1+e^{-u12}} \right) + 0.99$

where

$u1 = (-0.81) * K1 + (1.70) * K2 + (1.24) * K3 + (-1.86) * K4 + (0.37) * K5 + (0.50) * K6 + (-0.21) * K7 +$

$(-0.92) * K8 + (-3.10) * K9 + (1.83) * K10 + (-0.85) * K11 + (0.20) * K2 + (0.47)$

$u2 = (0.60) * K1 + (-1.29) * K2 + (-0.33) * K3 + (1.63) * K4 + (0.07) * K5 + (0.62) * K6 + (0.75) * K7 + (0.78) *$

$K8 + (0.80) * K9 + (0.07) * K10 + (0.51) * K11 + (0.58) * K2 + (0.06)$

$u3 = (0.19) * K1 + (-0.24) * K2 + (-2.11) * K3 + (-0.78) * K4 + (-0.09) * K5 + (-0.35) * K6 + (-0.66) * K7 +$

$(0.65) * K8 + (0.57) * K9 + (-0.08) * K10 + (0.88) * K11 + (0.58) * K2 + (-0.93)$

$u4 = (0.57) * K1 + (-1.63) * K2 + (-0.88) * K3 + (1.56) * K4 + (0.31) * K5 + (0.02) * K6 + (0.43) * K7 + (0.75) *$

$K8 + (1.57) * K9 + (-0.79) * K10 + (0.09) * K11 + (-1.16) * K2 + (0.51)$

$u5 = (0.30) * K1 + (-1.08) * K2 + (-0.87) * K3 + (1.94) * K4 + (0.89) * K5 + (0.44) * K6 + (0.70) * K7 + (0.82) *$

$K8 + (1.92) * K9 + (-0.03) * K10 + (0.73) * K11 + (0.89) * K2 + (0.43)$

$u6 = (-0.10) * K1 + (0.47) * K2 + (0.24) * K3 + (-0.39) * K4 + (0.25) * K5 + (-0.06) * K6 + (0.09) * K7 +$

$(-0.51) * K8 + (0.22) * K9 + (0.11) * K10 + (0.04) * K11 + (0.12) * K2 + (0.36)$

$u7 = (0.56) * K1 + (-0.33) * K2 + (-0.82) * K3 + (0.19) * K4 + (-0.69) * K5 + (0.07) * K6 + (0.65) * K7 +$

$(0.33) * K8 + (0.42) * K9 + (0.70) * K10 + (1.08) * K11 + (0.35) * K2 + (0.82)$

$u8 = (0.49) * K1 + (-1.50) * K2 + (-0.70) * K3 + (0.90) * K4 + (0.57) * K5 + (0.17) * K6 + (0.20) * K7 + (0.13) *$

$K8 + (0.91) * K9 + (0.03) * K10 + (0.69) * K11 + (0.82) * K2 + (0.53)$

$u9 = (0.37) * K1 + (-0.79) * K2 + (-0.10) * K3 + (0.88) * K4 + (-0.08) * K5 + (0.03) * K6 + (0.42) * K7 +$

$(-0.06) * K8 + (-0.74) * K9 + (0.26) * K10 + (-0.65) * K11 + (-0.36) * K2 + (0.10)$

$u10 = (1.21) * K1 + (1.26) * K2 + (-1.75) * K3 + (-1.79) * K4 + (-3.59) * K5 + (-1.40) * K6 + (-0.44) * K7 +$

$(-0.89) * K8 + (-0.68) * K9 + (0.22) * K10 + (-3.15) * K11 + (-2.48) * K2 + (-1.46)$

$u11 = (0.23) * K1 + (-0.20) * K2 + (-0.47) * K3 + (0.18) * K4 + (-0.09) * K5 + (0.07) * K6 + (0.38) * K7 +$

$(-0.37) * K8 + (-0.68) * K9 + (0.70) * K10 + (-0.22) * K11 + (-0.07) * K2 + (0.22)$

$u12 = (0.68) * K1 + (-0.33) * K2 + (-0.41) * K3 + (1.32) * K4 + (0.32) * K5 + (0.41) * K6 + (0.20) * K7 +$

$(0.74) * K8 + (0.49) * K9 + (-0.84) * K10 + (0.27) * K11 + (0.53) * K2 + (0.04)$

where K1, K2, K3, K4, K5, K56, K7, K8, K9, K10, K11 and K12 are normalized input data of Al, Mg, Si, Fe, Cu, Mn, Cr, Ti, Ni, Zn, P, Zr, Be, B, Li, Y and Na.

The training data set contains the weight percentage of all alloying elements and is used to determine the best set of neural network weights. Thus, the proposed equation can be used for determining the effects of all alloying elements on the strength of Al-Mg2Si composites. It is observed that the strength of Al-9.5Mg-5.5Si-0.01Cu-0.01Mn-0.02Cr alloy used in test set with 3 wt.% Cu increased from 252 to 317 ± 12 MPa using the derived formula. The multiple minor alloying elements (0.02P + 0.15Be + 0.2B + 0.2Y + 0.03Na wt.%) is theoretically added into

Al-16.2Mg-9.1Si-0.01Cu-0.03Mn-0.02Cr alloy. The strength of the alloy is increased from 108 to 189 ± 9 MPa. The main reason for this improvement can be attributed to the precipitation of more primary complex particles [34].

4. Conclusions

This work proposes an approach for ultimate tensile strength in prediction of Al-Mg$_2$Si composites containing different alloying elements. The back propagation NNs are used for the training process and the proposed NN model shows good agreement with experimental results. The results also demonstrate that all the data sets have quite high correlation and accuracy. Therefore, the mathematical function is derived in explicit form by using ANN. The outcomes of the study are very promising in general. The mean absolute error for predicted values does not exceed 5.8%. The sensitivity analysis of the model demonstrated that the addition of Mg has greater effect than the other elements on the UTS of Al-M$_2$Si composites. Hence it is concluded that considerable saving in terms of cost and time could be obtained by using the model and ANN is a successful analytical tool provided it is properly used.

Author Contributions

These authors contributed equally to this work. Kurt and Oduncuoglu analyzed and interpreted the data and prepared the manuscript. All authors discussed the conclusions and reviewed the manuscript.

Conflicts of Interest

The authors declare no conflict of interest.

References

1. Schwartz, M.M. *Composite Materials: Processing, Fabrication, and Applications*; Prentice Hall PTR: Upper Saddle River, NJ, USA, 1997.
2. Smith, W.F.; Hashemi, J. *Foundations of Materials Science and Engineering*; McGraw-Hill: New York, NY, USA, 2003.
3. Hassan, A.M.; Alrashdan, A.; Hayajneh, M.T.; Mayyas, A.T. Prediction of density, porosity and hardness in aluminum–copper-based composite materials using artificial neural network. *J. Mater. Process. Technol.* **2009**, *209*, 894–899.
4. Xu, C.L.; Wang, H.Y.; Yang, Y.F.; Jiang, Q.C. Effect of Al–P–Ti–TiC–Nd2O3 modifier on the microstructure and mechanical properties of hypereutectic Al–20 wt.%Si alloy. *Mater. Sci. Eng. A* **2007**, *452–453*, 341–346.
5. Akrami, A.; Emamy, M.; Mousavian, H. The effect of bi addition on the microstructure and tensile properties of cast Al-15%Mg2Si composite über den einfluss der zumischung von bi auf die mikrostruktur und zugeigenschaften von gegossenen al-15%mg2si verbunden. *Materialwissenschaft und Werkstofftechnik* **2013**, *44*, 431–435.
6. Jiang, Q.C.; Wang, H.Y.; Wang, Y.; Ma, B.X.; Wang, J.G. Modification of Mg2Si in Mg–Si alloys with yttrium. *Mater. Sci. Eng. A* **2005**, *392*, 130–135.

7. Qin, Q.D.; Zhao, Y.G.; Liu, C.; Cong, P.J.; Zhou, W.; Liang, Y.H. Effect of holding temperature on semisolid microstructure of Mg2Si/Al composite. *J. Alloys Compd.* **2006**, *416*, 143–147.

8. Liao, L.-H.; Wang, H.-W.; Li, X.-F.; Ma, N.-H. Research on the dislocation damping of Mg2Si/Mg–9Al composite materials. *Mater. Lett.* **2007**, *61*, 2518–2522.

9. Shabani, M.O.; Mazahery, A. Prediction of wear properties in A356 matrix composite reinforced with B4C particulates. *Synth. Metals* **2011**, *161*, 1226–1231.

10. Yong, L.Z. Supervised classification of multispectral remote sensing image using BP neural network. *J. Infrared Millim. Waves* **1998**, *2*, 153–156.

11. Durmuş, H.K.; Özkaya, E.; Meriç, C. The use of neural networks for the prediction of wear loss and surface roughness of AA 6351 Aluminium alloy. *Mater. Des.* **2006**, *27*, 156–159.

12. Altinkok, N.; Koker, R. Modelling of the prediction of tensile and density properties in particle reinforced metal matrix composites by using neural networks. *Mater. Des.* **2006**, *27*, 625–631.

13. Sun, Y.; Zeng, W.D.; Zhang, X.M.; Zhao, Y.Q.; Ma, X.; Han, Y.F. Prediction of tensile property of hydrogenated Ti600 titanium alloy using artificial neural network. *J. Mater. Eng. Perform.* **2011**, *20*, 335–340.

14. Amirjan, M.; Khorsand, H.; Siadati, M.H.; Eslami Farsani, R. Artificial neural network prediction of Cu–Al2O3 composite properties prepared by powder metallurgy method. *J. Mater. Res. Technol.* **2013**, *2*, 351–355.

15. *Matlab*, version 7.13; The MathWorks Inc.: Natick, MA, USA, 2011.

16. Rojas, R. *Neural Networks: A Systematic Introduction*; Springer: Berlin, Germany, 1996.

17. Hecht-Nielsen, R. *Neurocomputing*; Addison-Wesley Publishing Company: New York, NY, USA, 1990.

18. Rumelhart, D.E.; Hinton, G.E.; Williams, R.J. Learning internal representations by error propagation. Available online: http://psych.stanford.edu/~jlm/papers/PDP/Volume%201/Chap8_PDP86.pdf (accessed on 3 March 2015).

19. Cevik, A.; Kutuk, M.A.; Erklig, A.; Guzelbey, I.H. Neural network modeling of arc spot welding. *J. Mater. Process. Technol.* **2008**, *202*, 137–144.

20. Roiger, R.; Geatz, M. *Data mining: A Tutorial-Based Primer*; Addison Wesley: New York, NY, USA, 2003.

21. Larose, D.T. Neural networks. In *Discovering Knowledge in Data*; John Wiley & Sons, Inc.: New York, NY, USA, 2005; pp. 128–146.

22. Yeganeh, S.E.V.; Razaghian, A.; Emamy, M. The influence of Cu–15P master alloy on the microstructure and tensile properties of Al–25wt% Mg2Si composite before and after hot-extrusion. *Mater. Sci. Eng. A* **2013**, *566*, 1–7.

23. Azarbarmas, M.; Emamy, M.; Rassizadehghani, J.; Alipour, M.; karamouz, M. The influence of beryllium addition on the microstructure and mechanical properties of Al–15%Mg2Si *in situ* metal matrix composite. *Mater. Sci. Eng. A* **2011**, *528*, 8205–8211.

24. Ghorbani, M.R.; Emamy, M.; Nemati, N. Microstructural and mechanical characterization of Al–15%Mg2Si composite containing chromium. *Mater.Des.* **2011**, *32*, 4262–4269.

25. Azarbarmas, M.; Emamy, M.; karamouz, M.; Alipour, M.; Rassizadehghani, J. The effects of boron additions on the microstructure, hardness and tensile properties of *in situ* Al–15%Mg2Si composite. *Mater. Des.* **2011**, *32*, 5049–5054.

26. Ghorbani, M.R.; Emamy, M.; Khorshidi, R.; Rasizadehghani, J.; Emami, A.R. Effect of mn addition on the microstructure and tensile properties of Al–15%Mg2Si composite. *Mater. Sci. Eng. A* **2012**, *550*, 191–198.

27. Razaghian, A.; Bahrami, A.; Emamy, M. The influence of Li on the tensile properties of extruded *in situ* Al–15%Mg2Si composite. *Mater. Sci. Eng. A* **2012**, *532*, 346–353.

28. Nasiri, N.; Emamy, M.; Malekan, A. Microstructural evolution and tensile properties of the *in situ* Al–15%Mg2Si composite with extra Si contents. *Mater. Des.* **2012**, *37*, 215–222.

29. Emamy, M.; Jafari Nodooshan, H.R.; Malekan, A. The microstructure, hardness and tensile properties of Al–15%Mg2Si *in situ* composite with yttrium addition. *Mater. Des.* **2011**, *32*, 4559–4566.

30. Emamy, M.; Khorshidi, R.; Raouf, A.H. The influence of pure Na on the microstructure and tensile properties of Al-Mg2Si metal matrix composite. *Mater. Sci. Eng. A* **2011**, *528*, 4337–4342.

31. Emamy, M.; Nemati, N.; Heidarzadeh, A. The influence of cu rich intermetallic phases on the microstructure, hardness and tensile properties of Al–15% Mg2Si composite. *Mater. Sci. Eng. A* **2010**, *527*, 2998–3004.

32. Li, C.; Wu, Y.Y.; Li, H.; Liu, X.F. Morphological evolution and growth mechanism of primary Mg2Si phase in Al–Mg2Si alloys. *Acta Mater.* **2011**, *59*, 1058–1067.

33. Zhang, J.; Wang, Y.-Q.; Yang, B.; Zhou, B.-L. Effects of Si content on the microstructure and tensile strength of an *in situ* Al/Mg2Si composite. *J. Mater. Res.* **1999**, *14*, 68–74.

34. Norman, A.F.; Hyde, K.; Costello, F.; Thompson, S.; Birley, S.; Prangnell, P.B. Examination of the effect of Sc on 2000 and 7000 series aluminium alloy castings: For improvements in fusion welding. *Mater. Sci. Eng. A* **2003**, *354*, 188–198.

9

Effect of Milling Time and the Consolidation Process on the Properties of Al Matrix Composites Reinforced with Fe-Based Glassy Particles

Özge Balcı [1,2], Konda Gokuldoss Prashanth [2,†,*], Sergio Scudino [2], Duygu Ağaoğulları [1], İsmail Duman [1], M. Lütfi Öveçoğlu [1], Volker Uhlenwinkel [3] and Jürgen Eckert [2,4]

[1] Particulate Materials Laboratories (PML), Department of Metallurgical and Materials Engineering, İstanbul Technical University, 34469 İstanbul, Turkey; E-Mails: balciozg@itu.edu.tr (O.B.); bozkurtdu@itu.edu.tr (D.A.); iduman@itu.edu.tr (I.D.); ovecoglu@itu.edu.tr (M.L.O.)

[2] Institute for Complex Materials, IFW Dresden, 270116 Dresden, Germany; E-Mails: s.scudino@ifw-dresden.de (S.S.); j.eckert@ifw-dresden.de (J.E.)

[3] Institut für Werkstofftechnik, Universität Bremen, D-28359 Bremen, Germany; E-Mail: uhl@iwt.uni-bremen.de

[4] TU Dresden, Institut für Werkstoffwissenschaft, D-01062 Dresden, Germany

[†] Present Address: R&D Engineer, Additive manufacturing Center, Sandvik AB, 81181 Sandviken, Sweden; E-Mail: prashanth.konda_gokuldoss@sandvik.com.

[*] Author to whom correspondence should be addressed; E-Mail: kgprashanth@gmail.com

Academic Editors: K. C. Chan and Jordi Sort Viñas

Abstract: Al matrix composites reinforced with 40 vol% $Fe_{50.1}Co_{35.1}Nb_{7.7}B_{4.3}Si_{2.8}$ glassy particles have been produced by powder metallurgy, and their microstructure and mechanical properties have been investigated in detail. Different processing routes (hot pressing and hot extrusion) are used in order to consolidate the composite powders. The homogeneous distribution of the glassy reinforcement in the Al matrix and the decrease of the particle size are obtained through ball milling. This has a positive effect on the hardness and strength of the composites. Mechanical tests show that the hardness of the hot pressed samples increases from 51–155 HV, and the strength rises from 220–630 MPa by extending the milling time from 1–50 h. The use of hot extrusion after hot pressing reduces both the strength and hardness of the composites: however, it enhances the plastic deformation significantly.

Keywords: metallic glasses; composites; powder metallurgy; mechanical characterization

1. Introduction

Al-based metal matrix composites (MMCs) have attracted considerable interest due to their superior properties, including high strength and good fatigue and wear resistance [1–3]. They are advanced engineering materials able to meet the increasing demand for structural and thermal applications, particularly in the aerospace and automotive industries [4,5]. Al-based MMCs offer a unique combination of properties, including the ductility of the matrix and the strength of the reinforcement, that cannot be found in conventional unreinforced materials [4,6]. Various materials comprised of ceramics (e.g., SiO_2, Al_2O_3, SiC, TiC, TiB_2, ZrB_2 and AlN), metallic glasses and complex metallic alloys have been successfully used as reinforcements in Al-based MMCs in the form of fibers, flakes or particulates [7–15]. Amongst them, particulate-reinforced MMCs are particularly attractive due to their easier fabrication routes and lower costs compared to fibers or flakes [4,13].

Metallic glasses have been recently proposed as an effective type of reinforcement due to their exceptional mechanical properties, such as high strength, hardness and corrosion resistance [16–18]. Metallic glasses are believed to be more compatible with the metal matrix and may lead to improved interface strength between the matrix and reinforcement than their ceramic counterparts [16–18]. In addition, the sintering process conducted within supercooled liquid (SCL), where metallic glasses display a significant decrease of viscosity, can assist with the consolidation, resulting in bulk samples with reduced porosity [18,19]. Thus, in order to obtain highly-dense materials with increased mechanical properties, Al-, Zr-, Fe-, Ni- and Cu-based metallic glass particles or ribbons have been used as reinforcements in Al-based metal matrix composites [17–24]. Amongst the different types of metallic glasses, Fe-based glasses are of considerable interest, because of their ultrahigh strength, good corrosion resistance, low cost, excellent soft magnetic properties, good glass forming ability and large SCL region [25–29]. Aljerf et al. reported that the use of $[(Fe_{1/2}Co_{1/2})_{75}B_{20}Si_5]_{96}Nb_4$ glassy particles as reinforcement in the Al-6061 alloy leads to a remarkable combination of high strength and plasticity [19]. Similar results have been recently reported for Al-2024 matrix composites reinforced with $Fe_{73}Nb_5Ge_2P_{10}C_6B_4$ glassy particles [30].

Powder metallurgy is one of the methods successfully used for the preparation of MMCs [1,7,31]. The main advantage of powder metallurgy over other techniques is the low processing temperature, which may prevent unwanted interfacial reactions between the matrix and reinforcement and which permits the economic feasibility of large-scale production, thus allowing the commercial processing of MMCs [32,33]. Furthermore, it provides a homogeneous distribution of the reinforcements within the matrix, and it enables a high degree of control over the product microstructure (volume fraction, size, shape, etc.), which is comparatively limited in the casting or diffusion welding routes [7,24]. The current studies related to the metallic glass-reinforced Al-based MMCs produced via powder metallurgy are mainly focused on the effect of the glassy particle content on the properties of the composites [23–25]. On the other hand, the effect of microstructural modifications, such as particle shape and interparticle distance, induced by mechanical treatments, like ball milling, as well as the influence of the consolidation method on the properties of Al-based MMCs have received relatively little attention.

In this study, Al-based MMCs reinforced with Fe-based glassy particles have been produced via powder metallurgy and the effects of milling time (1, 10, 30 and 50 h) and the consolidation process (hot press or hot press followed by hot extrusion) on the microstructure and mechanical properties are investigated.

2. Experimental Section

2.1. Raw Materials

Glassy particles with a nominal composition of $Fe_{50.1}Co_{35.1}Nb_{7.7}B_{4.3}Si_{2.8}$ were produced by high-pressure N_2 gas atomization. The samples produced by this method are powders with sizes ranging from 38–112 μm. Al powders with a purity of 99.5% and an average particle size of 125 μm were used as the matrix material.

2.2. Mechanical Milling

Milling experiments on the powder mixtures comprised of pure Al and 40 vol% glassy particles were performed using a Retsch PM400 planetary ball mill (Retsch, Dusseldorf, Germany) equipped with hardened steel balls and vials and without any process control agents. The powders were milled at room temperature for 1, 10, 30 and 50 h using a ball-to-powder mass ratio (BPR) of 10:1 and at a milling speed of 150 rpm. Such a low milling speed was used only to have uniform dispersion of the reinforcement in the matrix, unlike conventional mechanical milling or alloying. Milling was carried out as a sequence of 15-min milling intervals interrupted by 15-min breaks to avoid a strong temperature rise during milling. All sample handling was carried out in a Braun MB 150B-G glove box under purified Ar atmosphere (less than 0.1 ppm O_2 and H_2O) in order to minimize atmospheric contamination.

2.3. Consolidation

Consolidation of the composite powders was done by uni-axial hot pressing (HP) or hot pressing followed by hot extrusion (HE) under Ar atmosphere at 673 K and 640 MPa. Hot pressing time and the hot extrusion ratio were 30 min and 4:1, respectively.

2.4. Characterization of the Powders and Consolidated Samples

Phase analysis of the powders and consolidated samples was performed by the X-ray diffraction technique (XRD) using a D3290 PANalytical X'pert PRO (PANalytical, Almelo, The Netherlands) with Co-Kα radiation (λ = 0.17889 nm) in the Bragg–Brentano configuration. After applying a series of metallographic treatments, microstructural characterization and elemental mapping of the consolidated samples were carried out by scanning electron microscopy (SEM, Zeiss, Oberkochen,

Germany) using a Gemini 1530 microscope (operated at 15 kV) equipped with energy-dispersive X-ray detection (EDX). The matrix ligament size ($\lambda = L/N$) was calculated from the arithmetic mean of ten measurements by superposing random lines on the high magnification SEM micrographs of the composites. It is determined by the total length falling in the matrix (L) and by counting the number of matrix region intercepts per unit length of test line (N). The thermal behavior of the powders was investigated by differential scanning calorimetry (DSC) with a Perkin-Elmer DSC7 calorimeter (Perkin Elmer, Waltham, MA, USA) at a heating rate of 20 K/min under a continuous flow of purified Ar.

The experimental densities of the consolidated samples were evaluated by the Archimedes principle, and the relative densities were calculated as the percent value from the ratio of the experimental to the theoretical density of Al matrix composite reinforced with 40 vol% $Fe_{50.1}Co_{35.1}Nb_{7.7}B_{4.3}Si_{2.8}$ (4.43 g/cm^3). Microhardness measurements were conducted by a computer-controlled Struers Duramin 5 Vickers hardness tester using a load of 10 g and indenter dwell time of 10 s. The microhardness test result of each sample includes the arithmetic mean of twenty successive indentations and standard deviations. Optical micrographs of the representative hardness indentations performed on the glassy phase distributed Al matrix were also given for each sample. Since the hardness comparison of the composites was aimed at a constant load, measurements were not directly applied on the glassy phase, whose indentation requires a higher value of load than the one utilized. Five different cylindrical specimens 2 mm diameter and 4 mm length were prepared from each of the hot pressed and hot extruded samples and tested at room temperature using an Instron 8562 testing facility under quasistatic compressive loading (strain rate 8×10^{-4} s^{-1}). Both ends of these specimens were carefully polished to make them parallel to each other prior to the compression tests. The strain during compression was measured directly on the specimens using a Fiedler laser-extensometer.

3. Results and Discussion

3.1. Processing and Characterization of the Composite Powders

Figure 1a–c illustrates the XRD patterns of the starting and composite powders milled for different times. Figure 1a shows the XRD patterns of the gas-atomized $Fe_{50.1}Co_{35.1}Nb_{7.7}B_{4.3}Si_{2.8}$ and pure Al (The International Centre for Diffraction Data (ICDD) Card No: 01-072-3440) powders. The glassy powder is not completely amorphous and shows the typical broad maxima characteristic for glassy materials at angles between $2\theta = 40°$ and $2\theta = 60°$ together with small diffraction peaks at about $2\theta = 41°$, $45°$, $50°$ and $53°$. As previously reported for gas-atomized Al-based glassy powders, the cooling rate may not be sufficient to completely suppress the formation of crystalline phases during gas atomization [34]. Figure 1b shows the XRD patterns of the composite powders with 40 vol% $Fe_{50.1}Co_{35.1}Nb_{7.7}B_{4.3}Si_{2.8}$ milled for 1, 10, 30 and 50 h, revealing sharp Bragg peaks, which correspond to the Al matrix. No peaks belonging to additional phases can be observed even after milling for 50 h. Moreover, the Al peaks are broadened, and their intensities gradually decrease with increasing milling time from 1–50 h, indicating a gradual decrease of the crystallite size, as well as the increase in the lattice strain. The weak and diffuse peak at about $2\theta = 50°$–$53°$ indicates the existence of the glassy phase in the Al matrix and proves the amorphous structure of the glassy phase after milling for 1 h. Further milling does not result in significant structural changes.

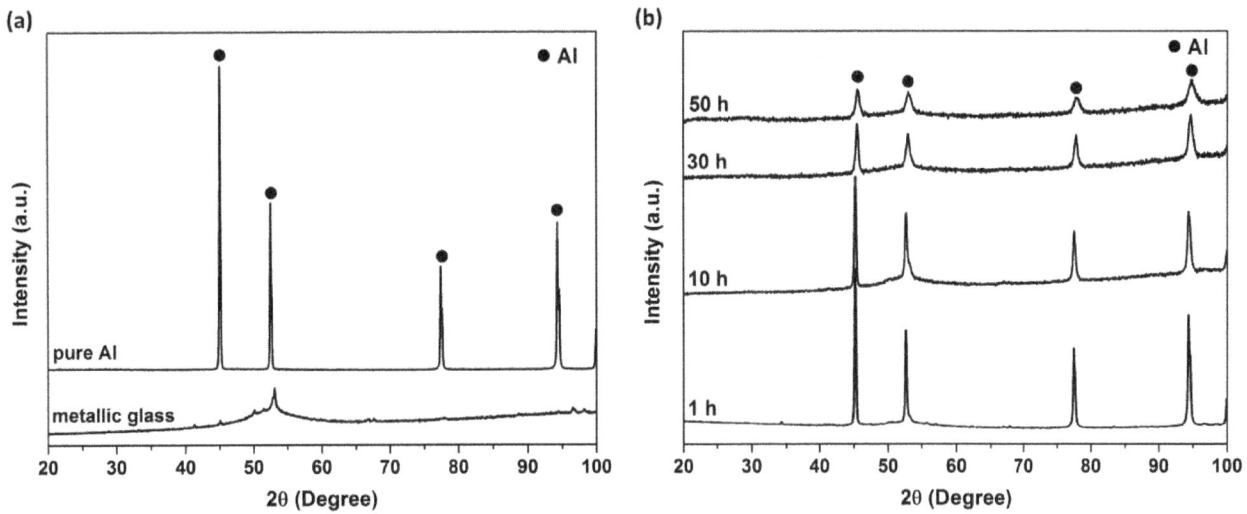

Figure 1. XRD patterns of the starting and composite powders milled for different times: (**a**) pure Al and metallic glass powders; and (**b**) composite powders milled for 1, 10, 30 and 50 h.

Figure 2a displays the DSC scan of the gas-atomized $Fe_{50.1}Co_{35.1}Nb_{7.7}B_{4.3}Si_{2.8}$ glassy powder. The scan displays a glass transition (T_g) at 838 K followed by a crystallization event with the onset (T_x) at 853 K. This is similar to what was observed for the $[(Fe_{1/2}Co_{1/2})_{75}B_{20}Si_5]_{96}Nb_4$ metallic glass, which has T_g and T_x values of 821 and 861 K, respectively [19]. DSC experiments were also carried out for the milled composite powders (Figure 2b) to analyze the effect of milling on T_g and T_x and to select the appropriate consolidation temperature. The DSC curves reveal a glass transition temperature (T_g) followed by the supercooled liquid (SCL) region (defined as $\Delta T_x = T_x - T_g$) before crystallization occurs at higher temperatures (T_x). The SCL regions lie below the melting temperature of the Al matrix, except for the powder milled for 1 h, which has T_x above 920 K. T_g values peaking with very weak endotherms are located at about 837, 800, 733 and 720 K, respectively, for the powders milled for 1, 10, 30 and 50 h. The prolonged milling time from 1–50 h results in a shift of T_g of about 117 K to lower temperatures. Continuous mechanical deformation of powder particles disturbs the bonds, creates dislocations, increases the fresh reactive surfaces of the particles and improves the chemical reactivity [32]. The shift of the glass transition temperature is larger (104 K) for the milling period between 1 and 30 h, whereas it is smaller for the powder milled between 30 and 50 h (13 K). Therefore, milling for 50 h is sufficient to get a desirable reduction in the T_g value of the composite powder, which correspondingly increases the temperature range of the SCL region. Figure 2b also displays broad exothermic events (corresponding to the crystallization of the glassy phase) with onset (T_x) at about 853, 840 and 837 K for the powders milled for 10, 30 and 50 h, therefore decreasing gradually with increasing the milling time. T_x determines the upper temperature limit for the sintering process, because at T_x, metallic glasses lose their liquid-like behavior, and the viscosity increases as crystallization starts [35].

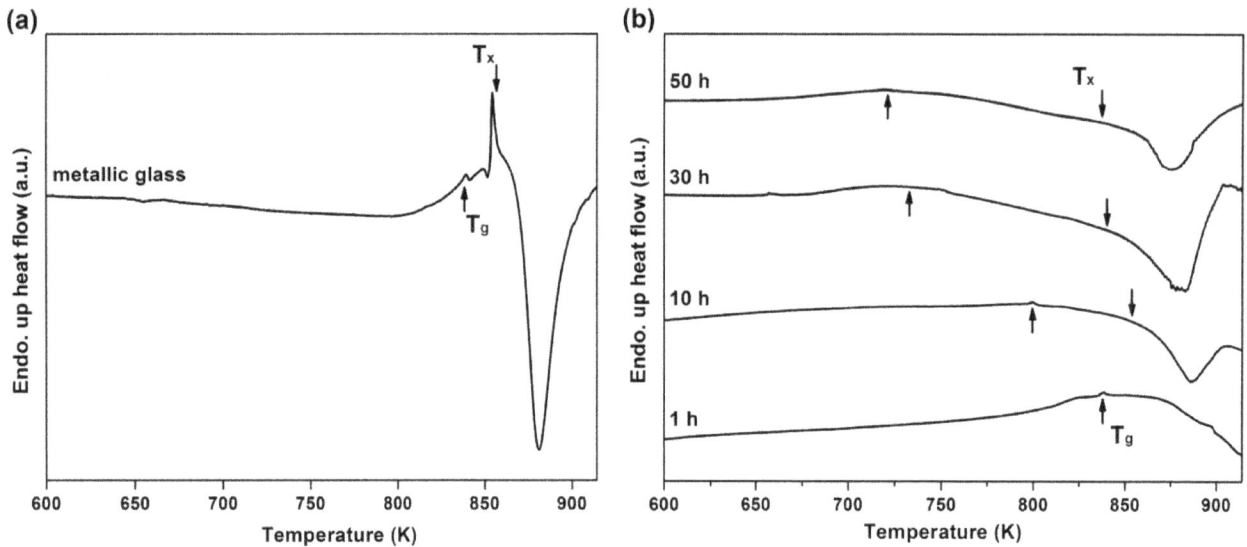

Figure 2. DSC scans **(a)** of the gas-atomized $Fe_{50.1}Co_{35.1}Nb_{7.7}B_{4.3}Si_{2.8}$ powder and **(b)** of the composite powders milled for 1, 10, 30 and 50 h.

The exothermic peaks in Figure 2b are rather broad, in contrast with what is generally observed for the crystallization of metallic glasses. This can be explained by the peak overlap for the peaks related to the crystallization of the glass and to the reaction of the Al matrix with the metallic glass induced by partial Al melting. When the crystallization temperature of the glassy phase in the composites is higher than that of Al melting, the corresponding DSC scan exhibits an endothermic event corresponding to the Al melting at about 835–897 K [21]. On the other hand, similar crystallization and Al melting temperatures were observed for Al-based composites reinforced with $Ni_{60}Nb_{40}$ metallic glass particles [18]. Similar to the present study, they showed broad exotherms in the DSC scans instead of the sharp peaks corresponding to the crystallization of the metallic glass [18]. The reason for this behavior was attributed to the amount of heat release originating from the $Ni_{60}Nb_{40}$ reinforcement, which was diluted in the Al matrix [18]. Furthermore, the reaction of the Al matrix with the metallic glass was observed, supported by the appearance of low-intensity diffraction peaks of aluminide intermetallics ($NiAl_3$ and $NbAl_3$) in the XRD patterns after annealing the composites at temperatures between 893 and 913 K [18].

3.2. Characterization of the Consolidated Samples

Figure 3a exhibits the XRD patterns of the hot pressed composites milled for different times. The patterns display the peaks of the Al matrix along with the diffuse amorphous halo at about $2\theta = 53°$. This indicates that, after consolidation, the glassy reinforcement remains amorphous, corroborating the DSC results showing the exothermic event due to crystallization of the glassy phase. Figure 3c shows the XRD patterns of the hot pressed and hot extruded composites milled for different times. The comparison of Figure 3a and Figure 3c reveals the effect of hot extrusion on the microstructure: after extrusion, the characteristic Al peaks become narrower, and their intensities increase with respect to the hot pressed samples, indicating an increase in the crystallite size, probably due to the stress-induced grain growth during secondary consolidation [36]. As seen in Figure 3c, a small amount of Al_5Fe_2 (ICDD Card No: 00-029-0043) is formed in the extruded specimens. The amount of the

Al_5Fe_2 phase increases with increasing the milling time, as expected from the contribution of deformation-induced crystallization.

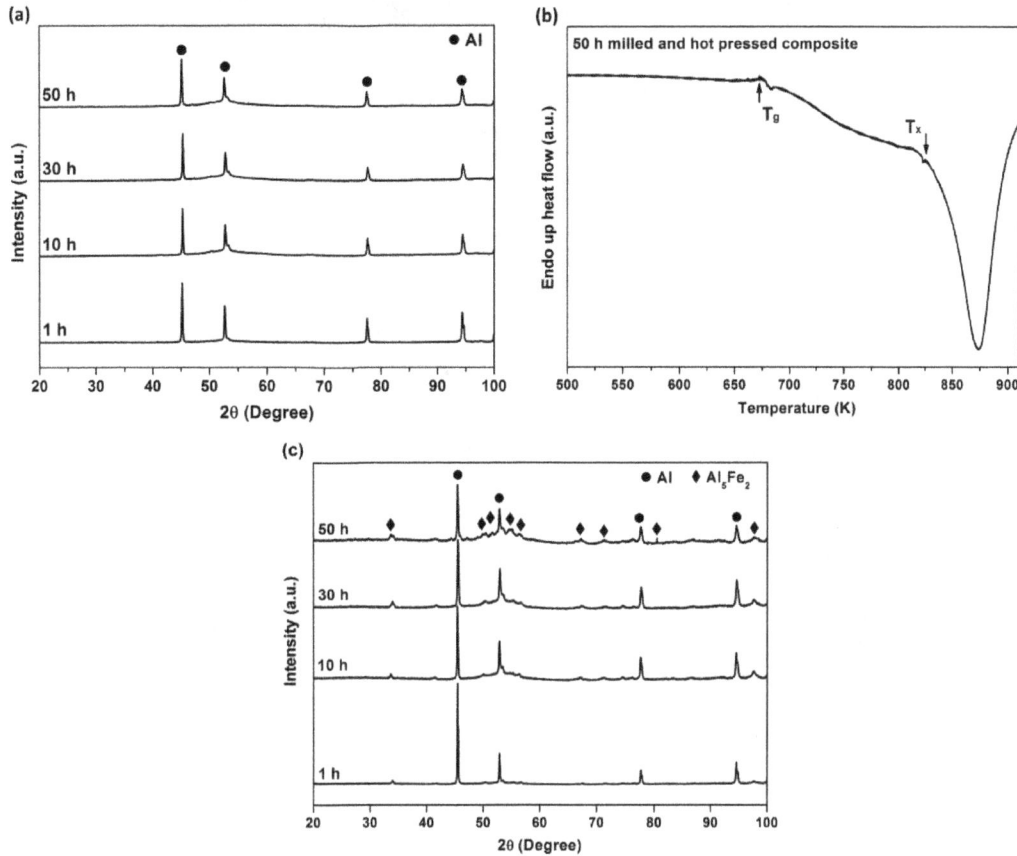

Figure 3. XRD patterns and DSC scan of the milled and consolidated samples: (**a**) XRD patterns of the hot pressed, (**b**) DSC scan of the 50-h milled and hot pressed and (**c**) XRD patterns of the hot pressed and hot extruded composites.

Figure 4a–d shows the SEM micrographs and the EDX elemental maps of the hot pressed composites. The SEM micrographs consist of dark and bright regions corresponding to the Al matrix and Fe-based glassy particles, as shown by the EDX analysis. The images also reveal the irregular shape and size of the Fe-based glassy particles embedded in the continuous Al matrix without observable cracks. The hot pressed composite produced from the powders milled for 1 h is characterized by a microstructure containing spherical glassy particles with an average size of 34 ± 8 μm and only a few pores (marked by arrows in Figure 4a). With increasing the milling time to 50 h (Figure 4d), repeated cold-welding, fracturing and re-welding [32] led to the formation of a heterogeneous microstructure consisting of large spheroidal particles along with small layered glassy particles. This gives rise to a smaller average size and to a broader particle size distribution (15 ± 12 μm). The decrease in the particle size of the Fe-based glassy particles resulting from the mechanically-induced fragmentation and the corresponding decrease of the inter-particle distance can be clearly seen in Figure 4a–d. The reduction of the distance between the particles can contribute considerably to the strength of the composites, because the matrix/particle interface can effectively inhibit dislocation movement [14].

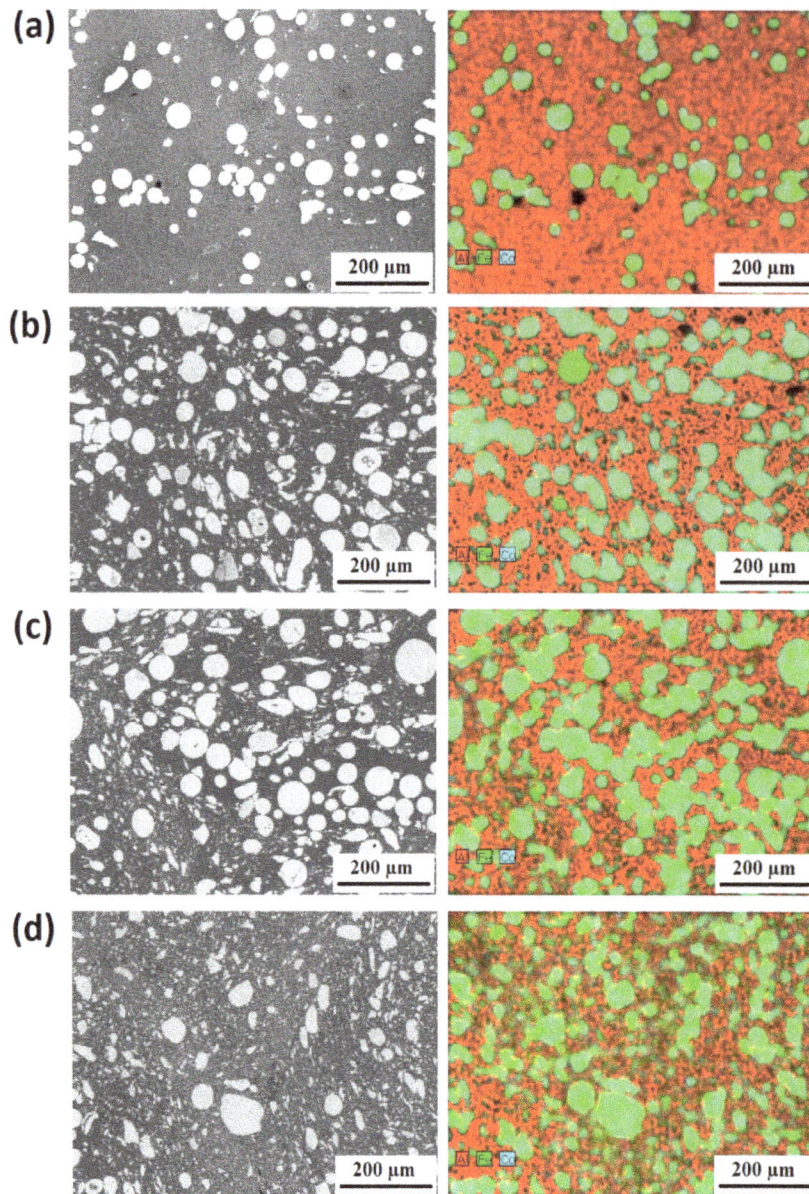

Figure 4. SEM micrographs and EDX maps of the hot pressed composites obtained from (**a**) 1-h, (**b**) 10-h, (**c**) 30-h and (**d**) 50-h milled powders.

Figure 5a–d illustrates the SEM micrographs of the hot extruded composites. The effects of milling on the microstructures of hot extruded samples are very similar to what was observed for the hot pressed samples. Compared to the hot pressed composites (Figure 4), hot extrusion leads to samples with enhanced density, as no pores are observed in the SEM micrographs (Figure 5). Although the XRD patterns in Figure 3c show the Al_5Fe_2 intermetallic product after hot extrusion, the matrix/particle interface of the hot extruded samples is clean, and no reaction zone is observed in Figure 5a–d. The Al_5Fe_2 phase cannot be detected by SEM, most likely because of the ultrafine dimension, as supported by the broad XRD peaks for this phase (Figure 3c). Furthermore, elongation of some large glassy particles occurs during hot extrusion with respect to the corresponding particles observed previously in the hot pressed composites (Figure 4a–d).

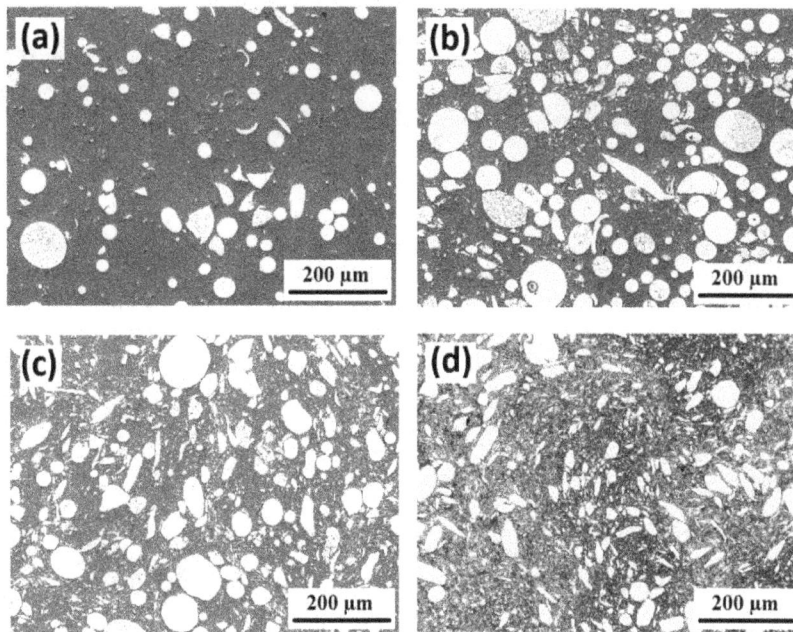

Figure 5. SEM micrographs of the hot pressed and hot extruded composites obtained from (**a**) 1-h, (**b**) 10-h, (**c**) 30-h and (**d**) 50-h milled powders.

The relative density (theoretical-experimental density) of the hot pressed and hot pressed and hot extruded composites are respectively ~98% and ~99%, as also supported by the SEM micrographs given in Figures 4 and 5, which show few or no pores. Therefore, hot pressing and hot extrusion provide nearly fully-dense specimens, since the combination of temperature and pressure stimulates the accelerated densification process and the elimination of residual porosity [37]. No significant effect of the milling time on the relative density of the composites is observed.

3.3. Mechanical Properties of the Consolidated Samples

Figure 6a,b displays the microhardness values of the composites as a function of milling time along with the optical micrographs of the indentations (under the same applied load). With increasing the milling time from 1–50 h, the microhardness increases from 51 ± 2.3–155 ± 6.5 HV for the hot pressed composites (Figure 6a) and from 45.1 ± 2.2–144 ± 10.5 HV for the hot pressed and hot extruded samples (Figure 6b). For a given milling time, the hardness of the hot pressed samples is higher than those of the hot pressed and hot extruded material. This can be explained by the smaller particle sizes (15 ± 12 µm) and more homogeneous distribution of the hard Fe-based glassy particles throughout the hot pressed microstructure than those consolidated by hot extrusion (22 ± 10 µm) [38]. A maximum average hardness value of 155 HV was measured for the 50-h milled and hot pressed sample, which is compatible with its microstructure, presenting a uniform distribution of fine glassy particles without significant clustering (Figure 4d). Moreover, the pyramidal indentations are not distorted, and no significant cracks are observed around the indentation, despite the difference in hardness between the soft matrix and the hard reinforcing phase (Figure 7). This indicates that the Fe-based glassy particles are strongly bonded with the Al matrix, which can play a significant role in enhancing the mechanical properties of the composites [30].

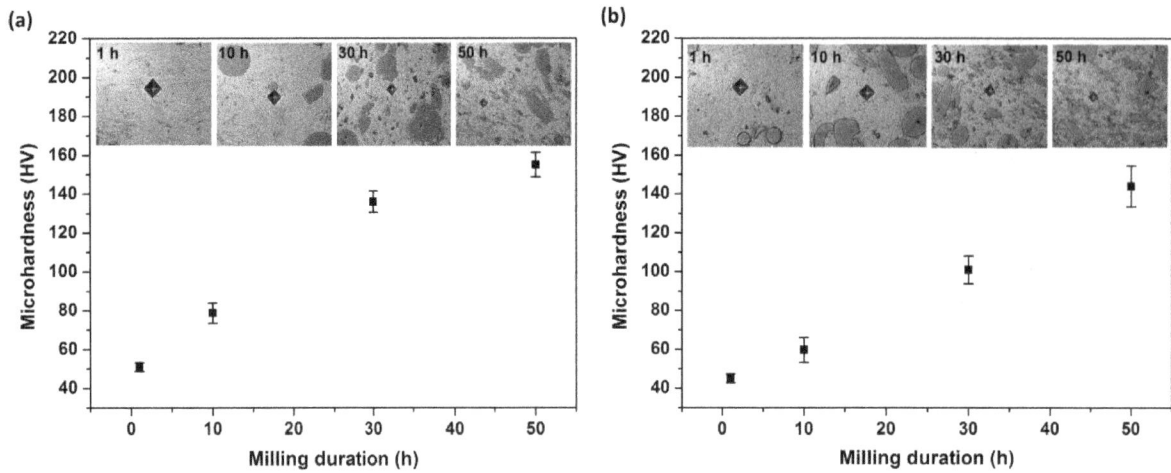

Figure 6. Microhardness values as a function of milling time and optical micrographs of the indentation marks obtained from: (**a**) hot pressed and (**b**) hot pressed and hot extruded composites.

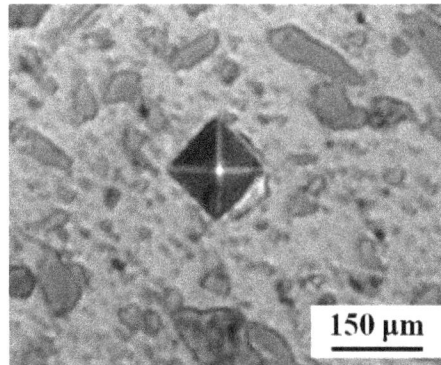

Figure 7. Optical microscopy image showing the indentation mark taken from the 50-h milled and hot pressed composite.

Room temperature compression true stress-true strain curves of the tests under quasistatic loading for the composite materials are shown in Figure 8. Data for the yield strength (0.2% offset), ultimate strength, strain at fracture and microhardness of the composites are summarized in Table 1. The yield and ultimate strength values of the composites are higher than those of the hot pressed and hot extruded pure Al (125 and 155 MPa, respectively), indicating the positive effect of Fe-based metallic glass reinforcements on the mechanical properties of Al [27,30]. These results also reveal the significant effect of milling and the consolidation process on the strength and plasticity of the composites. With increasing milling time from 1–50 h, the strength of the composites increases remarkably for both consolidation processes, showing a similar tendency as the hardness.

The ultimate strength of the hot pressed materials increases from 220 MPa for the composite milled for 1 h to 340, 440 and 630 MPa for the composites milled for 10, 30 and 50 h, respectively. In contrast, the plastic deformation of the hot pressed materials decreases from 35% for the composite milled for 1 h to 8.5%, 6.5% and 1.2% for the hot pressed composites produced from the powders milled for 10, 30 and 50 h. On the other hand, the ultimate strength of the hot extruded materials increases from 220 MPa for the composite milled for 1 h to 295 MPa for the composites milled for

10 h, while retaining appreciable plastic deformation, reaching an ultimate strain of 30% before fracture occurs. The 30 h milled, hot pressed and hot extruded composite gives a good combination of high strength and remarkable plasticity, exhibiting yield and ultimate strengths at 285 and 390 MPa and fracture strain at 21%.

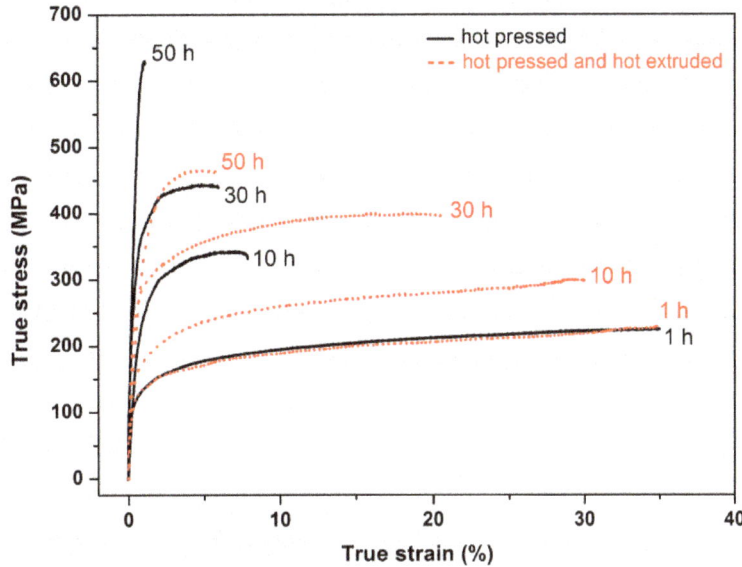

Figure 8. Room temperature compression true stress-true strain curves of the consolidated samples obtained from 1-, 10-, 30- and 50-h milled powders: (▬) hot pressed and (▬) hot pressed and hot extruded composites.

Table 1. Mechanical properties of Al metal matrix composites (MMCs) reinforced with 40 vol% $Fe_{50.1}Co_{35.1}Nb_{7.7}B_{4.3}Si_{2.8}$ metallic glass particles produced by mechanical milling for different times, hot pressing (HP) and hot pressing followed by hot extrusion (HP + HE) at 673 K and 640 MPa, as derived from the uniaxial compression and Vickers microhardness tests.

Milling time (h)	Consolidation process	Yield Strength (MPa)	Ultimate Strength (MPa)	Strain at Fracture (%)	Vickers Microhardness (HV)
1	HP	125	220	35.0	51 ± 2
	HP + HE	125	220	35.0	45 ± 2
10	HP	275	340	8.5	79 ± 5
	HP + HE	180	295	30.5	60 ± 6
30	HP	350	440	6.5	136 ± 5
	HP + HE	285	390	21.0	101 ± 7
50	HP	595	630	1.2	155 ± 7
	HP + HE	340	470	6.5	144 ± 11

These results show that the use of hot extrusion after hot pressing slightly reduces both the strength and hardness of the composites, while raising the plastic deformation, which ranges between 35 and 6.5%. The values of strength for the 50-h milled and consolidated samples are higher than those reported for Al/Al_2O_3 composites [39] and Al MMCs reinforced with different types of metallic glasses [21,25]. This indicates that Fe-based glassy particles may be a valid alternative to the conventional ceramic or other metallic glass reinforcements.

As revealed by the XRD patterns and SEM micrographs of the powders and consolidated samples (Figures 1 and 3–5), milling leads to the reduction of the reinforcement particle size (Figure 9), which is most likely the reason for the increase of the yield and ultimate strengths with increasing milling time (Table 2). The reduction of the particle size leads to an improvement in the mechanical properties if a clean particle-matrix interface is obtained [4]. The average inter-particle distance of the composites decreases to 3 µm after milling for 50 h. This observation can be translated into an effective grain size of the Al matrix in the composites smaller than 3.5 µm [30]. Besides the reduced particle size, the improvement of the strength and hardness of the composites resulting from milling can also be ascribed to the corresponding reduced distance between the glassy particles, as observed in the SEM micrographs (Figures 4 and 5). This phenomenon was reported for the Al and brass matrix composites reinforced with different types of metallic glasses [18,35].

Table 2. The average inter-particle distance for the hot pressed and hot pressed and hot extruded composites as a function of milling time.

Milling Time (h)	Average inter-particle distance (µm)	
	Hot pressed composites	Hot pressed and hot extruded composites
1	21 ± 7	26 ± 7
10	8 ± 1	10 ± 2
30	6 ± 2	6 ± 1
50	3 ± 1	4 ± 1

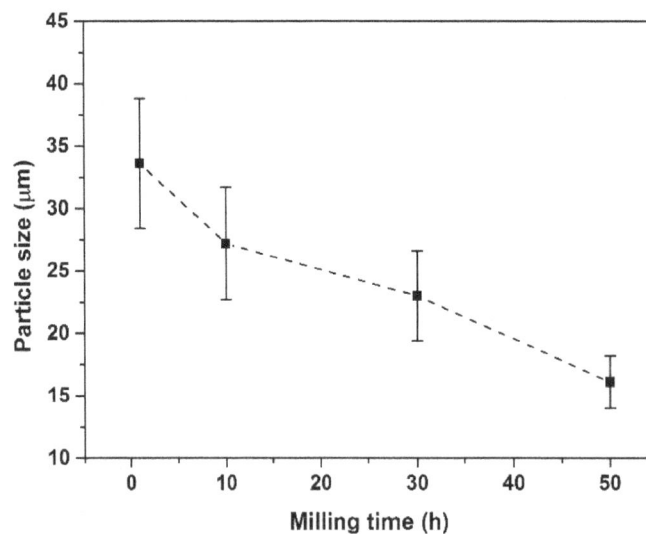

Figure 9. Average particle sizes of the hot pressed composites milled for different times.

3.4. Evaluation of the Mechanical Behavior by Theoretical Predictions

The prediction of the mechanical properties of a composite is an important prerequisite for material design and application. Recently, a model proposed by Gurland [40], which considers ($\sigma \propto V^{1/3} d^{-1/2}$)

the combined effects of the reinforcement volume fraction (V) and the particle size (d), has been rearranged to consider the effect of the matrix ligament size λ, as [35]:

$$\sigma \propto V^{1/3} \lambda^{-1/2} \tag{1}$$

Figure 10a,b shows the yield strength as a function of $V^{1/3}\lambda^{-1/2}$ for the present consolidated samples. The matrix ligament size depends on the milling time of the composite powders, and for the current volume fraction of reinforcement (40 vol%), the relationship between strength and $V^{1/3}\lambda^{-1/2}$ is linear for both hot pressed and hot extruded samples. The reduction of λ can give a considerable contribution to the strength of the composites because the matrix/particle interface can effectively reduce the movement of dislocations [15]. This corroborates the validity of the model (Equation (1)) for the present composites and further indicates that the strength can be accurately modeled by considering the effect of the matrix ligament size. Furthermore, the reason for the differences in strength and plasticity of the hot pressed and hot extruded composites can be explained by the grain growth of the Al matrix and the formation of particle clusters during hot extrusion, as previously observed in Figures 3c and 5. The grain refinement provides an additional strengthening effect resulting from the dislocation piling-up at the grain boundaries [41–43].

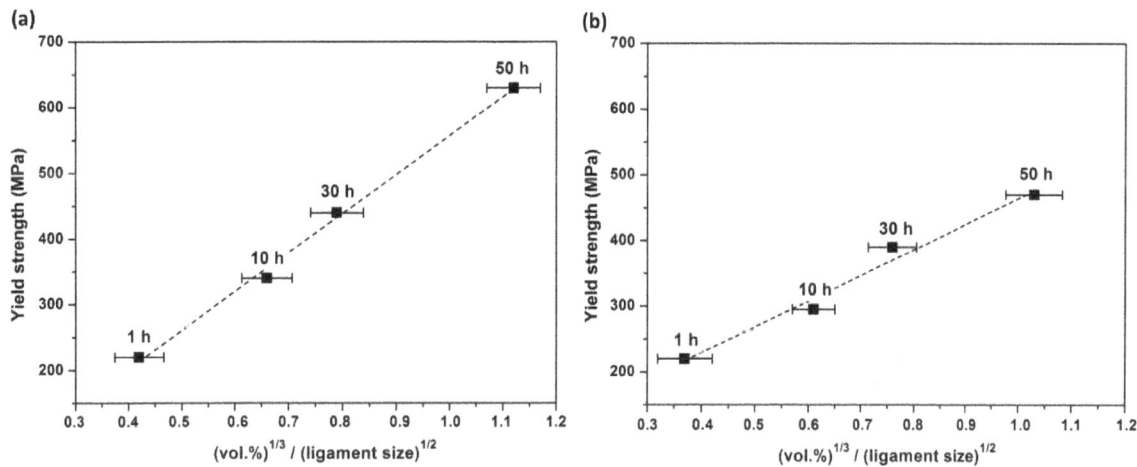

Figure 10. Yield strength as a function of volume fraction and matrix ligament size ($\sigma \propto V^{1/3} \lambda^{-1/2}$) for consolidated samples: (**a**) milled and hot pressed and (**b**) milled, hot pressed and hot extruded composites.

Eventually, composites fabricated by milling for different times, hot pressing and hot extrusion can be considered as potential candidates for applications demanding various mechanical properties. The hot pressed composites exhibit high strength and high hardness, which are of primary importance for structural applications [1]. On the other hand, hot extruded MMCs are potential materials for aerospace and automotive industries due to a superior balance of improved strength and plasticity [1].

4. Conclusions

Al matrix composites reinforced with 40 vol% $Fe_{50.1}Co_{35.1}Nb_{7.7}B_{4.3}Si_{2.8}$ glassy particles have been successfully produced via powder metallurgy routes using different milling times and consolidation processes. Based on the results of the present study, the following conclusions can be drawn:

1. Milled and hot pressed composites revealed no formation of intermetallic compounds even after milling for 50 h, within the sensitivity of the XRD measurements. Only a small amount of the Al_5Fe_2 intermetallic compound was observed in the XRD patterns of all milled, hot pressed and hot extruded samples. DSC scans revealed that milling time changes the overall crystallization behavior of the composite powders.

2. Ball milling resulted in the decrease of the grain sizes of the Al matrix, in the reduction of the particle size of the glassy reinforcements and in their homogeneous distribution in the Al matrix. This has a positive effect on the hardness and strength of the composites produced by both hot pressing and hot pressing followed by hot extrusion. With increasing the milling time from 1–50 h, the microhardness values of the hot pressed and hot extruded samples increase from 51 ± 2.26–155 ± 6.5 HV and from 45.1 ± 2.24–144 ± 10.5 HV, respectively. With increasing milling time from 1–50 h, the strength of the composites increases remarkably for both consolidation processes, showing a similar tendency as observed for the hardness.

3. The ultimate strength of the hot pressed materials increases from 220 MPa for the composite milled for 1 h to 340, 440 and 630 MPa for the composites milled for 10, 30 and 50 h, respectively. The 50-h milled and hot pressed composite exhibits small plastic deformation of 1.2% and a maximum strength of 630 MPa, which is in agreement with the highest Vickers microhardness of 155 ± 6.5 HV among all composites.

4. The use of hot extrusion after hot pressing slightly reduces both the strength and hardness of the composites, while raising the plastic deformation ranging between 35 and 6.5%. The 30-h milled, hot pressed and hot extruded composite gives a combination of high strength (390 MPa) and remarkable plasticity (21%).

The strength of both the hot pressed or hot pressed and hot extruded composites can be accurately described by a simple model considering the effect of matrix ligament size on the strengthening of the composites.

Acknowledgments

Özge Balcı would like to express her appreciation to German Academic Exchange Service (DAAD) for the financial support during her stay at IFW Dresden.

Author Contributions

Özge Balcı performed all the experiments and characterization studies and created the initial draft. Konda Gokuldoss Prashanth aided in SEM analyses and mechanical tests and conceived the final manuscript. Sergio Scudino formulated the idea of this research, designed the experiments and supervised the discussion of the results. Duygu Ağaoğulları, İsmail Duman and M. Lütfi Öveçoğlu aided the review of the paper and provided critical comments. Volker Uhlenwinkel prepared the gas-atomized glassy powders used in the experiments. Jürgen Eckert contributed to the overall development of the main concepts of this study.

Conflicts of Interest

The authors declare no conflict of interest.

References

1. Epple, M. *Biomaterialien und Biomineralisation—Eine Einführung für Naturwissenschaftler, Mediziner und Ingenieure*; Springer: Wiesbaden, Germany, 2003.
2. Miracle, D.B. Metal matrix composites—From science to technological significance. *Compos. Sci. Technol.* **2005**, *65*, 2526–2540.
3. Miracle, D.B.; Donaldson, S.L. Composites. In *ASM Handbook*; ASM International: Materials Park, OH, USA, 2001.
4. Davis, J.R. Aluminum and Aluminum alloys. In *ASM Specialty Handbook*; ASM International: Materials Park, OH, USA, 1993.
5. Clyne, T.W.; Withers, P.J. *An Introduction to Metal Matrix Composites*; Cambridge University Press: New York, NY, USA, 1993.
6. Kainer, K.U. *Metal Matrix Composites: Custom-Made Materials for Automotive and Aerospace Engineering*; WILEY-VCH: Weinheim, Germany, 2006.
7. Christman, T.; Needleman, A.; Suresh, S. An experimental and numerical study of deformation in metal-ceramic composites. *Acta Metall.* **1998**, *37*, 3029–3050.
8. Slipenyuk, A.; Kuprin, V.; Milman, Y.; Goncharuk, V.; Eckert, J. Properties of P/M processed particle reinforced metal matrix composites specified by reinforcement concentration and matrix-to-reinforcement particle size ration. *Acta Mater.* **2006**, *54*, 157–166.
9. Song, M.S.; Zhang, M.X.; Zhang, S.G.; Huang, B.; Li, L.G. *In situ* fabrication of TiC particulates locally reinforced aluminum matrix composites by self-propagating reaction during casting. *Mater. Sci. Eng. A* **2008**, *473*, 166–171.
10. Wang, J.; Yi, D.; Su, X.; Yin, F.; Li, H. Properties of submicron AlN particulate reinforced aluminium matrix composite. *Mater. Des.* **2009**, *30*, 78–81.
11. Feng, C.F.; Froyen, L. *In situ* synthesis of Al_2O_3 and TiB_2 particulate mixture reinforced aluminium matrix composites. *Scr. Mater.* **1997**, *36*, 467–473.
12. Arsenault, R.J. The strengthening of aluminum 6061 by fiber and platelet silicon carbide. *Mater. Sci. Eng. A* **1984**, *64*, 171–181.
13. Balcı, Ö.; Ağaoğulları, D.; Gökçe, H.; Duman, İ.; Öveçoğlu, M.L. Influence of TiB_2 particle size on the microstructure and properties of Al matrix composites prepared via mechanical alloying and pressureless sintering. *J. Alloys Compd.* **2013**, *586*, S78–S84.
14. Ibrahim, I.A.; Mohammed, F.A.; Lavernia, E.J. Particulate reinforce metal matrix composites: Review. *J. Mater. Sci.* **1991**, *26*, 1137–1156.
15. Scudino, S.; Liu, G.; Sakaliyska, M.; Surreddi, K.B.; Eckert, J. Powder metallurgy of Al-based metal matrix composites reinforced with β-Al_3Mg_2 intermetallic particles: Analysis and modeling of mechanical properties. *Acta Mater.* **2009**, *57*, 4529–4538.
16. Inoue, A. Bulk amorphous and nanocrystalline alloys with high functional properties. *Mater. Sci. Eng. A* **2001**, *304–306*, 1–10.

17. Ashby, M.F.; Greer, A.L. Metallic glasses as structural materials. *Scr. Mater.* **2006**, *54*, 321–326.

18. Yu, P.; Zhang, L.C.; Zhang, W.Y.; Das, J.; Kim, K.B.; Eckert, J. Interfacial reaction during the fabrication of $Ni_{60}Nb_{40}$ metallic glass particles-reinforced Al based MMCs. *Mater. Sci. Eng. A* **2007**, *444*, 206–213.

19. Aljerf, M.; Georgarakis, K.; Louzguine-Luzgin, D.; le Moulec, A.; Inoue, A.; Yavari, A.R. Strong and light metal matrix composites with metallic glass particulate reinforcement. *Mater. Sci. Eng. A* **2012**, *532*, 325–330.

20. Dudina, D.V.; Georgarakis, K.; Aljerf, M.; Li, Y.; Braccini, M.; Yavari, A.R.; Inoue, A. Cu-based metallic glass particle additions to significantly improve overall compressive properties of an Al alloy. *Compos. A* **2010**, *41*, 1551–1557.

21. Lee, M.H.; Kim, J.H.; Park, J.S.; Kim, J.C.; Kim, W.T.; Kim, W.T.; Kim, D.H. Fabrication of Ni–Nb–Ta metallic glass reinforced Al-based alloy matrix composites by infiltration casting process. *Scr. Mater.* **2004**, *50*, 1367–1371.

22. Yu, P.; Kim, K.B.; Das, J.; Baier, F.; Xu, W.; Eckert, J. Fabrication and mechanical properties of Ni–Nb metallic glass particle-reinforced Al-based metal matrix composite. *Scr. Mater.* **2006**, *54*, 1445–1450.

23. Scudino, S.; Surreddi, K.B.; Sager, S.; Sakaliyska, M.; Kim, J.S.; Löser, W.; Eckert, J. Production and mechanical properties of metallic glass-reinforced Al-based metal matrix composites. *J. Mater. Sci.* **2008**, *43*, 4518–4526.

24. Scudino, S.; Liu, G.; Prashanth, K.G.; Bartusch, B.; Surredi, K.B.; Murty, B.S.; Eckert, J. Mechanical properties of Al-based metal matrix composites reinforced with Zr-based glassy particles produced by powder metallurgy. *Acta Mater.* **2009**, *57*, 2029–2039.

25. Prashanth, K.G.; Kumar, S.; Scudino, S.; Murty, B.S.; Eckert, J. Fabrication and response of $Al_{70}Y_{16}Ni_{10}Co_4$ glass reinforced metal matrix composites. *Mater. Manuf. Processes* **2011**, *26*, 1242–1247.

26. Shen, T.D.; Schwarz, R.B. Bulk ferromagnetic glasses prepared by flux melting and water quenching. *Appl. Phys. Lett.* **1999**, *75*, 49–51.

27. Fujii, H.; Sun, Y.; Inada, K.; Ji, Y.; Yokoyama, Y.; Kimura, H.; Inoue, A. Fabrication of Fe-based metallic glass particle reinforced Al-based composite materials by Friction Stir processing. *Mater. Trans.* **2011**, *52*, 1634–1640.

28. Suryanarayana, C.; Inoue, A. Iron-based bulk metallic glasses. *Int. Mater. Rev.* **2013**, *58*, 131–166.

29. Kaban, I.; Jovari, P.; Waske, A.; Stoica, M.; Bednarcik, J.; Beuneu, B.; Mattern, N.; Eckert, J. Atomic structure and magnetic properties of Fe–Nb–B metallic glasses. *J. Alloys Compd.* **2014**, *586*, S189–S193.

30. Zheng, R.; Yang, H.; Liu, T.; Ameyama, K.; Ma, C. Microstructure and mechanical properties of aluminum alloy matrix composites reinforced with Fe-based metallic glass particles. *Mater. Des.* **2014**, *53*, 512–518.

31. Murty, B.S.; Ranganathan, S. Novel materials synthesis by mechanical alloying/milling. *Int. Mater. Rev.* **1998**, *43*, 101–141.

32. Suryanarayana, C. Mechanical alloying and milling. *Prog. Mater. Sci.* **2001**, *46*, 1–184.

33. Harrigan, W.C., Jr. Commercial processing of metal matrix composites. *Mater. Sci. Eng. A* **1998**, *244*, 75–79.

34. Surreddi, K.B.; Scudino, S.; Sakaliyska, M.; Prashanth, K.G.; Sordelet, D.J.; Eckert, J. Crystallization behavior and consolidation of gas-atomized $Al_{84}Gd_6Ni_7Co_3$ glassy powder. *J. Alloys Compd.* **2010**, *491*, 137–142.

35. Kim, J.Y.; Scudino, S.; Kühn, U.; Kim, B.S.; Lee, M.H.; Eckert, J. Production and characterization of Brass-matrix composites reinforced with $Ni_{59}Zr_{20}Ti_{16}Si_2Sn_3$ glassy particles. *Metals* **2012**, *2*, 79–94.

36. German, R.M. *Sintering Theory and Practice*; Wiley-Interscience: Weinheim, Germany, 1996.

37. Prashanth, K.G.; Murty, B.S. Production, kinetic study and properties of Fe-based glass and its composites. *Mat. Manuf. Processes* **2010**, *25*, 592–597.

38. Keryvin, V.; Hoang, V.H.; Shen, J. Hardness, toughness, brittleness and cracking systems in an iron-based bulk metallic glass by indentation. *Intermetallics* **2009**, *17*, 211–217.

39. San Marchi, C.; Cao, F.; Kouzeli, M.; Mortensen, A. Quasistatic and dynamic compression of aluminum-oxide particle reinforced pure aluminum. *Mater. Sci. Eng. A* **2002**, *337*, 202–211.

40. Gurland, J. The fracture strength of sintered tungsten carbide-cobalt alloys in relation to composition and particle spacing. *Trans. Metall. Soc. AMIE* **1963**, *227*, 1146–1149.

41. Chawla, K.K. *Composite Materials: Science and Engineering*; Springer-Verlag: New York, NY, USA, 1987.

42. Kim, H.S. On the rule of mixtures for the hardness of particle reinforced composites. *Mater. Sci. Eng. A* **2000**, *289*, 30–33.

43. Hirth, J.P. *Physical Metallurgy*, Cahn, R.W., Haasen, P., Eds.; North-Holland: Amsterdam, The Netherlands, 1996.

10

Damage Behavior of Sintered Fiber Felts

Nicolas Lippitz * and Joachim Rösler

Institute for Materials, TU Braunschweig, Langer Kamp 8, Braunschweig D-38106, Germany;
E-Mail: j.roesler@tu-bs.de

* Author to whom correspondence should be addressed; E-Mail: n.lippitz@tu-bs

Academic Editor: Hugo F. Lopez

Abstract: The reduction of aircraft noise is important due to a rising number of flights and the growth of urban centers close to airports. During landing, a significant part of the noise is generated by flow around the airframe. To reduce that noise porous trailing edges are investigated. Ideally, the porous materials should to be structural materials as well. Therefore, the mechanical properties and damage behavior are of major interest. The aim of this study is to show the change of structure and the damage behavior of sintered fiber felts, which are promising materials for porous trailing edges, under tensile loading using a combination of tensile tests and three dimensional computed tomography scans. By stopping the tensile test after a defined stress or strain and scanning the sample, it is possible to correlate structural changes and the development of damage to certain features in the stress-strain curve and follow the damage process with a high spatial resolution. Finally, the correlation between material structure and mechanical behavior is demonstrated.

Keywords: porosity; fibers; tensile testing; X-ray; 3D tomography

1. Introduction

As the exposure to significant noise loads, caused by an increasing number of flights and airports close to residential areas, impairs physical health and productivity, a reduction of aircraft noise has become important in engineering [1,2]. A major factor for aircraft noise during landing is the noise generated by flow around the airframe. One way to lower that noise is to use porous surfaces, e.g., a porous trailing edge [3–7]. Ideally, the porous materials are supposed to bear mechanical loads. As the structural changes influence the aeroacoustic performance, not only the mechanical properties, but also the damage behavior is of major interest. X-ray tomography is a non-destructive tool that enables material characterization and in situ experiments to reveal structural changes and damage during mechanical tests and, therefore, it has been used in several studies [8–13]. In this study, the damage behavior is analyzed using a combination of tensile tests and three dimensional computed tomography, which allows determination of the damage location with a high resolution. To detect fiber rotation a line segmentation technique, wich allows to chartacterize the morphology and orientation of pores and ligaments, is applied [14–16]. For these experiments the sintered fiber felts SFF 3 and SFF 100, obtained from GKN Sinter Metals (GKN Sinter Metals GmbH, Dahlienstraße, Radevormwald, Germany), that showed a good acoustic performance were chosen and characterized extensively in previous studies [16,17]. The test sample was scanned to create a reference and afterwards used in a tensile test. The tensile test was then stopped at defined strains or significant changes in stress and the sample was analyzed for structural changes and damage.

2. Material Characterisation

The sintered fiber felts selected for this study consist of one functional layer (SFF 100), three functional layers (SFF 3), and a support grid with significantly bigger dimensions. Both fiber felts were manufactured from the CrNi alloy AISI 316L. The materials were characterized using 3D computed tomography as described in [17,18]. CT images of both fiber felts are shown in Figure 1. The material properties can be found in Table 1. The fibers of SFF 3 are finer and the porosity is lower compared to SFF 100. Additionally the wires of the support grid of SFF 3 are around 30% thinner than those of SFF 100. In relation to the overall thickness, SFF 3 has a relatively thick functional layer and SFF 100 a relatively thin one. The voxel size achieved with the samples used to characterize the sintered fiber felts was $V \approx 1 \ \mu m$. This resolution is four times higher than the resolution achieved with the tensile samples ($V \approx 4 \ \mu m$) due to an increasing sample size. Therefore all data in Table 1 results from scans and samples separate from the tensile tests.

Table 1. Material properties of sintered fiber felts (SFF), including porosity (ϕ), thickness of functional layer (d_i), fiber diameter of functional layers (D_i) and diameter of the wires of the support grid (D_{grid}); the layer designated as $i = 1$ corresponds to the outer layer which is not in contact with the support grid

Material	Layer 1			Layer 2			Layer 3				
	ϕ_1	d_1	D_1	ϕ_2	d_2	D_2	ϕ_3	d_3	D_3	ϕ_{total}	D_{grid}
	(%)	(μm)	(μm)	(%)	(μm)	(μm)	(%)	(μm)	(μm)	(%)	(μm)
SFF 3	85 ± 24	260	14 ± 2	78 ± 62	340	5 ± 2	83 ± 42	140	8 ± 2	81 ± 36	174 ± 8
SFF 100	92 ± 15	430	25 ± 2	-	-	-	-	-	-	-	253 ± 8

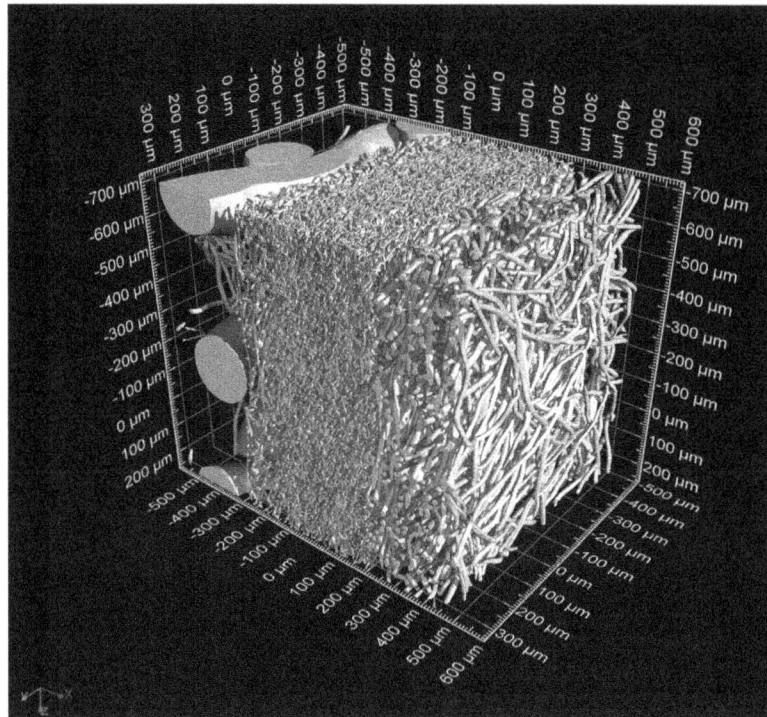

(a)

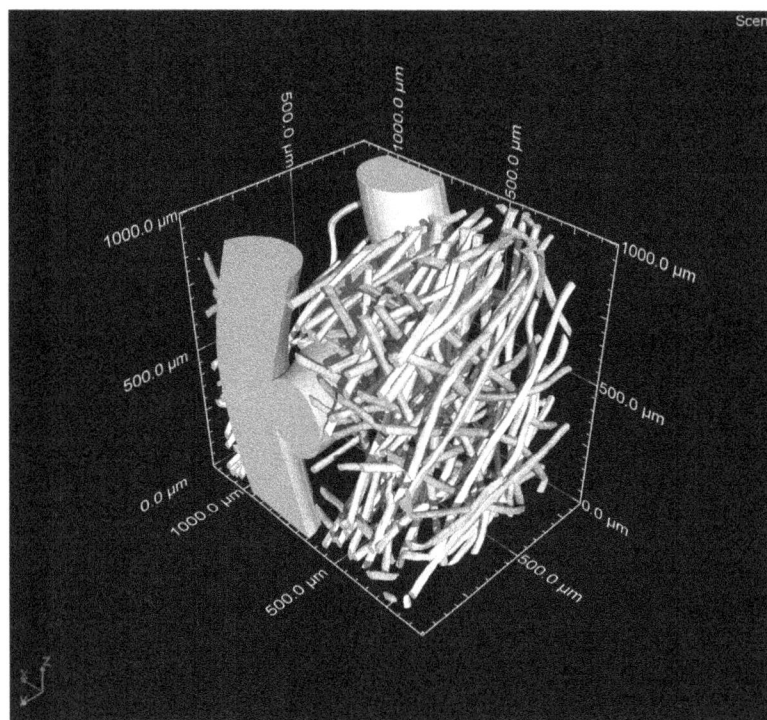

(b)

Figure 1. CT image of sintered fiber felts. (**a**) SFF 3; (**b**) SFF 100.

3. Experimental Procedure

To analyze the damage behavior of sintered fiber felts, tensile test samples were repeatedly examined at defined strains and before the tensile tests to create a reference using computed tomography. Because the resolution of CT scans decreases with increasing field of view, the samples were kept small and the scan was divided into four scans along the gage length, which were subsequently merged to one reconstruction. Samples with a width of 7 mm turned out to be suitable for tensile testing and enabled a voxel size of $V \approx 4 \, \mu m$ m in the CT scan, which is sufficient for the selected sintered fiber felts. To support the relatively soft sintered fiber felts during the scan, they were scanned in a carbon fiber tube (see Figure 2a). As the carbon fiber tube has a low density and is nearly transparent to X-rays, it does not impair the scan quality and was neglected in the reconstruction. To ensure that after every tensile step the same part of the sample was scanned it was marked using copper strips. The investigated area lies between the two copper strips as shown in Figure 2b. The distance between the copper strips corresponds to the initial separation distance of the strain extensometer clips used for the tensile tests. The stress given in Figure 3 refers to the cross-sectional area of the sample without taking the porosity into account. For the first step of the tensile tests a maximum strain of $\epsilon = 2\%$ was selected. For all following steps the tensile test was stopped after additional strains of $\epsilon = 5\%$.

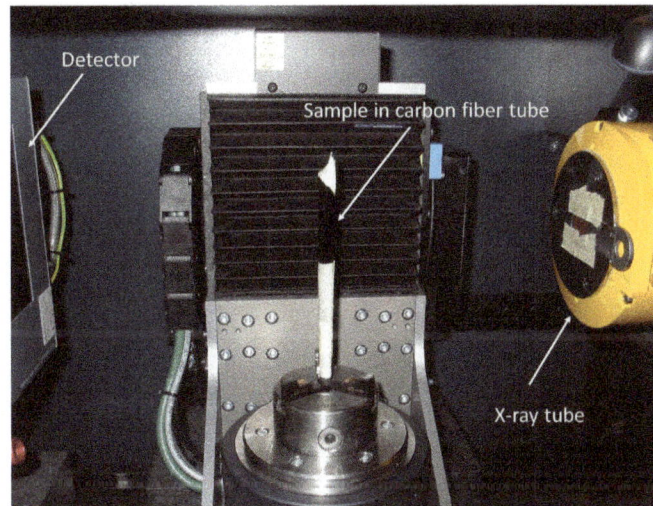

(a)

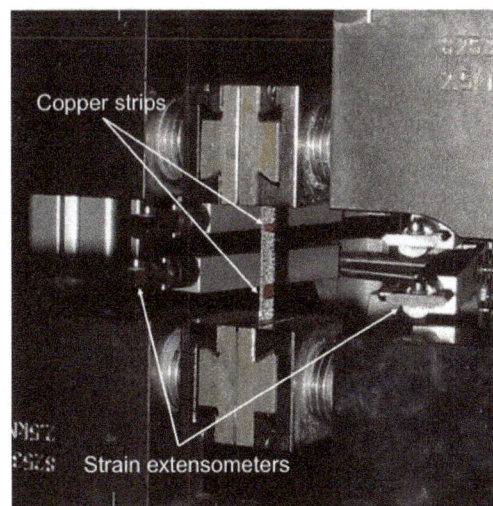

(b)

Figure 2. Experimental setup. (a) sample in CT; (b) sample in tensile test machine.

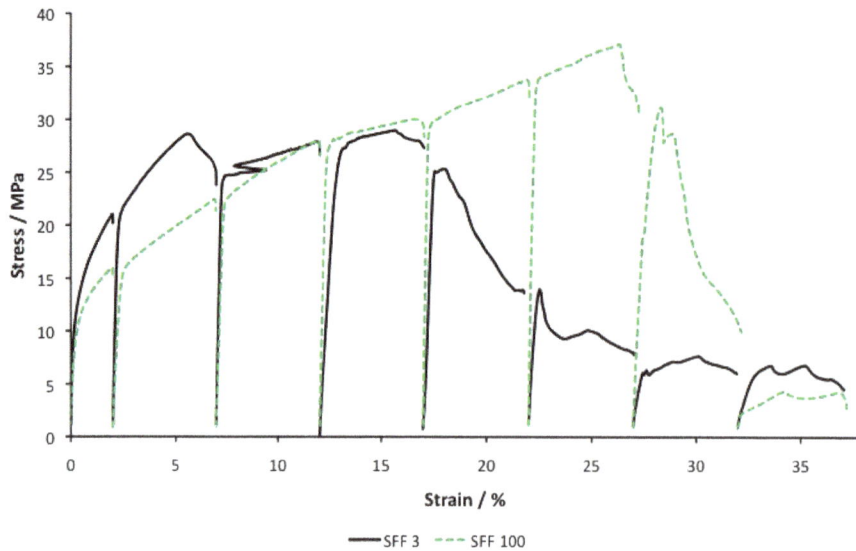

Figure 3. Stress-strain curves of SFF 3 and SFF 100 after eight steps of tensile testing.

3.1. Structure Analysis

As the sintered fiber felts deform during the tensile test, the fibers of the functional layers are expected to be rotating. X-ray tomography is a powerful tool to measure the fiber orientations in fibrous materials, as demonstrated in several works [11,13,19]. Usually, the fibers are idealized as cylinders and the orientation distribution of the cylinder axes is plotted. In this paper however, not the fiber orientations, but their evolution during the tensile test is analyzed. This is done using a line segmentation technique, which is mainly used to measure the sizes of pores and ligaments [14–16]. For this a lattice of parallel lines with a defined distance is superimposed to cross-sectional binarized black and white images of the sintered fiber felts, which were extracted from the CT scans. The lattice is then rotated 180° in steps of 1° and the lengths of the black and white segments is measured. The mean length of the black and white segments for all angles can then be plotted in structural ellipses [14].

The structural ellipses of a cross-sectional area perpendicular to the surface of the fiber felts and without a preferred fiber orientation would be circular. When the fibers rotate during the tensile test into a preferred orientation, the structural ellipses will change their appearance from a circular to a more elliptic shape. The line segmentation technique was applied to images perpendicular to the surface of the sintered fiber felt and parallel to the direction of loading (YZ plane in Figures 4 and 5). For every tensile test step five planes were analyzed and the results were averaged. Additionally, the longest and shortest mean segment length from the line segmentation analysis were used to define an aspect ratio. This aspect ratio can be used as an indicator for a preferred fiber orientation and fiber rotation. Note that the longest and shortest mean segment length are not equivalent to the long and short axis of the ellipses.

Due to the small fiber diameters of the inner functional layer of SFF 3 and the rather coarse resolution during the interrupted tensile tests, the line segmentation analysis was only applied to SFF 100.

4. Results and Discussion

The damage behavior analyses performed for SFF 3 and SFF 100 consist of eight tensile test steps accompanied by nine CT scans. The stress strain curves of both sintered fiber felts are shown in Figure 3. Selected CT images are given in Figure 4 for SFF 3 and Figure 5 for SFF 100.

SFF 3   After a strain of $\epsilon = 2\%$, SFF 3 does not show any damage or changes of the material structure in the CT images. The first structural changes can be observed after a strain of $\epsilon = 7\%$ as shown in Figure 4b,c. While the outer functional layer (indicated as layer 1) and the layer directly on the support grid (indicated as layer 3) are intact, the middle layer shows a tear perpendicular to the direction of loading (see arrow in Figure 4b). The CT images clearly show how the tear opens in layer 2 and closes again in layer 3. This structural change corresponds to a drop of stress observed in the stress-strain curve of SFF 3. After the middle layer of SFF 3 fails, the stress remains essentially constant (strains between $\epsilon = 7\%$ and $\epsilon \approx 10\%$) while the two remaining functional layers fail. In this strain range the increase of stress in the wires of the support grid, caused by strain hardening of the metal, compensates for the reduction of the cross-sectional area. However, the tear does not expand into the adjacent layers, but new tears appear in locally weakened areas of the sample. The distance between the failure locations of layer 1 and layer 2 is approximately 4 mm in the Z-direction of the reconstruction as indicated by the arrows shown in Figure 4d. Between these two failure locations the interface between layer 1 and layer 2 delaminated parallel to the direction of loading. Because one clip of the strain extensometer was located near the tear in the outer layer it slipped on the surface resulting in the loop shown in the stress-strain curve. After complete failure of the functional layers ($\epsilon > 10\%$) only the support grid is bearing the mechanical load. Then individual wires of the support gird begin to fail in various regions of the sample resulting in multiple tears rather than one tear propagating from a single failure location. It can be observed that wires that were originally oriented perpendicular to the direction of loading rotate into the direction of loading and enable a very high strain after the functional layers and all wires originally in the direction of loading had failed (see Figure 4f). The support grid finally tears apart by a stepwise failure of single wires and failure at the intersections between the rotating wires and the wires that were originally oriented in the direction of loading. This stepwise failure is reflected by the small fluctuations in the last segment of the stress-strain curve.

SFF 100   Like SFF 3, SFF 100 does not show any damage or changes of the material structure in the CT images for $\epsilon = 2\%$. While SFF 3 shows a clear drop of stress corresponding to damage of the functional layer for strains smaller than $\epsilon = 12\%$, SFF 100 only shows a slightly decreasing slope in the stress-strain curve and local necking of the functional layer along with first signs of a tearing functional layer (see Figure 5b) for strains between $\epsilon = 12\%$ and $\epsilon = 17\%$. A complete failure of the functional layer, shown in Figure 5c, was observed after $\epsilon = 22\%$. The first wires of the support gird failed at $\epsilon = 27\%$, corresponding with a drop in stress. The support grid then finally fails similarly to SFF 3 as shown in Figure 5d.

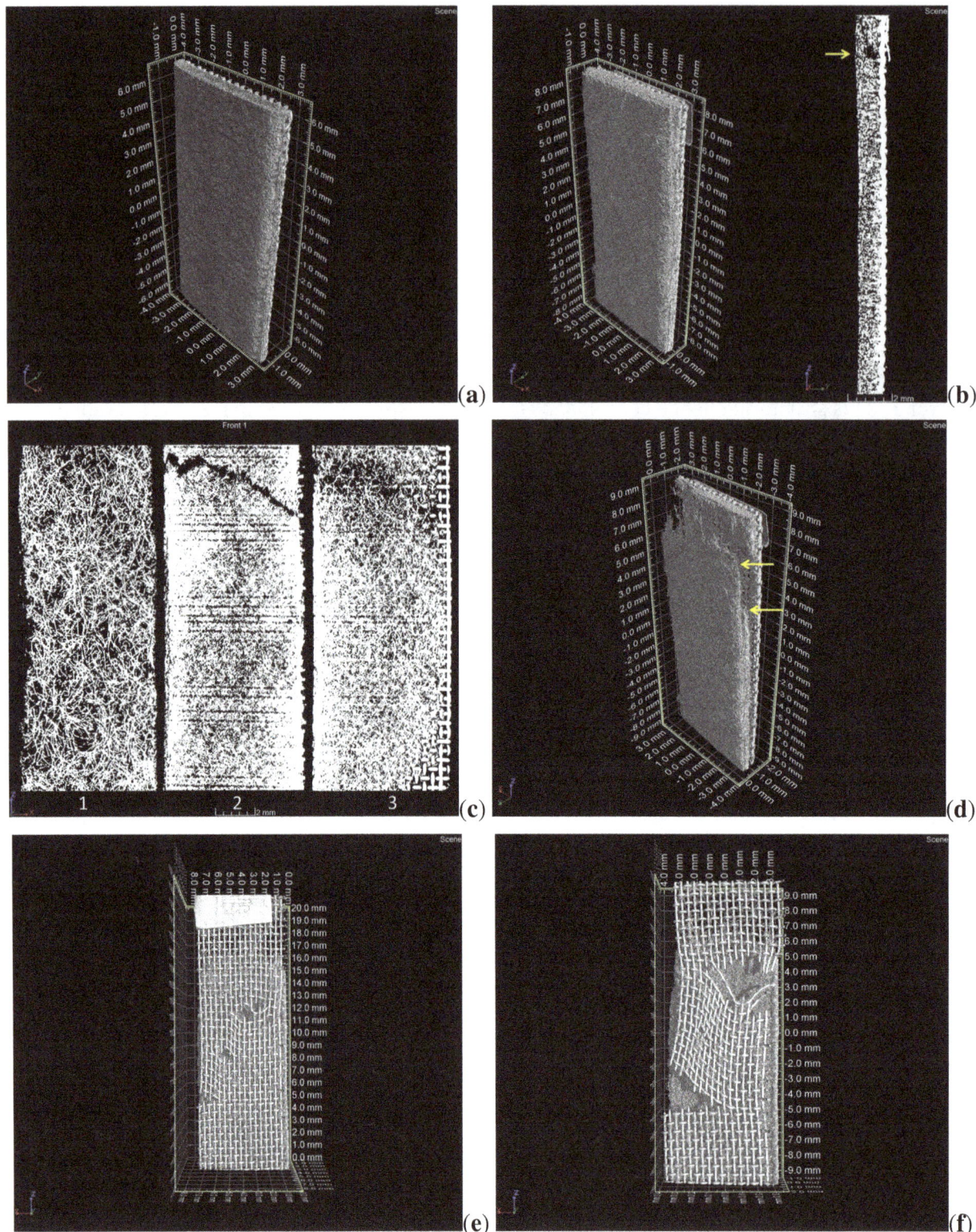

Figure 4. CT images of the SFF 3 tensile test sample. (**a**) three dimensional view before tensile tests; (**b**) three dimensional view and sectional view of the y-z-plane oriented parallel to the direction of loading after $\epsilon = 7\%$; (**c**) sectional views of layer 1, 2 and 3 after $\epsilon = 7\%$; (**d**) three dimensional view after $\epsilon = 12\%$; (**e**) three dimensional view of the support grid after $\epsilon = 22\%$; (**f**) three dimensional view of the support grid after $\epsilon = 37\%$.

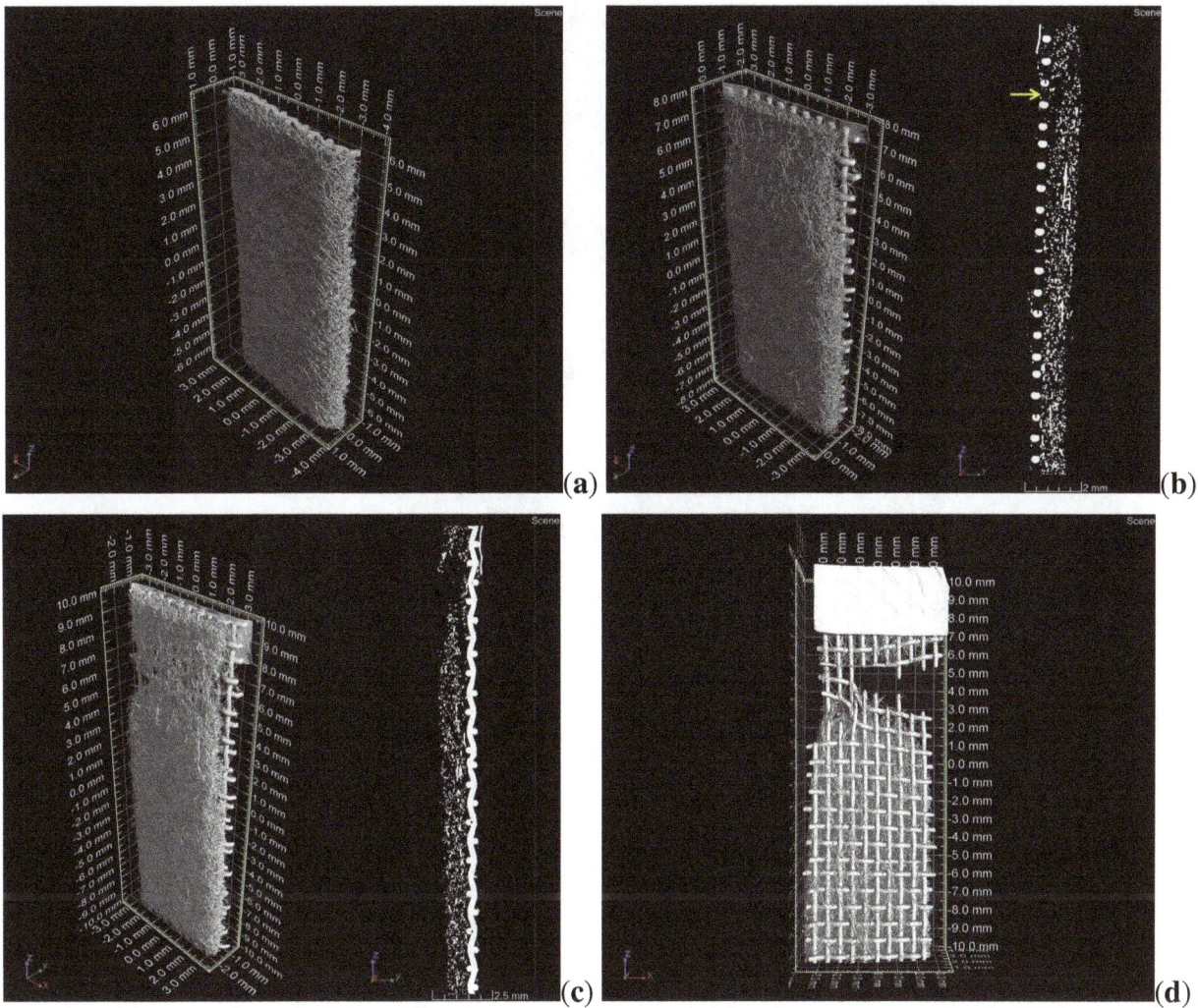

Figure 5. CT images of the SFF 100 tensile test sample. (**a**) three dimensional view before tensile tests; (**b**) three dimensional view and sectional view of the y-z-plane oriented parallel to the direction of loading after $\epsilon = 17\%$; (**c**) three dimensional view after $\epsilon = 27\%$; (**d**) three dimensional view of the support grid after $\epsilon = 37\%$.

To analyze the fiber rotation in the functional layer, the line segmentation technique was applied to binarized images, which were extracted from the CT-scans of SFF 100. The images were extracted between wires oriented in the direction of loading at the position indicated in Figure 6a. As the three-dimensional images show local deformation and tearing of the functional layer, the line segmentation analysis was split into two sections. The upper section of the reconstruction, including the tear formation is indicated as section A. The lower section, without tear formation, is indicated as section B (compare Figure 6a).

Along with the necking and initial tear formation in the functional layer (between $\epsilon = 12\%$ and $\epsilon = 17\%$), the line segmentation reveals a rotation of the fibers in the functional layer into the direction of loading. The structural ellipses for $\epsilon = 0\%$ (section A and B) have a roughly circular shape, whereas the ellipses for $\epsilon = 17\%$ are clearly elliptical. Although the fibers all over the gage length (section A and B) are rotating, the rotation is stronger in the vicinity of the tear (section A). The long axes of the ellipses in Figure 6b correspond to the direction of loading in the tensile test.

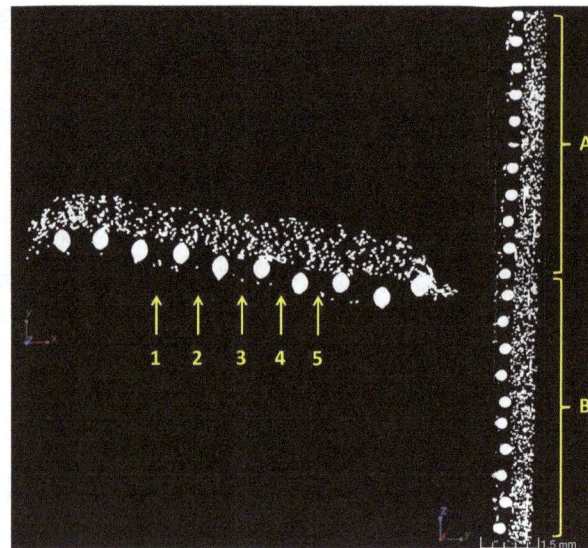

(a)

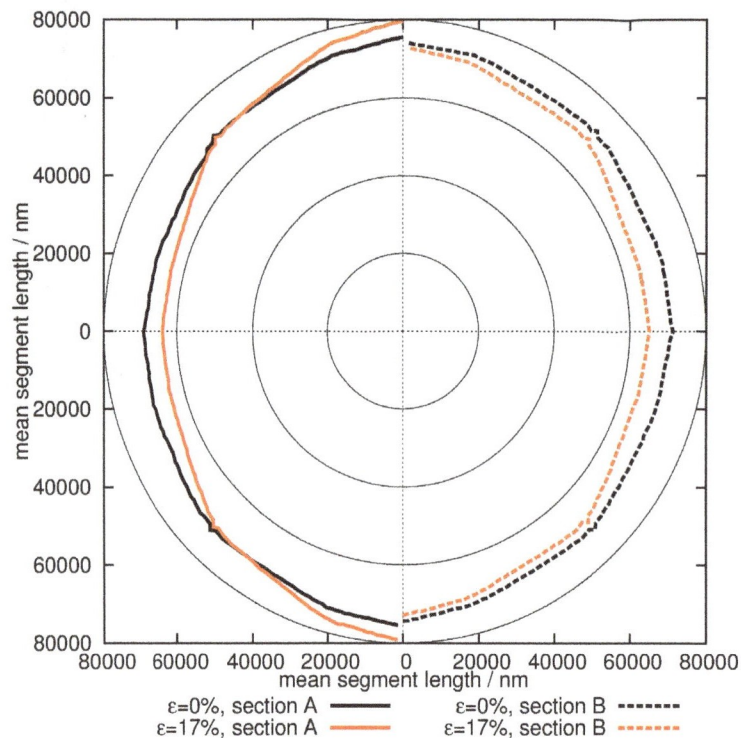

(b)

Figure 6. (a) extracted planes of SFF 100 at $\epsilon = 0\%$; (b) structural ellipses of SFF 100 at $\epsilon = 0\%$ and $\epsilon = 17\%$.

The aspect ratio given in Table 2 for section A is increasing up to a strain of $\epsilon = 17\%$ and reaches a maximum of 1,272. The aspect ratio for section B stays roughly constant for strains higher than $\epsilon = 7\%$ and reaches a maximum of 1,166. That indicates that the fiber rotation in section B stops when the tear starts to form in Section A. The functional layer delaminates from the support grid (compare Figure 5c) and its deformation is concentrated on the vicinity of the tear. After a complete tear formation and delamination of the functional layer, it is decoupled from the support grid and its fiber orientation is not influenced by the last tensile steps.

Table 2. Aspect ratios for SFF 100 in section A and B.

Section	$\epsilon = 0\%$	$\epsilon = 2\%$	$\epsilon = 7\%$	$\epsilon = 12\%$	$\epsilon = 17\%$	$\epsilon = 22\%$	$\epsilon = 27\%$	$\epsilon = 32\%$	$\epsilon = 37\%$
A	1,138	1,087	1,151	1,204	1,272	1,241	1,260	1,272	1,262
B	1,099	1,085	1,155	1,166	1,148	1,160	1,149	1,150	1,163

Comparison SFF 3, which has a complex structure with a relatively thick functional layer and fine fibers, shows a drop in stress around $\epsilon = 6\%$. This drop in stress is caused by a tearing of the inner functional layer (layer 2) with fine fibers and low porosity. At that point the outer functional layer (layer 1), with thicker fibers and higher porosity, is still intact. Because of this, the initial damage that occurs in layer 2 cannot be observed from the surface. The functional layer of SFF 100, which has the thickest fibers and highest porosity observed, fails at higher strains than the functional layers of SFF 3. The fibers of a functional layer with fine fibers and relatively low porosity have many connections, whereas the number of connections for thicker fibers and higher porosity is lower. As a result the functional layer of SFF 3 is stiffer than the one of SFF 100 and fails at lower strains. While the functional layer of SFF 100 fails, the fibers are rotating into the direction of loading. Here the fiber rotation is concentrated to the vicinity of the tear. As the functional layer of SFF 100 is thin in relation to the total thickness of the sintered fiber felt, it bears only a small portion of the applied load. Consequently, its failure is merely reflected by a decreasing slope in the stress-strain curve. In contrast, a drop in stress is observed in the stress-strain curve when the functional layers of SFF 3 start to fail. Which demonstrates that the functional layers of SFF 3 bear a significant part of the applied load.

5. Conclusions

The combination of tensile tests and computed tomography allows observation of the damage process and structural changes in sintered fiber felts with a high spatial resolution. It is possible to detect damage in the inner layers of a sintered fiber felt, which cannot be seen from the surface, and correlate the damage to certain features in the stress-strain curve. The application of a line segmentation technique to images extracted from the CT scans, makes it possible to detect fiber rotation in the functional layer of sintered fiber felts. For SFF 3, which has fine fibers, low porosity and a thick functional layer in relation to the total thickness of the sintered fiber felt, the functional layer has a significant influence on the mechanical strength at small strains. For SFF 100, which has a relatively thin functional layer, thicker fibers and higher porosity, the functional layer has no considerable effect on the mechanical strength. At high strains, only the support grid is bearing the mechanical load while the functional layer has already failed completely. Therefore, the functional range for aeroacoustic applications of sintered fiber felts is limited to strains lower than the failure strain of the functional layer, which depends on the material structure.

Acknowledgments

Financial support of the Deutsche Forschungsgemeinschaft in the frame of the SFB 880-Fundamentals of High-Lift for Future Commercial Aircraft is greatly appreciated.

Author Contributions

Nicolas Lippitz and Joachim Rösler conceived and designed the experiments; Nicolas Lippitz performed the experiments and analyzed the data; Nicolas Lippitz and Joachin Rösler wrote the paper.

Conflicts of Interest

The authors declare no conflict of interest.

References

1. Haines, M.M.; Stansfeld, S.A.; Job, R.F.S.; Berglund, B.; Head, J. Chronic aircraft noise exposure, stress responses, mental health and cognitive performance in school children. *Psychol. Med.* **2001**, *31*, 265–277.

2. Franssen, E.A.M.; van Wiechen, C.M.A.G.; Nagelkerke, N.J.D.; Lebert, E. Aircraft noise around a large international airport and its impact on general health and medication use. *Occup. Environ. Med.* **2004**, *61*, 405–413.

3. Revell, J.; Revell, J.; Kuntz, H.; Balena, F.; Horne, C.; Storms, B.; Dougherty, R.; Kuntz, H.; Balena, F.; Horne, C.; *et al.* Trailing-edge flap noise reduction by porous acoustic treatment. In *Aeroacoustics Conferences*; American Institute of Aeronautics and Astronautics: Reston, VA, USA, 1997.

4. Herr, M.; Dobrinsky, W. Experimental investigation in low-noise trailing edge design. *AIAA J.* **2005**, *43*, 1167–1175.

5. Herr, M. Design criteria for low-noise trailing-edges. In *Aeroacoustics Conferences*; American Institute of Aeronautics and Astronautics: Reston, VA, USA, 2007.

6. Delfs, J.; Faßmann, B.; Lippitz, N.; Lummer, M.; Mößner, M.; Müller, L.; Rurkowska, K.; Uphoff, S. SFB 880: Aeroacoustic research for low noise take-off and landing. *CEAS Aeronaut. J.* **2014**, *5*, 1–15.

7. Herr, M.; Rossignol, K.S.; Delfs, J.; Lippitz, N.; Mößner, M. Specification of porous materials for low-noise trailing-edge applications. In *AIAA Aviation*; American Institute of Aeronautics and Astronautics: Reston, VA, USA, 2014.

8. Salvo, L.; Cloetens, P.; Maire, E.; Zabler, S.; Blandin, J.; Buffière, J.; Ludwig, W.; Boller, E.; Bellet, D.; Josserond, C. X-ray micro-tomography an attractive characterisation technique in materials science. In *Nuclear Instruments and Methods in Physics Research Section B: Beam Interactions with Materials and Atoms*; Elsevier: Amsterdam, The Netherlands, 2003; Volume 200, pp. 273–286.

9. Clyne, T.; Markaki, A.; Tan, J. Mechanical and magnetic properties of metal fibre networks, with and without a polymeric matrix. *Compos. Sci. Technol.* **2005**, *65*, 2492–2499.

10. Adrien, J.; Maire, E.; Gimenez, N.; Sauvant-Moynot, V. Experimental study of the compression behaviour of syntactic foams by in situ X-ray tomography. *Acta Mater.* **2007**, *55*, 1667–1679.

11. Tsarouchas, D.; Markaki, A. Extraction of fibre network architecture by X-ray tomography and prediction of elastic properties using an affine analytical model. *Acta Mater.* **2011**, *59*, 6989–7002.

12. Maire, E.; Zhou, S.; Adrien, J.; Dimichiel, M. Damage quantification in aluminium alloys using in situ tensile tests in X-ray tomography. *Eng. Fract. Mech.* **2011**, *78*, 2679–2690.

13. Neelakantan, S.; Bosbach, W.; Woodhouse, J.; Markaki, A. Characterization and deformation response of orthotropic fibre networks with auxetic out-of-plane behaviour. *Acta Mater.* **2014**, *66*, 326–339.

14. Rösler, J.; Näth, O. Mechanical behaviour of nanoporous superalloy membranes. *Acta Mater.* **2010**, *58*, 1815–1828.

15. Hinze, B.; Rösler, J.; Schmitz, F. Production of nanoporous superalloy membranes by load-free coarsening of γ'-precipitates. *Acta Mater.* **2011**, *59*, 3049–3060.

16. Hinze, B.; Rösler, J.; Lippitz, N. Noise reduction potential of cellular metals. *Metals* **2012**, *2*, 195–201.

17. Lippitz, N.; Rösler, J.; Hinze, B. Potential of metal fibre felts as passive absorbers in absorption silencers. *Metals* **2013**, *3*, 150–158.

18. Lippitz, N.; Hinze, B.; Rösler, J. *SFB 880-Fundamentals of High-Lift for Future Commercial Aircraft: Biennial Report*; Technical University Campus Forschungsflughafen: Braunschweig, Germany, 2013; Chapter Characterization of Materials, pp. 29–39.

19. Tan, J.C.; Elliot, J.A.; Clyne, T.W. Analysis of tomography images of bonded fibre networks to measure distributions of fibre segment length and fibre orientation. *Adv. Eng. Mater.* **2008**, *8*, 495–500.

Measurement and Determination of Friction Characteristic of Air Flow through Porous Media

Wei Zhong [1,2,*]**, Xin Li** [3]**, Guoliang Tao** [3] **and Toshiharu Kagawa** [4]

[1] School of Mechanical Engineering, Jiangsu University of Science and Technology, Zhenjiang 212003, China

[2] Wuxi Pneumatic Technology Research Institute, Wuxi 214072, China

[3] State Key Laboratory of Fluid Power Transmission and Control, Zhejiang University, Hangzhou 310027, China; E-Mails: vortexdoctor@zju.edu.cn (X.L.); gltao@zju.edu.cn (G.T.)

[4] Precision and Intelligence Laboratory, Tokyo Institute of Technology, Yokohama 226-8503, Japan; E-Mail: kagawa.t.aa@m.titech.ac.jp

* Author to whom correspondence should be addressed; E-Mail: zhongwei@just.edu.cn

Academic Editors: Jordi Sort and Eva Pellicer

Abstract: Sintered metal porous media currently plays an important role in air bearing systems. When flowing through porous media, the flow properties are generally represented by incompressible Darcy-Forchheimer regime or Ergun regime. In this study, a modified Ergun equation, which includes air compressibility effects, is developed to describe friction characteristic. Experimental and theoretical investigations on friction characteristic are conducted with a series of metal-sintered porous media. Re = 10 is selected as the boundary for a viscous drag region and a form drag region. Experimental data are first used to determine the coefficient α in the viscous drag region, and then the coefficient β in the form drag region, rather than both simultaneously. Also, the theoretical mass flow rate in terms of the modified Ergun equation provides close approximations to the experimental data. Finally, it is also known that both the air compressibility and inertial effects can obviously enhance the pressure drop.

Keywords: porous media; friction characteristic; modified Ergun; air flow; compressibility

1. Introduction

Air flow through porous media is universal in a broad range of industrial equipments, e.g., filters, heat exchanges, catalytic reactors and pneumatic silencers. In such porous media the distribution of pores with respect to shape and size is irregular [1]. Therefore, the porous media easily provide impedance when they are connected in a pneumatic circuit. There is a crucial need for an accurate description of the friction characteristic in order to evaluate their performance. In pneumatics, it is generally known that flow rate characteristics of normal pneumatic components are characterized in terms of sonic conductance and critical pressure ratio by the ISO 6358 standard [2], which is deduced from isentropic flow through a perfect nozzle, or by an ISO expanded expression proposed by Oneyama in 2003 [3]. Unfortunately, the flow pathway in porous media is much more complicated than that of the ordinary pneumatic components, and there is no such a generalized specification for characterizing the flow throughout porous medium.

A number of studies [4–7] have demonstrated the physics of flow through porous media under the assumption that the internal structure is isotropic and homogeneous. The results revealed that the linear Darcy regime [4] can properly predict the flow behavior when the flow velocity is sufficiently small, or in other words, when the regime is dominated by viscous effects. However, when inertial effects become dominant, the pressure gradient *versus* flow velocity exhibit a nonlinear relation. On this issue, probably the two most known equations including the effects of viscosity and inertia should be the Forchheimer equation and the Ergun equation [8,9]. Many researchers have attempted to correlate experimental data to assess the flow characteristics of porous media using the two fundamental equations. Early experimental works by Beavers *et al.* [10,11] indicate that the flow characteristics of fibrous porous media can be clarified by Forchheimer equation with appropriate permeability and inertial coefficient. Montillet *et al.* [12] proposed a correlating equation to predict pressure drops through packed beds and provided experimental evidence for validation. Antohe *et al.* [13] presented a study that uses the Forchheimer extended flow model to compute the permeability and inertial coefficient for the compressed aluminum matrices. Dukhan and Minjeur [14] suggested the use of two permeabilities, one for the Darcy regime and another for the Forchheimer regime, to depict pressure drop properties. Moreover, Dukhan and Patel [15,16], Dukhan and Ali [17] experimentally investigated wall, size, entrance and exit effects on permeability and form drag coefficient. In addition to the use of Forchheimer equation, Dukhan [18] also verified the validity of the Ergun-type relationship to characterize pressure drop with respect to flow velocity and confirmed that both permeability and inertial coefficient correlate well with porosity. Du Plessis [19] theoretically derived a momentum transport equation for fully-developed flow in porous media which is based on Ergun equation. J.F. Liu [20] described the friction characteristic based on the measured pressure drop of air through foam matrixes using an empirical equation which is similar to Ergun equation.

According to the literature survey it is noted that porous media with a high range of porosities (over 80%) have been frequently used in earlier applications, and, existing references that use either the Forchheimer regime or the Ergun equation, commonly neglect the velocity gradient in the flow direction. When unidirectional airflow passes through those loose media, air compressibility is negligible due to slight pressure drops, and velocity in the flow direction is generally considered constant [21–23]. However, in contemporary air-bearing feeding systems, porous media fabricated by sintered metal that have a small porosity also receive great attention. Belforte *et al.* [24,25] showed a considerable pressure drop between the upstream and downstream, and verified the capability of porous media to eliminate pressure peak compared with the case involving an orifice. Amano *et al.* [26] and Oiwa *et al.* [27] developed air conveyor systems that use sintered metal porous air supply pads. In these applications, air compressibility cannot be ignored, and therefore, the conventional correlation of friction characteristics needs to be reconstructed. The authors [28] have developed a modified Forchheimer equation to represent the pressure drop characteristic for a series of metal sintered porous media, but the feasibility in terms of the Ergun equation is not yet known and also needs to be confirmed.

In this study, we determined the friction characteristic in terms of a modified Ergun equation. Several sintered metal porous resistances were tested, and the effectiveness of the proposed correlation was experimentally verified.

2. Theoretical Section

Porous media consist of thousands of irregular pores, and its microstructure is characterized by several factors, such as pore size, pore shape, and porosity. The physical model of airflow through porous media is schematically shown in Figure 1, and the following assumptions are made to enable a theoretical derivation.

(1) The flow is steady and fully developed. (2) The porous medium is isotropic and homogeneous, so its property parameters are constant. (3) The flow is in the horizontal direction only. (4) The flow through the porous medium remains isothermal.

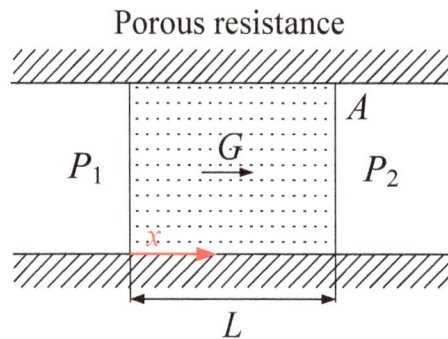

Figure 1. Schematic of flow through porous medium.

Consequently, the mean flow velocity changes with the air density along the length direction and is written in the following form [25]

$$v = \frac{G}{\rho A} \tag{1}$$

where v is the flow velocity, G is the mass flow rate, ρ is the density and A is the cross-sectional area.

Assuming that the air is a perfect gas, density ρ is expressed in the following form according to the ideal gas state equation

$$\rho = \frac{P}{RT} \tag{2}$$

where P is the pressure, R is the gas constant and T is the temperature.

Ergun [8] published an empirical relation for describing the pressure drop through porous media based on the porosity and a geometrical length scale as

$$-\frac{dP}{dx} = \alpha \frac{(1-\varepsilon)^2 \mu}{\varepsilon^3 \varphi^2 D_p^2} v + \beta \frac{(1-\varepsilon)\rho}{\varepsilon^3 \varphi D_p} v^2 \tag{3}$$

where α is a factor for the viscous drag portion of the pressure drop, β is a factor for the form drag portion, ε is the porosity and D_p is the particle diameter. In addition, a coefficient of sphericity φ, which multiplies the particle diameter, is employed in the equation in order to correct the pressure losses.

The construction of Ergun equation was based on modeling the space between packed beds of spheres as parallel capillaries, with the first term accounting for viscous effects and the other term for kinetic effects. The multipliers α and β can be viewed as correction factors to account for the geometrical difference between flow paths through packed spheres and parallel capillaries, Dukhan et al. [29]. Substituting Equations (1) and (2) into Equation (3), the Ergun equation takes the following form for an incompressible flow

$$\rho \frac{\Delta P}{L} = \frac{\alpha (1-\varepsilon)^2 \mu}{\varepsilon^3 A \varphi^2 D_p^2} G + \frac{\beta (1-\varepsilon)}{\varepsilon^3 \varphi D_p A^2} G^2 \tag{4}$$

The Reynolds number based on porosity ε and particle diameter D_p is given by

$$\mathrm{Re} = \frac{\rho v \varphi D_p}{\mu (1-\varepsilon)} \tag{5}$$

A dimensionless friction factor f is defined as

$$f = -\frac{dP}{dx} \frac{\varphi D_p}{\rho v^2} \left(\frac{\varepsilon^3}{1-\varepsilon} \right) \tag{6}$$

Substituting Equations (5) and (6) into Equation (3), we obtain the following correlation for the friction factor as a function of the Reynolds number

$$f = \frac{\alpha}{\mathrm{Re}} + \beta \tag{7}$$

For traditional Ergun, coefficients α and β are valued 150 and 1.75, respectively. However, by experiments, we found that the coefficients differ from these values. Experimental method and results will be clarified in the next two sections.

3. Experimental Section

A series of commercial porous resistances with different dimensions were selected from many specimens, and are designated as #1 to #6 in Figure 2. Each sample was a cylinder having a length of 2, 5, 2, 4, 10, and 3 mm and a diameter of 4, 4, 8, 8, 8, and 12 mm. All of these samples were fabricated by sintering SUS316L powder into a cylindrical shape at a temperature above 1000 °C. During sintering, complicated flow paths were formed by interconnected and dead-end pores, as shown in the microscope photograph of the porous surface (Figure 2). The average particle diameter D_p is approximately 75 μm for #1 and #3, approximately 90 μm for #2 and #5, and approximately 50 μm for

#4 and #6. Porosity ε, defined as the fraction of the total volume that is occupied by void space, is calculated as follows

$$\varepsilon = 1 - \frac{4m}{\pi\gamma LD^2}.$$

(8)

where $\gamma = 8.03 \times 10^3$ kg/m^3 is the density of the steel alloy, m is the mass of the resistance, L is the length, and D is the diameter. Dimensions, weights, porosities, average particle diameters, coefficient of sphericity and hydraulic diameter of the test resistances are tabulated in Table 1.

Table 1. Dimensions, weights, porosities, average particle diameters, coefficient of sphericity and hydraulic diameter of the test resistances.

Porous media	Diameter (mm)	Length (mm)	Weight (g)	Porosity	Particle diameter (µm)	Hydraulic diameter (µm)
#1	4	2	0.122	0.397	75	16.5
#2	4	5	0.277	0.451	90	24.6
#3	8	2	0.494	0.388	75	15.8
#4	8	4	0.974	0.379	50	10.2
#5	8	10	2.301	0.430	90	22.6
#6	12	3	1.476	0.458	50	14.1

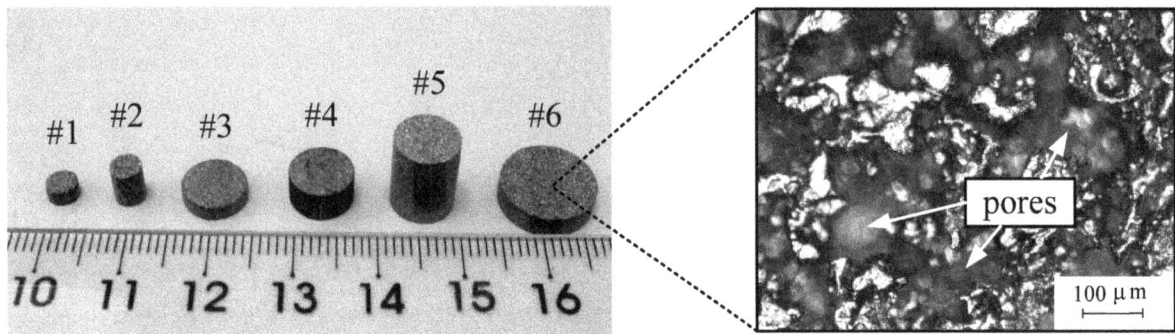

Figure 2. Test porous resistances.

The arrangement of the pneumatic circuit is shown in Figure 3. The porous resistance is sealed in a metal sleeve using a cylindrical rubber to prevent possible leakage between the side face and the sleeve. Compressed air is supplied and regulated to the selected pressure. The buffer tank with a volume of 30 L, located after the pressure regulator, is used to stabilize the supply pressure. Two calibrated pressure gauges are respectively placed upstream and downstream of the porous resistance, and an adjustable valve is connected to change the downstream pressure. The upstream and downstream pressure gauges are installed at a distance not less than three times the resistance length before and after the medium to avoid the entrance and exit effect [16]. It should be noted that the pressure drops in the connecting pipes are negligibly small in comparison with that in the porous media and it is believable the pressure data can be measured with sufficient accuracy and do not need any correction. The total air volume through the porous resistance is collected by a dry test gas meter which has a range of 0.04~6 m^3/s, with a stopwatch used to record the time and then to calculate the mean flow rate. An attempt was also made in advance to clarify the effects of flow direction by reversing the resistance position relative to the flow direction. No apparent variation in flow rate measured under the same pressure condition was observed before and after the position changed. This result validates that the friction characteristic is independent of the flow direction.

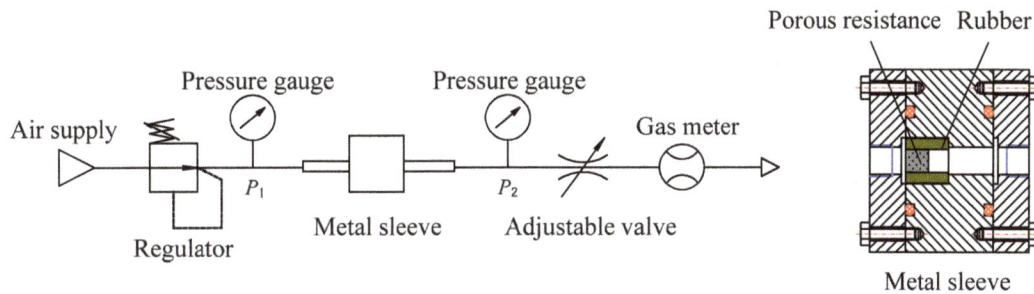

Figure 3. Schematic of experimental setup.

During the experiments, the procedure is performed as follows: (1) maintain a steady flow rate and keep the upstream pressure unchangeable at 584.9 kPa. (2) Use the adjustable valve to change the pressure ratio P_2/P_1, and record the paired pressure and flow rate for more than 20 points. To reduce the uncertainty, steady flow is maintained at least 30 s for procuring each data point, and the measurements are repeated three times for each sample to get an average result. In the case of small pressure drop, a differential pressure transducer (KL17 [30], Nagano Keiki Co., Tokyo, Japan), with a measuring range of 0~2 kPa, is used to measure the pressure difference, and correspondingly, the gas meter is replaced by a wet type one with a measuring range of 0.016~5 L/min.

4. Results and Discussion

The first term on the right-hand side of Equation (4) represents the viscous drag portion of the pressure drop, and therefore, the coefficient α can be determined by the experimental data under small pressure drop. The pressure drop results (<1.5 kPa) are considered negligibly small in comparison with the supply pressure, and therefore air is treated as incompressible fluid, without density variation. Figure 4 shows that the pressure drop (0 to 1 kPa) is in close linearity to the flow rate, a result implying that form drag effects are negligible. If we visualize the porous medium as a bundle of tubes, it is not difficult to understand the linear relationship with the knowledge of equivalent hydraulic diameter. In the meantime, it should be noted that the medium size also plays an important role in the relationship between the pressure drop and the flow rate. Therefore, a least squares straight line is passed through the data points, with the slope proportional to the coefficient α. Then, all the data are used to gain the coefficient β by Equation (4). The R^2 value for the linear fitting line exceeds 0.99 for all the test media so that the error is thought to be acceptable for the calculation.

Figure 5 shows the relation between coefficient β and flow rate in terms of incompressible Ergun equation. The β is not an average coefficient for all the test media, and, it is just for a specific case and varies with the properties of the medium. The β is obtained with the resulting coefficient α using Equation (9) with the paired pressure drop and flow rate. Observation shows that β is a negative value and appears to continuously change as the flow rate increases. However, these results are theoretically unreasonable. First, the negative value is physically meaningless because the form drag effects have positive impacts on the pressure drop. Second again, in general, when form drag effects dominate the flow pattern, β should have the tendency to be a universal constant because it is completely dependent on the properties of the media. The negative and increasing β indicates that it is the compressibility effects enhance the pressure drop and thus cannot be ignored. Actually, in theory, the coefficient β should be a constant that only depends on the internal geometry at high Reynolds number. However, numerous studies indicate that a transitional regime exists between the viscous and inertial regimes. The visible variations in β at small Reynolds number are mainly because the data points in the transitional regime are mistakenly used for the calculation. As a result, the optimal β should be selected for the converged value under the maximum flow rate.

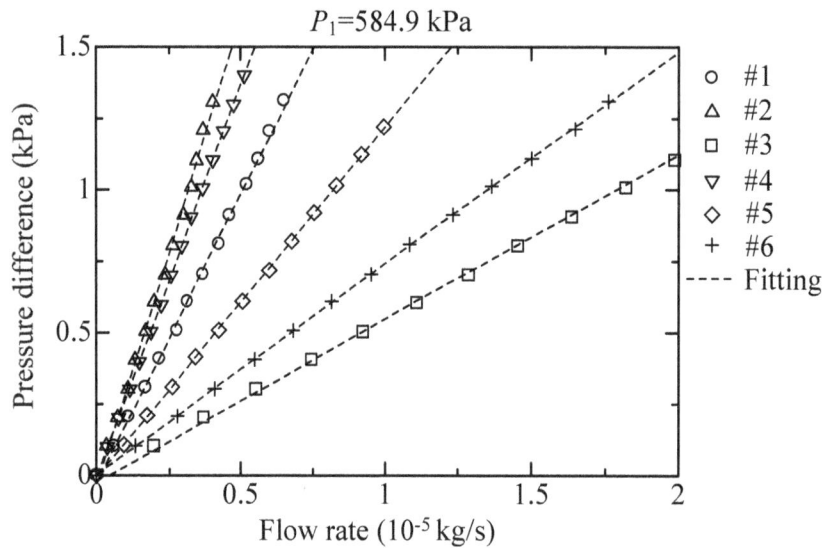

Figure 4. Pressure difference *versus* flow rate.

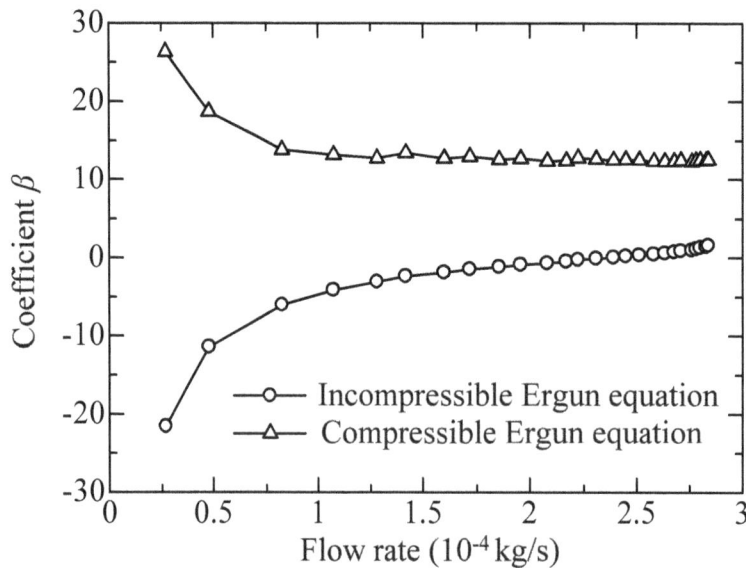

Figure 5. Coefficient β *versus* flow rate (resistance #1).

Accordingly, substituting Equations (1) and (2) into Equation (3) and integrating the pressure with respect to length, with the boundary conditions $P = P_1$ ($x = 0$ [m]), $P = P_2$ ($x = L$ [m]), obtain the following modified Ergun equation

$$\frac{\beta(1-\varepsilon)RT}{\varepsilon^3 \varphi D_p A^2}G^2 + \frac{\alpha(1-\varepsilon)^2 \mu RT}{\varepsilon^3 A\varphi^2 D_p^2}G + \frac{P_2^2 - P_1^2}{2L} = 0 \qquad (9)$$

Equation (9) includes the contribution of air compressibility to the pressure drop. The relation of β with respect to flow rate in terms of the modified Ergun equation is also presented in Figure 5, which shows that β is a positive value and indeed converges to constant as the flow rate increases. As a result, considering compressibility effects is necessary.

According to Equation (9), the coefficients α and β are determined non-simultaneously by, α first in viscous drag region, then β in form drag region. This method is named non-simultaneous method hereinafter. Table 2 lists the determined coefficients for each test medium. The values of α and β were generally thought to be universal; however some early experiments proved otherwise. In [31], Fand et al. tested spheres of various diameters and found that α and β exhibit some variations with diameter. Comiti and Renaud [32] obtain an average value of 141 for α and 1.63 for β for all of the packed spheres they tested. In this work, the authors contend to first use different α and β to describe the friction characteristic for each of the test media for a comparison. Noted that as the flow rate increases, small changes in coefficient β will induce an obvious variation in the pressure drop, and therefore it is corrected to the second decimal place to improve accuracy. Moreover, an alternative method is to determine the coefficients simultaneously according to the Gauss–Newton fitting method, which is hereinafter referred to as the simultaneous method. The simultaneous method is simple in comparison with the non-simultaneous method. Coefficients obtained by the simultaneous method are also listed in Table 2. Clearly, the coefficients determined by different methods have different values.

Table 2. Results of coefficients α and β using non-simultaneous and simultaneous method.

Porous medium	Non-simultaneous method		Simultaneous method	
	α	β	α	β
#1	466.2	12.43	515.9	12.12
#2	458.0	12.84	550.4	12.01
#3	475.8	8.73	419.3	9.08
#4	457.9	14.82	502.4	13.73
#5	460.8	10.01	599.8	8.81
#6	468.4	14.99	332.0	18.59

A presentation of the data in the form of friction factor versus Reynolds number is given in Figure 6. The linear dashed curve is for $f = 464.5/Re$, which leaves out the second term of the Equation (4) and serves as a reference line for medium #1. The experimental Reynolds number ranges for the medium #1 to #6 are 0.35~154, 0.26~153, 0.28~169, 0.04~42, 0.17~118 and 0.06~67, respectively. Dybbs and Edwards [33] reported that the boundary for viscous flow regime and inertial flow regime is Re = 1. However, the authors contend that the region with Re from 1 to 10 would involve the transition regime. For this class of porous resistances, the form drag effects prevail over the viscous effects when Reynolds number is approximately beyond 10. Therefore, Re = 10 can be seen as the boundary to distinguish the viscous drag region and the form drag region. Following Dybbs's classification the experimental range only cover the laminar regime. Actually, both viscous and inertial regimes, along with transition between the two, are laminar in nature. Moreover, friction factor decreases with the increase of Reynolds number, and gradually converges to an asymptotic value depended on the coefficient β as the Reynolds number becomes adequately large. The curve, which represents the Ergun equation with coefficients α and β respectively valued 150 and 1.75, is also given in Figure 6 for a comparison with the experimental data. Quite clearly, the traditional Ergun equation cannot be directly used to describe the friction characteristic. Table 2 shows that the values of α and β varies little for each of the materials by the non-simultaneous method. So, the optimal α and β can be obtained as the average value for all the media, and therefore, the generalized Ergun equation is now represented by 464.5/Re + 12.30, which shows a good agreement with all of the experimental data. However, even if the coefficients α and β are constants, Equation (9) indicates that the pressure losses do not only depend on the particle diameter and porosity but also the diameter and length of the medium. Taking the media #1 and #3 as an example, the two have almost the same particle diameter and porosity, but provide totally different pressure changes under the same flow rate due to the size difference (Figure 4). In addition to the size difference, as for media #2 and #5, the difference in coefficient of sphericity also greatly affects the pressure changes.

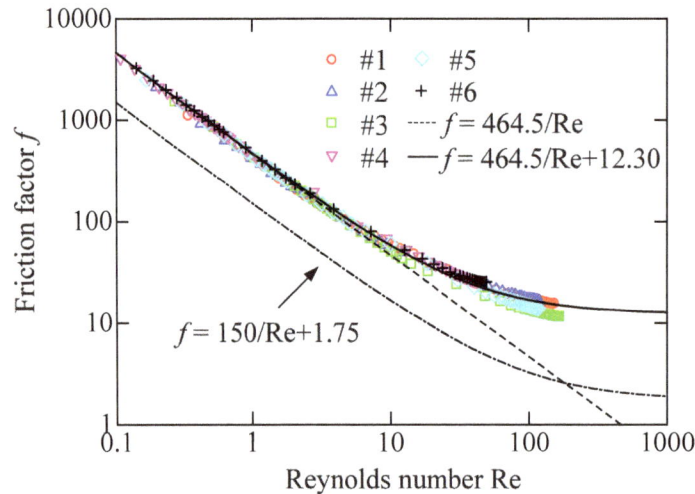

Figure 6. Relation between friction factor and Reynolds number.

Taking the resistance #2 as an example, Figure 7 plots the experimental data and the calculated results by the non-simultaneous method and simultaneous method. It is observed that both calculated curves accord with the experimental data when Re > 50. However, in the viscous region, the black curve, by the simultaneous method, shows significant departures from the experiment data. Then, to evaluate the applicability of the two methods, flow rate satisfying Equation (9) are calculated, and the errors E between the experimental data and theoretical results are considered as

$$E = \sqrt{\frac{\sum\left(G_{(\exp)i} - G_{(\text{cal})i}\right)^2}{\sum G_{(\exp)i}^2}} \times 100\% \tag{10}$$

where $G_{(\exp)i}$ represents the experimental data, and $G_{(\text{cal})i}$ represents the calculated result.

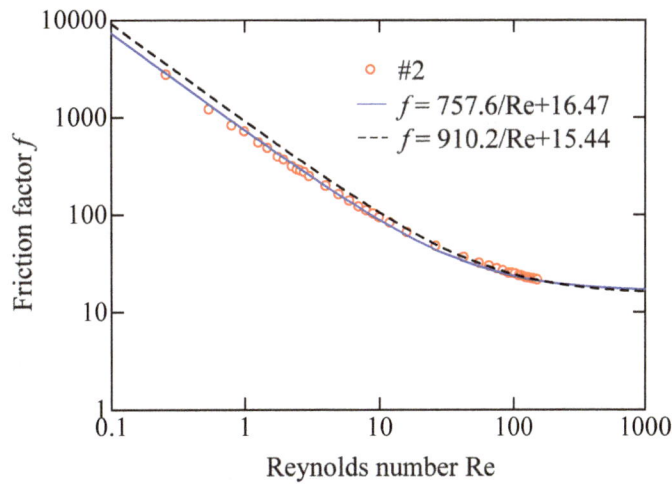

Figure 7. Relation between friction factor and Reynolds number (#2).

Table 3 lists a comparison of the errors for the two methods in different regions. It is made clear that in the form drag region, the simultaneous method provides a slightly better accuracy. However, in the viscous drag region, it is found that such approach is subject to considerable errors, even sometimes over 20%, which is unacceptable in applications. That is to say, although in some parts the simultaneous method can achieve a higher accuracy, it is just a curve fitting method, and those determined coefficients may lose physical meanings. Inversely, the representation with coefficients by the non-simultaneous method shows good accordance with the experimental data over the whole range.

Table 3. Errors E.

Tested resistances	Viscous region		Form drag region	
	Non-simultaneous method (E)	Simultaneous method (E)	Non-simultaneous method (E)	Simultaneous method (E)
#1	4.6%	13.3%	1.1%	0.8%
#2	5.0%	20.2%	1.0%	0.4%
#3	2.8%	11.0%	0.4%	0.4%
#4	0.8%	9.5%	1.0%	0.4%
#5	1.4%	24.1%	1.5%	0.5%
#6	0.5%	39.4%	1.1%	1.6%

Air density is variable in the porous media, and Expression (2) is another form of the continuity equation that includes air compressibility. When flowing through the porous media, air expands, accelerates, and the pressure decreases along the length direction. Figure 8 gives a comparison of the experimental data and the calculated results. It can be easily figured out that the proposed modified Ergun equation considering air compressibility accords well with the experimental data. As the pressure drop becomes visible, air density changes greatly throught the porous media, and of course, compressibility effects cannot be ignored. The dashed line, which depicts the incompressible case, expresses a less conspicuous pressure drop, indicating that air compressibility can lead to a more conspicuous pressure drop. In addition, the calculated curve for the incompressible case appears nonlinear because the inertial effects also play an important role.

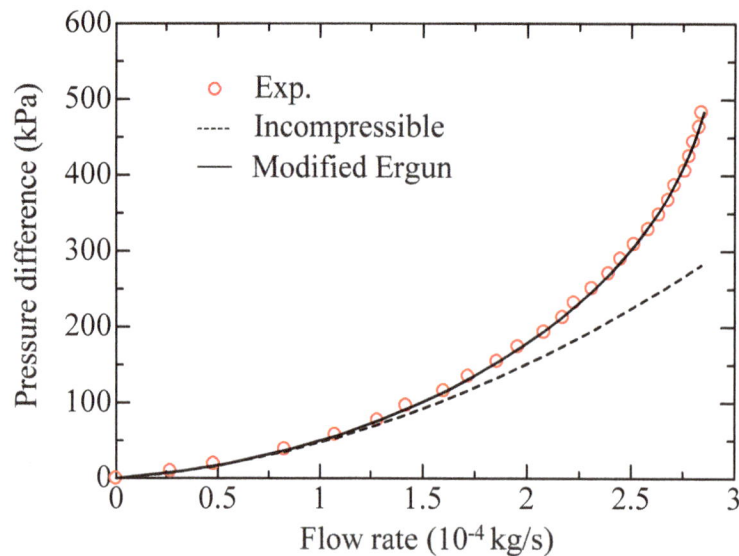

Figure 8. Pressure difference *versus* flow rate (#1).

5. Conclusions

This study provides an insight into the determination of friction characteristic for airflow through sintered metal porous media. An experimental setup was established to measure the pressure drop versus flow rate through porous media, and a few samples were tested within the laminar region. The traditional Ergun equation is proven inappropriate for the description of the friction characteristic. Therefore, a modified Ergun equation taking into account the air compressibility effects is proposed. The coefficients α and β in the modified equation are factors for the viscous drag portion and form drag portion of the pressure drop. Air is treated as an incompressible fluid when the pressure drop is sufficiently small. As the pressure drop becomes visible, air density changes greatly and the compressibility effects cannot be ignored. Two methods, the simultaneous method and the non-simultaneous method, are respectively used to determine the coefficients. The simultaneous method gives a perfect accordance with the experimental data when the flow rate becomes adequately large, but brings unacceptable errors even over 30% at small flow ranges. To completely avoid the transitional region, Re = 10 is selected as the boundary for the viscous drag region and form drag region. The non-simultaneous method is used to determine the coefficients by, α first in the viscous drag region, then β in the form drag region, and, the optimal β is selected for the converged value under maximum flow rate. Moreover, theoretical mass flow rate in terms of the modified Ergun equation can provide close approximations over the entire range within 5% uncertainty, which is sufficient for most applications.

It should be noted that, in addition to the experimental study, the proposed modified Ergun equation can also be of importance when dealing with numerical studies (e.g., [34,35], *etc.*). Efforts are needed to develop some numerical model that incorporates the basic fluid dynamics features in porous media in the future.

Acknowledgments

The authors wish to thank the financial support of the National Natural Science Foundation of China (Grant No. 51205174), the Postdoctoral Science Foundation of China (Grant No. 2014M550309), the Open Foundation of the State Key Laboratory of Fluid Power Transmission and Control (Grant No. GZKF-201407) and the Youth Funds of the State Key Laboratory of Fluid Power Transmission and Control (SKLoFP_QN_1304).

Author Contributions

Wei Zhong designed experiments, analyzed data and wrote the paper; Xin Li gave technical support and conceptual advice; Guoliang Tao gave technical support and discussed the results; Toshiharu Kagawa discussed the results and commented on the manuscript.

Conflicts of Interest

The authors declare no conflict of interest.

References

1. Nield, D.A.; Bejan, A. *Convection in Porous Media*, 3rd ed.; Springer: New York, NY, USA, 2006.
2. ISO 6358-1: 2013. Pneumatic fluid power-Determination of flow-rate characteristics of components using compressible fluids-Part 1: General rules and test methods for steady-state flow. Available online: http://www.iso.org/iso/home/store/catalogue_ics/catalogue_detail_ics.htm?ics1= 23&ics2=100&ics3=01&csnumber=56612 (accessed on 4 March 2015).
3. Oneyama, N.; Zhang, H.P.; Senoo, M. Study and suggestions on flow-rate characteristics of pneumatic components. In Proceedings of the Fourth International Symposium on Power Transmission and Control, Wuhan, China, 8–10 April 2003; pp. 326–331.
4. Lage, J.L.; Antohe, B.V.; Nield, D.A. Two types of nonlinear pressure-drop versus flow-rate relation observed for saturated porous media. *J. Fluids Eng.* **1997**, *119*, 700–706.
5. Andrade, J.S.; Costa, U.M.S.; Almeida, M.P.; Makse, H.A.; Stanley, H.E. Inertial effects on fluid flow through disordered porous media. *Phys. Rev. Lett.* **1999**, *82*, 5249–5252.
6. Boomsma, K.; Poulikakos, D. The effects of compression and pore size variations on the liquid flow characteristics in metal foams. *J. Fluids Eng.* **2002**, *124*, 263–272.
7. Medraj, M.; Baril, E.; Loya, V.; Lefebvre, L.P. The effect of microstructure on the permeability of metallic foams. *J. Mater. Sci.* **2007**, *42*, 4372–4383.
8. Ergun, S. Fluid flow through packed columns. *Chem. Eng. Prog.* **1952**, *48*, 89–94.
9. Macdonald, I.F.; EI-Sayed, M.S.; Mow, K.; Dullien, F.A.L. Flow through porous media-the Ergun Equation Revisited. *Ind. Eng. Chem. Fundam.* **1979**, *18*, 199–208.
10. Beavers, G.S.; Sparrow, E.M. Non-Darcy flow through fibrous porous media. *J. Appl. Mech.* **1969**, *36*, 711–714.
11. Beavers, G.S.; Sparrow, E.M.; Rodenz, D.E. Influence of bed size on the flow characteristics and porosity of randomly packed beds of spheres. *J. Appl. Mech.* **1973**, *40*, 655–660.

12. Montillet, A.; Akkari, E.; Comiti, J. About a correlating equation for predicting pressure drops through packed beds of spheres in a large range of Reynolds numbers. *Chem. Eng. Process.* **2007**, *46*, 329–333.

13. Antohe, B.V.; Lage, J.L.; Price, D.C.; Weber, R.M. Experimental determination of permeability and inertia coefficients of mechanically compressed aluminum porous matrices. *J. Fluids Eng.* **1997**, *119*, 404–412.

14. Dukhan, N.; Minjeur, C.A. A two-permeability approach for assessing flow properties in metal foam. *J. Porous Mater.* **2011**, *18*, 417–424.

15. Dukhan, N.; Patel, K. Entrance and exit effects for fluid flow in metal foam. *AIP Conf. Proc.* **2010**, *1254*, 299–304.

16. Dukhan, N.; Patel, K. Effect of sample's length on flow properties of open-cell metal foam and pressure-drop correlations. *J. Porous Mater.* **2011**, *18*, 655–665.

17. Dukhan, N.; Ali, M. Effect of confining wall on properties of gas flow through metal foam: An experimental study. *Transp. Porous Media* **2012**, *91*, 225–237.

18. Dukhan, N. Correlations for the pressure drop for flow through metal foam. *Exp. Fluids* **2006**, *41*, 665–672.

19. Du Plessis, J.P. Analytical quantification of coefficients in the Ergun equation for fluid friction in a packed bed. *Transp. Porous Media* **1994**, *16*, 189–207.

20. Liu, J.F.; Wu, W.T.; Chiu, W.C.; Hsieh, W.H. Measurement and correlation of friction characteristic of flow through foam matrixes. *Exp. Therm. Fluid Sci.* **2006**, *30*, 329–336.

21. Kim, T.; Lu, T.J. Pressure drop through anisotropic porous medium like cylinderbundles in turbulent flow regime. *J. Fluids Eng.* **2008**, *130*, 104501:1–104501:5.

22. Jin, L.W.; Kai, C.L. Pressure drop and friction factor of steady and oscillating flows in open-cell porous media. *Transp. Porous Media* **2008**, *72*, 37–52.

23. Mancin, S.; Zilio, C.; Cavallini, A.; Rossetto, L. Pressure drop during air flow in aluminum foams. *Int. J. Heat Mass Tran.* **2010**, *53*, 3121–3130.

24. Belforte, G.; Raparelli, T.; Viktorov, V.; Trivella, A. Feeding system of aerostatic bearings with porous media. Available online: http://www.jfps.jp/proceedings/tukuba2005/pdf/100193.pdf (accessed on 2 March 2015).

25. Belforte, G.; Raparelli, T.; Viktorov, V.; Trivella, A. Metal woven wire cloth feeding system for gas bearings. *Tribol. Int.* **2009**, *42*, 600–608.

26. Amano, K.; Yoshimoto, S.; Miyatake, M.; Hirayama, T. Basic investigation of noncontact transportation system for large TFT-LCD glass sheet used in CCD inspection section. *Precis. Eng.* **2001**, *35*, 58–64.

27. Oiwa, N.; Masuda, M.; Hirayama, T.; Matsuoka, T.; Yabe, H. Deformation and flying height orbit of glass sheets on aerostatic porous bearing guides. *Tribol. Int.* **2012**, *48*, 2–7.

28. Zhong, W.; Li, X.; Liu, F.H.; Tao, G.L.; Lu, B.; Kagawa, T. Measurement and correlation of pressure drop characteristics for air flow through sintered metal porous media. *Transp. Porous Media* **2014**, *101*, 53–67.

29. Dukhan, N.; Bağci, Ö.; Özdemir, M. Experimental flow in various porous media and reconciliation of Forchheimer and Ergun relations. *Exp. Therm. Fluid Sci.* **2014**, *57*, 425–433.

30. KL 17, Differential Pressure Transmitter. Available online: http://products.naganokeiki.co.jp/assets/files/3030/E-KL17J130401.pdf (accessed on 2 March 2015).

31. Fand, R.M.; Kim, B.Y.K.; Lam, A.C.C.; Phan, R.T. Resistance to the flow of fluids through simple and complex porous media whose matrices are composed of randomly packed spheres. *J. Fluids Eng.* **1987**, *109*, 268–273.

32. Comiti, J.; Renaud, M. A new model for determining mean structure parameters of fixed beds from pressure drop measurements: application to beds packed with parallelepipedal particles. *Chem. Eng. Sci.* **1989**, *44*, 1539–1545.

33. Dybbs, A.; Edwards, R.V. A new look at porous media fluid mechanics—Darcy to Turbulent. *Fundam. Transp. Phenom. Porous Media* **1984**, *82*, 199–256.

34. Machado, R. Numerical simulations of surface reaction in porous media with lattice Boltzmann. *Chem. Eng. Sci.* **2012**, *69*, 628–643.

35. Rong, L.W.; Dong, K.J.; Yu, A.B. Lattice-Boltzmann simulation of fluid flow through packed beds of uniform spheres: Effect of porosity. *Chem. Eng. Sci.* **2013**, *99*, 44–58.

12

Ultrafine-Grained Precipitation Hardened Copper Alloys by Swaging or Accumulative Roll Bonding

Igor Altenberger [1], Hans-Achim Kuhn [1,*], Mozhgan Gholami [2], Mansour Mhaede [2] and Lothar Wagner [2]

[1] Central Laboratory, Research & Development, Wieland-Werke AG, Graf-Arco-Str. 36, 89079 Ulm, Germany; E-Mail: igor.altenberger@wieland.de
[2] Institute of Materials Science and Engineering, TU Clausthal, Agricola-Str. 6, 38678 Clausthal-Zellerfeld, Germany; E-Mails: mozhgan.gholami.kermanshahi@tu-clausthal.de (M.G.); mansour.mhaede@tu-clausthal.de (M.M.); lothar.wagner@tu-clausthal.de (L.W.)

* Author to whom correspondence should be addressed; E-Mail: achim.kuhn@wieland.de

Academic Editor: Heinz Werner Höppel

Abstract: There is an increasing demand in the industry for conductive high strength copper alloys. Traditionally, alloy systems capable of precipitation hardening have been the first choice for electromechanical connector materials. Recently, ultrafine-grained materials have gained enormous attention in the materials science community as well as in first industrial applications (see, for instance, proceedings of NANO SPD conferences). In this study the potential of precipitation hardened ultra-fine grained copper alloys is outlined and discussed. For this purpose, swaging or accumulative roll-bonding is applied to typical precipitation hardened high-strength copper alloys such as Corson alloys. A detailed description of the microstructure is given by means of EBSD, Electron Channeling Imaging (ECCI) methods and consequences for mechanical properties (tensile strength as well as fatigue) and electrical conductivity are discussed. Finally the role of precipitates for thermal stability is investigated and promising concepts (e.g. tailoring of stacking fault energy for grain size reduction) and alloy systems for the future are proposed and discussed. The relation between electrical conductivity and strength is reported.

Keywords: Cu-Ni-Si alloys; swaging; accumulative roll bonding; precipitation hardening

1. Introduction

Precipitation hardening can provide a combination of high strength and high (thermal or electrical) conductivity in copper alloys. By concentrating the alloying elements in fine precipitates, the Cu-matrix remains relatively pure with only few interstitial or substitutional atoms left in the Cu-matrix. Consequently conductivity is not detrimentally affected by solid solution impurities while maintaining high yield strength by finely dispersed precipitates which effectively impede dislocation movement.

The industrially most relevant precipitation hardened copper alloys combining high strength with high electrical conductivity are essentially Corson-alloys [1] which are often (but not exclusively) based on the ternary system Cu-Ni-Si. The Cu-Ni-Si-system has been thoroughly studied already in 1927 [1]. Microstructurally, the high yield strength (up to 800–900 MPa after precipitation hardening and cold working) is caused by finely dispersed semi-coherent Ni-Si-precipitates [2,3] with a diameter lower than 20 nm. As explained by the phase diagram (Figure 1), the temperature of solution annealing strongly depends on the amount of alloyed Ni and Si. Today, preferred and standardized alloys such as C7025 or C7035 contain ~3% silicides.

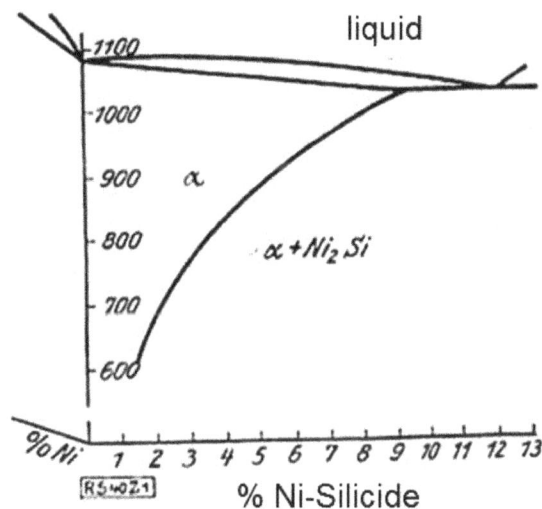

Figure 1. Pseudo-binary phase diagram of the Cu-Ni$_2$Si-system after Corson, 1927 [1].

The traditional approach for the development of high strength copper alloys is focused on chemical variation of precipitation hardened alloys [4]. In addition, microstructural control, e.g., generation of very fine grained or even ultra fine grained copper alloys, opens the door for tailored copper alloys combining optimized precipitate- as well as grain- or subgrain structure. Both approaches are presently used to generate high-performance components for industrial practice. Common methods for generating ultra fine grained metals by Severe Plastic Deformation (SPD), such as Equal Channel Angular Pressing (ECAP) or Accumulative Roll Bonding (ARB) [5–7], are well known and established, especially for pure copper. In contrast to SPD-related studies on pure copper, the archival literature sources dealing with severe plastic deformation of copper *alloys* are significantly more rare [8,9].

In the present study the authors seek to investigate and discuss the applicability of swaging (as a continuous method) as well as ARB for achieving very fine grained to ultra fine grained microstructures in classical Cu-Ni-Si alloys. A key feature of the research presented here is the stabilization of the microstructure by optimized aging treatments after swaging or ARB.

2. Experimental Section

The investigated copper alloy is the Corson-type alloy CuNi3Si1Mg (UNS designation C70250), which has experienced wide-spread use as connector-, leadframe- and high-strength wire material. Traces of Mg are alloyed to enhance the stress relaxation stability. Mg contributes to solid solution hardening as well as to precipitation hardening since Mg atoms may also form mixed (Ni,Mg)-silicides. The material investigated in our present study was hot extruded at 900 °C, then solution annealed at 800 °C/2 h (or alternatively at 950 °C/10 min). This condition was then rotary swaged and finally precipitation hardened. It should be noted that a complete dissolution of coarse Ni-silicides is not possible at these temperatures [10]. In the present study, the precipitation hardening was carried out at 450 °C.

Backscatter (ECCI, electron channeling contrast imaging) electron microscopy [11,12] was carried out using an AsB (Angle Selective Detector, Zeiss, Oberkochen, Germany) [13] in a Zeiss ULTRA scanning electron microscope (SEM, Zeiss, Oberkochen, Germany) equipped with a thermal field emission cathode. Typically, an aperture lens of 120 μm and acceleration voltages of 15–20 kV at a working distance of 2–6 mm were used.

For the Electron Backscatter Diffraction (EBSD) investigations, an EBSD-unit by Oxford was used. The EBSD patterns were recorded using a 4 × 4 binning, data acquisition and calculation of the patterns were performed by a Nordlys camera and AZTEC software by Oxford (UK), respectively. Prior to EBSD and ECCI-characterization in the SEM, the samples were carefully mechanically ground up to 2400 grid (SiC paper) and then polished up to 1 μm. Finally, samples were vibration polished for 3 h with dispersed magnesium oxide to aim for a sample surface with as little preparational cold work as possible. Two sets of experiments were carried out: swaging (which was carried out at TU Clausthal, Clausthal-Zellerfeld, Germany) of solution annealed bars from an initial diameter of 24 mm to a diameter of 7 mm (phi = −2.4) as well as swaging of a solution annealed wire with a diameter of 5.3 mm to a diameter of 2.7 mm (phi = −1.39). In both cases, precipitation annealing after swaging was done at 450 °C at different aging times. In the following elaborations we will use the terminology "peak-aged" for samples which were precipitation hardened at 450 °C for 1–6 h and "over-aged" for samples which were precipitation hardened for >16 h. Further details concerning the aging kinetics for the 2.7 mm wire are given in [14]. Finally, the swaged and subsequently precipitation hardened samples were mechanically characterized by tensile- and hardness tests. Moreover, the electrical conductivity of all the samples, before and after artificial aging, was measured using a SIGMATEST@-probe (eddy current method, Foerster, Reutlingen, Germany).

3. Results and Discussion

CuNi3Si1Mg was hot extruded at 800–900 °C, then solution treated at 800 °C/2 h (or alternatively 950 °C/10 min) and subsequently water-quenched. After this treatment the alloy exhibited a coarse grained microstructure with grain sizes of 100–150 μm and a few coarse silicides (typical diameter of a few hundred nm) which were not completely dissolved during homogenization.

The strain hardening curve for the solution treated and swaged condition can be seen in Figure 2. The highest hardness increase was observed after swaging up to a logarithmic strain of 2.5, then some saturation or even slight softening takes place. Therefore, further investigations focused on the condition swaged to a strain of 2.5, corresponding to a hardness increase of 100 HV as compared to the solution annealed state.

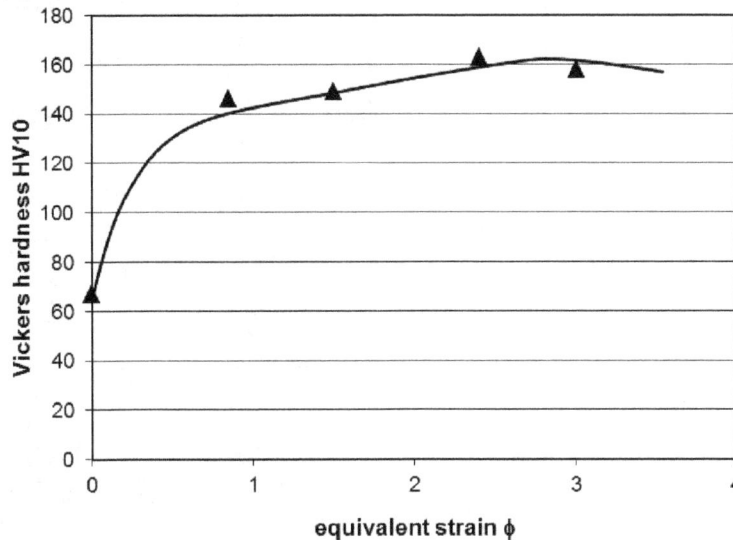

Figure 2. Hardness evolution of solution treated CuNi3Si1Mg by swaging.

After the extruded and solution treated bars were swaged from a diameter of 24 mm down to a diameter of 7 mm, an artificial aging treatment at 450 °C for aging times ranging from 30 min to 16 h was carried out. Figure 3 exhibits the resulting ultra fine grained microstructure at high resolution (magnification ~10000 times) before and after aging (distance from surface 100 μm). The swaging treatment significantly reduced the grain- or subgrain size to 200–800 nm. In addition, the formed ultra-fine grains show some pronounced elongation. With increasing distance from surface the grains become more equiaxed and their size increased up to ~2 μm in the center of the swaged bar. With increasing aging time, characteristic structural changes can be observed in the grain- and precipitate size- and arrangement (Figure 3). More and more precipitates (presumably largely Ni_2Si [10]) are formed as the aging continues, until, after 5–6 h at 450 °C a maximum hardness increase (peak-aging) occurs (see also Figures 4 and 6). By then, the grains have become more equiaxed in shape, however the grain coarsening is still not very pronounced. The precipitates are mainly concentrated on grain boundaries (Figure 3, 5 h). Only after extended aging of 16 h at 450 °C (Figure 3, 16 h) the microstructure becomes over-aged with clearly coarser grains as well as coarse precipitates, leading to a decline of hardness also. In this context, the characteristic sizes of the precipitates in the peak-aged and over-aged condition have to be pointed out.

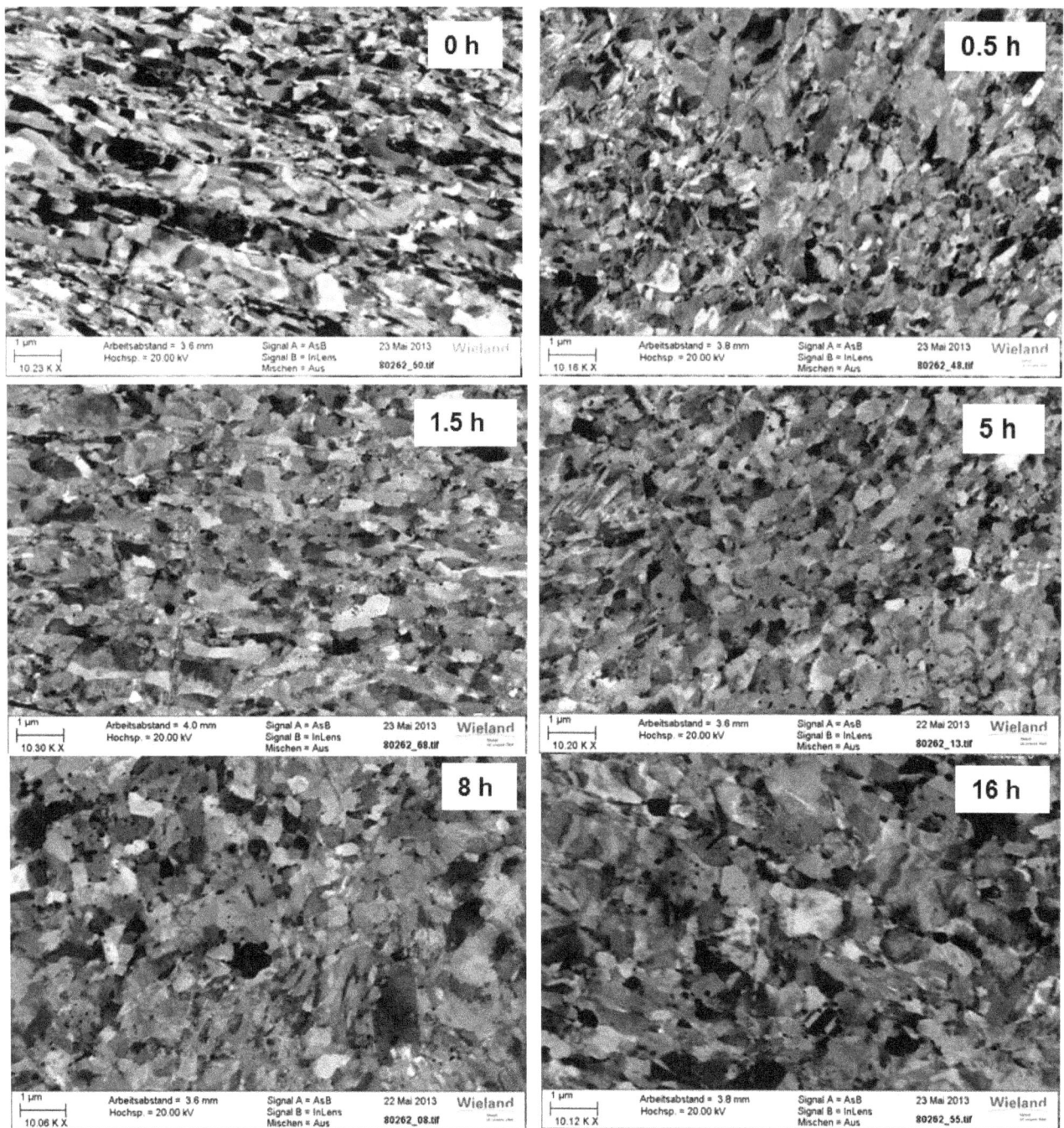

Figure 3. Microstructure of swaged CuNi3Si1Mg after different aging times (T = 450 °C) (ECCI-micrographs).

Figures 4 and 5 illustrate the size distribution of the precipitates in the peak-aged and over-aged condition, as derived from manual counting and measuring of precipitates in Electron Channeling Contrast Imaging (ECCI)–SEM micrographs. At the hardness peak (peak-aging) the average precipitate diameter is around 14 nm, whereas in the over-aged condition, after 16 h, the average precipitate diameter increased to around 47 nm. These aging kinetics are 3 to 5 times faster than in the non-swaged condition where a hardness maximum is found after 16 hours accompanied by typical mean precipitate diameters of 5 nm (see also [2] for comparison). Interestingly, for both conditions, peak-aged as well as over-aged, a bimodal precipitate size distribution was detected (Figures 4 and 5). By means of Energy Dispersive Spectroscopy (EDS) microanalysis we can not distinguish between nanoscale orthorhombic Ni_2Si and the possible hexagonal minority phase $Ni_{31}Si_{12}$ (which is thermodynamically expected for low Ni-content of ~2%). From measurements using EBSD, it appeared that there is also a fraction of hexagonal Ni-silicides. If this observation is any evidence of $Ni_{31}Si_{12}$-phase in addition to Ni_2Si, it is likely that the bimodal size distribution is also driven by this second type of precipitation. Alternatively, it can also be speculated, that the bimodal size distribution is possibly caused by different aging kinetics of precipitates at or near grain/subgrain boundaries and within the grains where diffusion is drastically different.

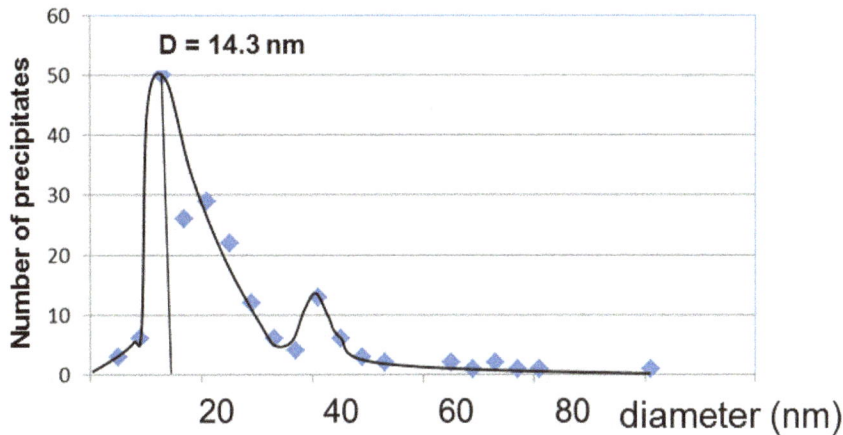

Figure 4. Precipitate size distribution in the peak-aged condition.

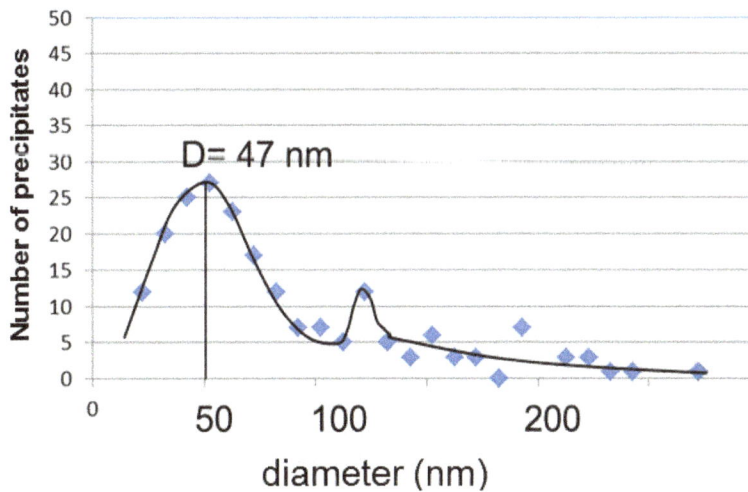

Figure 5. Precipitate size distribution in the over-aged condition.

Figure 6 exhibits the aging curves (hardness *vs.* aging time) of swaged CuNi3Si1Mg for isothermal aging in the temperature range 300–500 °C. Hardness values of up to 245 HV can be reached in the peak-aged conditions at aging temperatures of 400, 450 or 500 °C. With increasing temperature the hardness peak is shifted to smaller aging times. At an aging temperature of 350 °C, only a maximum hardness of ~230 HV appears to be possible.

Figure 7 shows the change of electrical conductivity of the swaged and non-swaged condition during aging at 450 °C. The diffusion of the alloying elements Ni and Si from the solid solution into the precipitates decreases the scattering of electrons by the strain fields of solute atoms. As a result, the electrical conductivity increases.

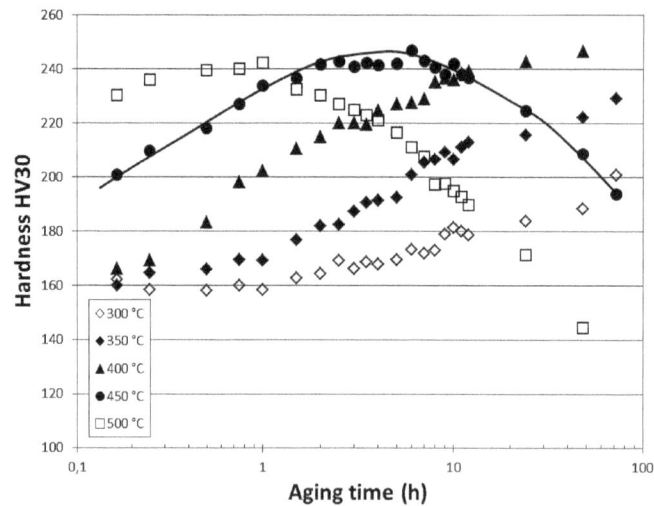

Figure 6. Aging curves of swaged CuNi3Si1Mg for different aging temperatures.

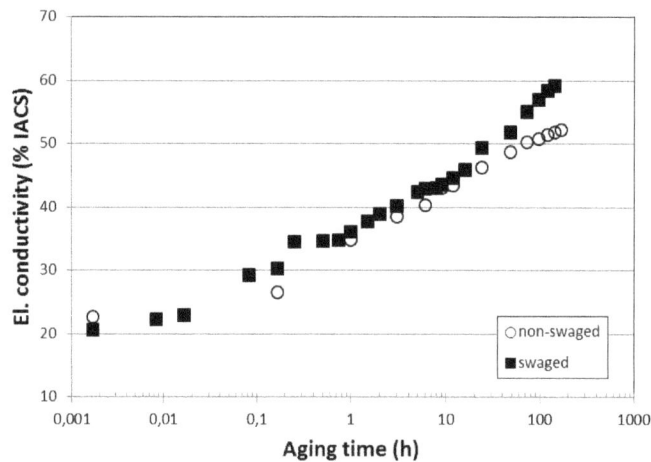

Figure 7. Conductivity *vs.* aging time for swaged (UFG) and coarse-grained CuNi3Si1Mg ($T = 450$ °C).

It is noteworthy, that already after 10 min the conductivity of the swaged condition is slightly higher than the conductivity of the non-deformed condition at this temperature. Obviously, the diffusion of solute elements is significantly accelerated by fast diffusion paths such as high- and low angle boundaries which are prevalent in the swaged condition. Throughout the further aging process the electrical conductivity of the swaged condition stays superior to the conductivity of the non-deformed condition. This difference amounts up to 8% IACS (International Annealed Copper Standard, 58 MS/m) in severely over-aged specimens, possibly being caused also by recrystallization which drastically reduces the grain boundary area in the swaged and severely over-aged condition. For comparison, standardized commercial CuNi3Si1Mg strips typically have electrical conductivities of 35%–45% IACS.

At an aging temperature of 450 °C, the microstructure of the swaged condition was not stable for long aging times. Nevertheless, at a lower aging temperature of 300 °C an aging effect can be induced without pronounced over-aging. At this temperature, a drop of hardness is not observed within several hundred hours thermal exposure (Figure 8). For the application of electromechanical connectors, this is a significant finding, since electromechanical connectors (in the presented conductivity range) are usually not exposed to temperatures higher than 150–200 °C during service. At this moment, implications for stress relaxation behavior of swaged or severely deformed precipitation hardened Corson-alloys remain speculative, however it can be assumed, that finely precipitated Ni-silicides at grain boundaries as well as within grains may also serve to effectively diminish stress relaxation or creep.

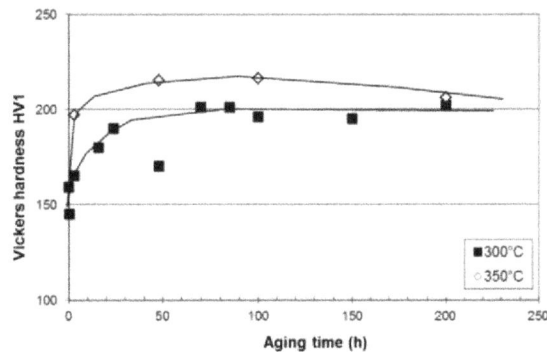

Figure 8. Hardness evolution of swaged CuNi3Si1Mg during thermal exposure at 300 °C (specimens swaged from 5.3 to 2.7 mm diameter).

As aforementioned, no homogeneous ultrafine grained (UFG) structure was achieved by swaging CuNi3Si1Mg bars from diameters of 24 mm to 7 mm. However, for wire which was swaged from 5.3 to 2.7 mm, a fully ultra fine grained structure was observed in the whole cross section after swaging and subsequent aging. Figures 9 and 10 show ECCI- as well as EBSD results for the obtained microstructure after aging at 450 °C for one hour. This microstructure (pictured in the center of the specimen) is characterized by strong orientation contrast (Figure 9), rather low dislocation density within the grains and a high percentage of high-angle boundaries (as seen by mapping grain orientations (Figure 10a) and grain boundary misorientations (Figure 10b) using EBSD). The typical resulting grain size in this UFG structure is about 350 nm.

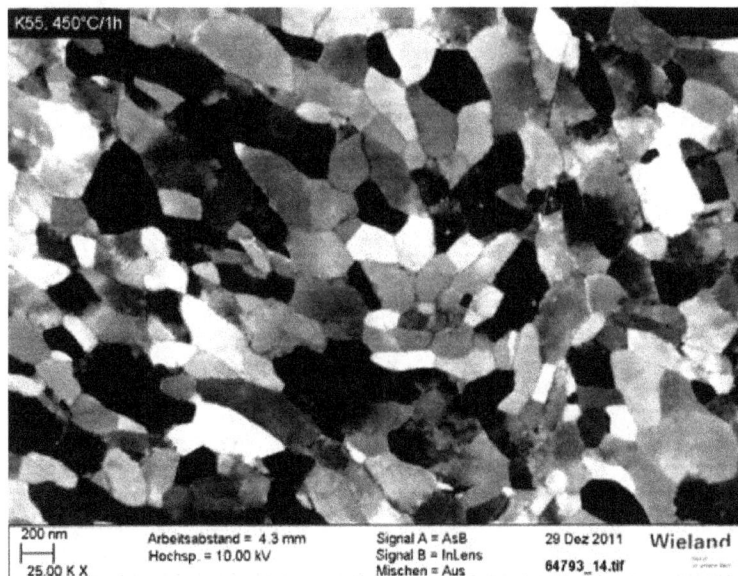

Figure 9. ECCI-picture of ultra-fine grained CuNi3Si1MgSi (swaged and aged at 450 °C/1 h, swaging from diameter of 5.3 to 2.7 mm).

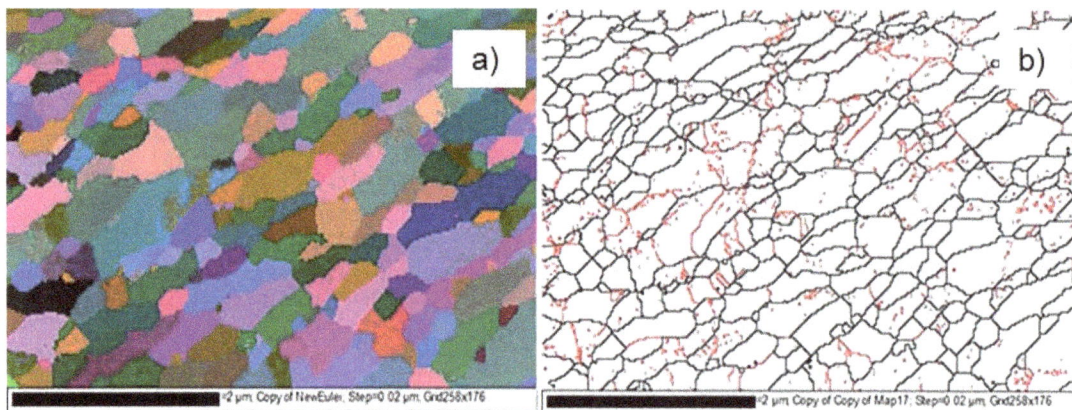

Figure 10. Fully ultra-fine grained CuNi3Si1Mg Electron Backscatter Diffraction Pattern (EBSD) orientation mapping showing Euler-orientations (**a**) and grain boundaries (**b**). High-angle grain boundaries are depicted in black, Low-angle grain boundaries are depicted in red).

Ultra fine grained microstructures can be obtained also in CuNi3Si1Mg by accumulative roll bonding (ARB) [15]. Figure 11 shows a high resolution ECCI micrograph of ARB-processed CuNi3Si1Mg. (for details see [15]). The typical microstructure of ARB-processed CuNi3Si1Mg is characterized by elongated grains perpendicular to rolling direction, having grain diameters lower than 200 nm. An analysis of high angle grain boundaries by EBSD reveals a grain width of 100 nm. This corresponds to other findings [16], where for the same logarithmic strain of ~5, ARB-processed pure copper showed similar grain widths. Also here, aging experiments after severe plastic deformation were carried out. In analogy to the swaged condition, precipitates were formed preferentially at grain boundaries, thus reducing the grain boundary mobility during further thermal exposure.

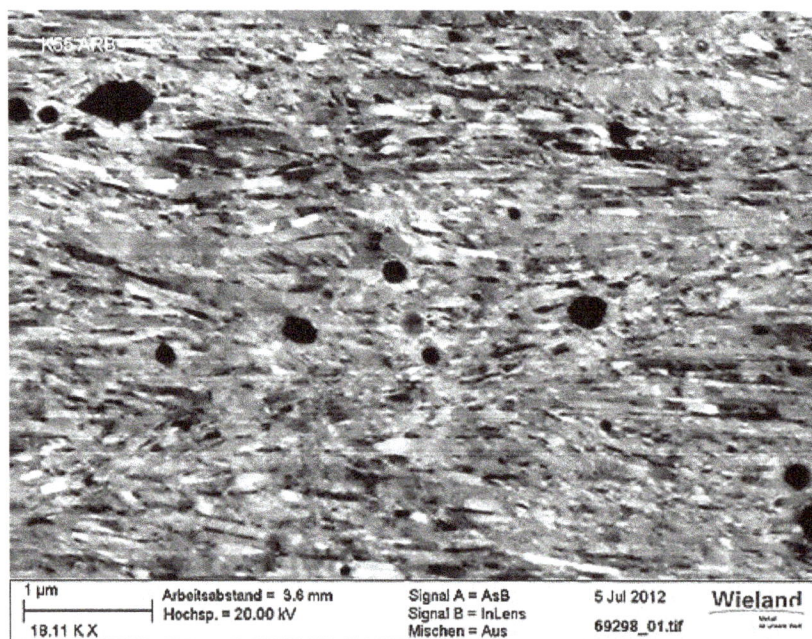

Figure 11. Microstructure of solution treated ARB-processed CuNi3Si1Mg. The larger precipitates were not dissolved by the solution treatment and are remains from the prematerial as produced by continuous casting.

Figure 12 exhibits a thermal softening curve (hardness vs. temperature) of differently processed CuNi3Si1Mg after 1 hour annealing at 300 to 600 °C. If we define the onset of softening as a loss of 10% in initial hardness, the conventionally processed spring hard strip softens at 500 °C whereas the swaged and ARB conditions start softening at 475 °C. This thermal stability is significantly higher compared to pure copper, where softening in the ARB condition already starts at 200 °C [17]. By using high resolution ECCI, grain boundary pinning by Ni$_2$Si-precipitates is confirmed as the underlying mechanism for the high thermal stability (Figure 13).

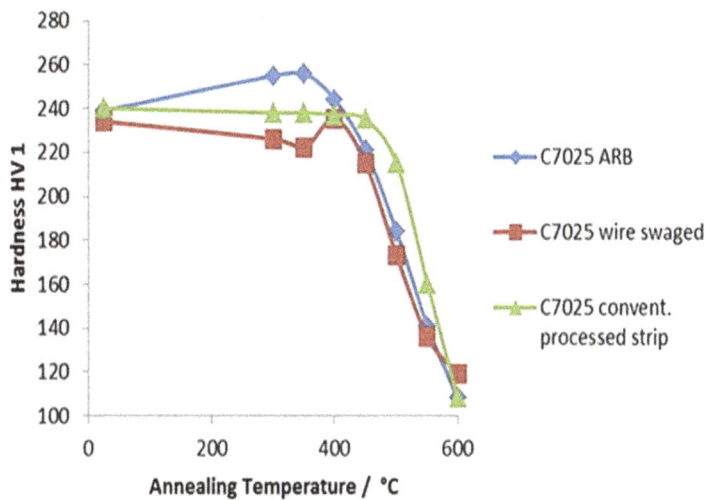

Figure 12. Thermal stability of swaged and ARB-processed CuNi3Si1Mg as compared to conventional strip.

Figure 13. ECCI-micrograph of swaged and under-aged CuNi3Si1Mg showing a low-angle grain boundary which is pinned by Ni-silicides (arrows).

The resulting mechanical (quasistatic as well as cyclic) properties after severe plastic deformation and optimized aging treatments are significantly superior to conventionally processed strip. As an example, thin CuNi3Si1Mg wires (diameter 0.1 mm) which were processed from swaged and optimized aged bars show tensile strengths of up to 1050 MPa and yield strengths higher than 1000 MPa. However, this strength increase is achieved at the expensive of electrical conductivity which is reduced to less than 25% IACS.

In addition to excellent quasistatic strength, optimized swaging plus consecutive aging leads to a marked increase of the 10^7-fatigue endurance strength from 250 to 300 MPa (Figure 14). By combining these processes with a final mechanical surface treatment (such as shot peening, laser peening or deep rolling) even higher fatigue strengths of about 400 MPa can be achieved [18]. Higher fatigue strengths for copper alloys are only reported for high-alloyed spray-formed copper alloys and Cu-Be alloys [19].

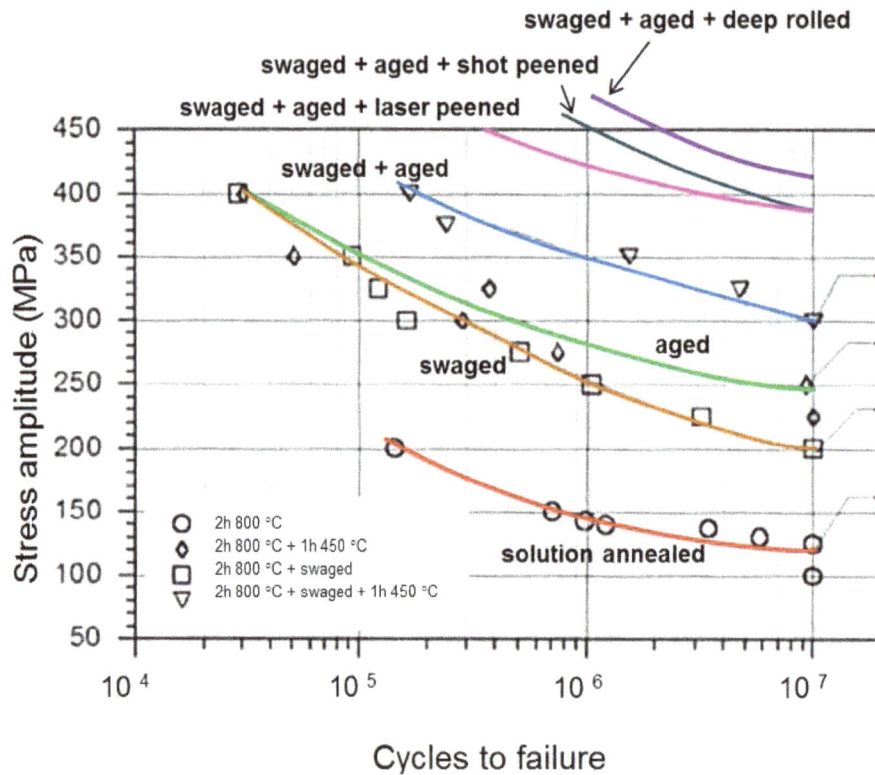

Figure 14. Wöhler-curves (rotation bending, R = −1) of swaged and non-swaged optimized precipitation hardened CuNi3Si1Mg for different mechanical surface treatments (part of this data from [18]).

As an outlook, in context with swaging the following further approaches should be addressed:

Firstly, CuNi3Si1Mg is low alloyed, therefore the achievable strength is limited by the volume fraction of the precipitates. To enable highest strengths of CuNi3Si1Mg, significant cold work is necessary (e.g., by drawing) which in turn lowers the conductivity. Higher alloyed Cu-Ni-Si alloys or more complex alloy systems such Cu-Ni-Si-Cr or Cu-Ni-Co-Si allow higher volume fractions of precipitates or multiphase hardening, respectively. First results on swaged and aged CuNi7Si2Cr are promising and indicate possible conductivities of 34% IACS at tensile strengths of ~1100 MPa after swaging and optimized aging, which is in the strength regime of Cu-Be-alloys.

Another approach to generate UFG copper alloys is to introduce an extremely high twinning density by severe deforming of single-phased copper alloys which exhibit very low stacking fault energies (such as Cu-Al, Cu-Al-Si, Cu-Zn or Cu-Zn-Si). By finely spaced twinning a very fine grain size can be achieved after swaging leading to a hardness of more than 420 HV (which corresponds to tensile strengths higher than 1300 MPa). Figure 15 shows the refinement of the microstructure by twinning in CuAl5.8Si2 (estimated stacking fault energy around 6 mJ/m² or lower [20,21]) with increasing logarithmic deformation degree during swaging. However, CuAl5.8Si2 is non-age hardenable, therefore there are no precipitates stabilizing the grain boundaries during thermal exposure. As a consequence, a hardness loss of more than 10% is observed at 300 °C after just one hour.

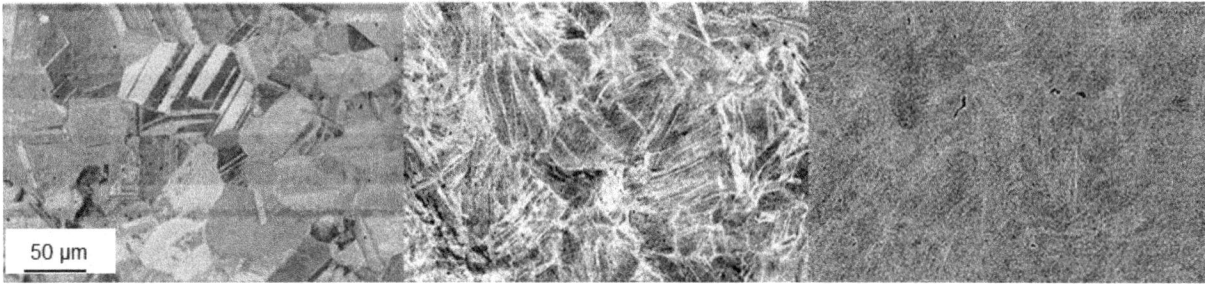

Figure 15. Evolution of an ultrafine-grained structure in single phased low-stacking fault alloy CuAl5.8Si2 by swaging (**Left**: non-swaged; **Middle**: swaged to a logarithmic strain of −0.8; **Right**: swaged to a logarithmic strain of −1.4).

4. Conclusions

The combination of swaging and subsequent optimized precipitation hardening is a simple method to produce Cu-Ni-Si materials with very fine grain size in the range 0.2–2 μm. For strip material accumulative roll bonding (ARB) is a suitable method to generate ultra fine grained microstructures in Cu-Ni-Si alloys. A crucial role is ascribed to the artificial aging treatment after the severe plastic deformation. The thermal stability of ultra fine grained CuNi3Si1Mg is significantly enhanced as compared to pure copper, owing to nanoscopically small precipitates which effectively pin the grain boundaries during aging or annealing. At 300 °C no over-aging was detected within 200 h. For short time exposure (1 h) the grain structure is fairly stable up to 400 °C. In addition to enhanced ultimate tensile- and yield strength (with possible strength >1000 MPa), also the fatigue behavior in the High Cycle Fatigue (HCF)-regime was significantly improved by the UFG-structure in the swaged plus peak-aged condition close to the surface.

Acknowledgments

Experimental help during wire drawing of Cu-Ni-Si wire by D. Vucic-Seele is kindly acknowledged.

Author Contributions

I.A. is the primary author of the paper and performed analysis of the experimental data. M.G. and M.M. contributed to the experimental research work. H.A.K. and L.W. discussed the results and analysis with the other authors.

Conflicts of Interest

The authors declare no conflict of interest.

References

1. Corson, M.G. Copper hardened by a new method. *Z. Metallkunde* **1927**, *19*, 370–371.

2. Lockyer, S.A.; Noble, F.W. Precipitate structure in a Cu-Ni-Si alloy. *J. Mater. Sci.* **1994**, *29*, 218–226.

3. Wang, C.; Zhu, J.; Lu, Y.; Guo, Y.; Liu, X. Thermodynamic description of the Cu-Ni-Si system. *J. Phase Equilib. Diffus.* **2014**, *35*, 93–104.

4. Kuhn, H.-A.; Altenberger, I.; Käufler, A.; Hölzl, H.; Fünfer, M. Properties of high performance alloys for electromechanical connectors. In *Copper Alloys—Early Applications and Current Performance—Enhancing Processes*; Collini, L., Ed.; InTech: Rijeka, Croatia, 2012; p. 52.

5. Valiev, R.Z.; Islamgaliev, R.K.; Alexandrov, I.V. Bulk nanostructured materials from severe plastic deformation. *Prog. Mater. Sci.* **2000**, *45*, 103–189.

6. Höppel, H.W.; May, J.; Göken, M. Enhanced strength and ductility in ultrafine-grained aluminium produced by accumulative roll bonding. *Adv. Eng. Mater.* **2004**, *6*, 781–784.

7. Mughrabi, H.; Höppel, H.W.; Kautz, M. Fatigue and microstructure of ultrafine-grained metals produced by severe plastic deformation. *Scr. Mater.* **2004**, *51*, 807–812.

8. Neishi, K.; Horita, Z.; Langdon, T.G. Achieving superplasticity in a Cu–40%Zn alloy through severe plastic deformation. *Scr. Mater.* **2001**, *45*, 965–970.

9. Wang, J.; Zhang, P.; Duan, Q.; Yang, G.; Wu, S.; Zhang, Z. Tensile deformation behaviors of Cu-Ni alloy processed by equal channel angular pressing. *Adv. Eng. Mater.* **2010**, *12*, 304–311.

10. Kinder, J.; Huter, D. TEM-Untersuchungen an höherfesten und elektrisch hochleitfähigen CuNi2Si-Legierungen. *Metall* **2009**, *63*, 298–303.

11. Altenberger, I.; Kuhn, H.-A.; Gholami, M.; Mhaede, M.; Wagner, L. Characterization of ultrafine grained Cu-Ni-Si alloys by electron backscatter diffraction. *IOP Conf. Ser. Mater. Sci. Eng.* **2014**, *63*, 012135.

12. Altenberger, I.; Kuhn, H.-A.; Hölzl, H. Mikrostrukturelle Charakterisierung von hochfesten Cu-Ni-Si-Legierungen mittels Electron-Channeling-Rückstreukontrast im Rasterelektronenmikroskop. *Sonderband d. Prakt. Metallographie* **2012**, *44*, 79–84. (In German)

13. Jaksch, H. Strain related contrast mechanisms in crystalline materials imaged with AsB detection. In *EMC 2008-14th European Microscopy Congress 2008*; Luysberg, M., Tillmann, K., Weirich, T., Eds.; Springer: Berlin/Heidelberg, Germany, 2008; pp. 553–554.

14. Altenberger, I.; Kuhn, H.-A.; Mhaede, M.; Gholami, M.; Wagner, L. Wie viel NANO steckt in Kupfer–ein klassischer Werkstoff im 21. Jahrhundert. *Metall* **2012**, *66*, 500–504. (In German)

15. Kuhn, H.-A.; Altenberger, I.; Riedle, J.; Hölzl, H. Microstructure and mechanical properties of ultra fine grained high performance copper alloys. In *Proceedings of Copper 2013*; Leibbrandt, J., Ignat, M., Sanchez, M., Eds.; The Chilean Institute of Mining Engineers: Santiago, Chile, 2013; pp. 129–138.

16. Murata, Y.; Nakaya, I.; Morinaga, M. Assessment of strain energy by measuring dislocation density in copper and aluminium prepared by ECAP and ARB. *Mat. Trans.* **2008**, *49*, 20–23.

17. Li, Y.J.; Zeng, X.H.; Blum, W. Transition from strengthening to softening by grain boundaries in ultrafine-grained Cu. *Acta Mater.* **2004**, *52*, 5009–5018.

18. Gholami, M.; Altenberger, I.; Mhaede, M.; Sano, Y.; Wagner, L. Surface treatments to improve fatigue performance of age-hardenable CuNi3Si1Mg. In Proceedings of the 12th International Conference on shot Peening, Goslar, Germany, 15–18 September 2014; Wagner, L., Ed.; pp. 208–213.

19. Altenberger, I.; Kuhn, H.-A.; Müller, H.R.; Mhaede, M.; Gholami, M.; Wagner, L. Material properties of high-strength Beryllium-free copper alloys. *Int. J. Mater. Prod. Technol.* **2015**, *50*, 124–146.

20. Rohatgi, A.; Vecchio, K.S.; Gray, G.T., III. The influence of stacking fault energy on the mechanical behavior of Cu and Cu-Al alloys: Deformation twinning, work hardening, and dynamic recovery. *Metall. Mater. Trans A* **2001**, *32*, 135–145.

21. An, X.H.; Lin, Q.Y.; Wu, S.D.; Zhang, Z.F.; Figueiredo, R.B.; Gao, N.; Langdon, T.G. The influence of stacking fault energy on the mechanical properties of nanostructured Cu and Cu–Al alloys processed by high-pressure torsion. *Scr. Mater.* **2011**, *64*, 954–957.

13

Manufacturing Ultrafine-Grained Ti-6Al-4V Bulk Rod Using Multi-Pass Caliber-Rolling

Taekyung Lee [1], Donald S. Shih [2], Yongmoon Lee [3] and Chong Soo Lee [3,*]

[1] Department of Mechanical Engineering, Northwestern University, Evanston, IL 60208, USA; E-Mail: taekyung.lee@northwestern.edu

[2] Boeing Research & Technology, St. Louis, MO 63166, USA; E-Mail: donald.s.shih@boeing.com

[3] Graduate Institute of Ferrous Technology, Pohang University of Science and Technology (POSTECH), Pohang 790-784, Korea; E-Mail: ymlee0725@postech.ac.kr

* Author to whom correspondence should be addressed; E-Mail: cslee@postech.ac.kr

Academic Editor: Heinz Werner Höppel

Abstract: Ultrafine-grained (UFG) Ti-6Al-4V alloy has attracted attention from the various industries due to its good mechanical properties. Although severe plastic deformation (SPD) processes can produce such a material, its dimension is generally limited to laboratory scale. The present work utilized the multi-pass caliber-rolling process to fabricate Ti-6Al-4V bulk rod with the equiaxed UFG microstructure. The manufactured alloy mainly consisted of alpha phase and showed the fiber texture with the basal planes parallel to the rolling direction. This rod was large enough to be used in the industry and exhibited comparable tensile properties at room temperature in comparison to SPD-processed Ti-6Al-4V alloys. The material also showed good formability at elevated temperature due to the occurrence of superplasticity. Internal-variable analysis was carried out to measure the contribution of deformation mechanisms at elevated temperatures in the manufactured alloy. This revealed the increasing contribution of phase/grain-boundary sliding at 1073 K, which explained the observed superplasticity.

Keywords: multi-pass caliber-rolling; grain refinement; internal-variable theory; Ti-6Al-4V

1. Introduction

Titanium and its alloys have attracted attention from various fields such as structural-material, biomedical, munitions, and information-technology industries. In particular, Ti-6Al-4V alloy has been a key material in aerospace industries since it was developed in 1954 [1]. The alloy possesses a superior strength-to-weight ratio that increases the fuel efficiency of rocket and aircraft. It also exhibits excellent corrosion resistance and mechanical stability at various temperatures. Finally, the alloy has good formability at elevated temperatures after applying a certain thermomechanical process due to the superplasticity [2–4].

Many researchers have focused on the fact that grain refinement can improve mechanical properties of titanium alloys. The increasing fraction of grain boundaries act as barriers against dislocation slip, leading to grain-boundary strengthening. In addition, the grain refinement provides the increasing sources of grain-boundary sliding and hence induces the superplastic behavior at elevated temperatures. It is thus natural that ultrafine-grained (UFG) titanium alloys have been actively studied for decades [5–9]. To attain the UFG structure, most of previous works have utilized a severe plastic deformation (SPD) process, such as equal-channel angular pressing (ECAP) and high-pressure torsion (HPT) [10]. However, the manufactured samples were generally 10s of centimeters in length, which were limited to laboratory scale.

To produce a UFG bulk rod applicable to the industries, the authors have introduced a multi-pass caliber-rolling process as an alternative to the conventional SPD processes. Caliber-rolling machine includes several calibers with various sizes and shapes (e.g., oval and circular) in its rolls by which a multi-axial deformation is imposed on a workpiece during the process. Since Kimura *et al.* [11] reported the considerable mechanical improvement in a caliber-rolled low-alloy steel, related studies have been actively carried out with various materials [12–19]. Nevertheless, studies on caliber-rolled titanium alloys have just begun in spite of its importance both in academia and industry [20–22]. A UFG bulk rod was successfully manufactured by caliber-rolling Ti-6Al-4V alloy in this work. The researchers investigated microstructures and tensile properties of the manufactured material and discussed mechanisms of grain refinement and superplasticity.

2. Materials and Methods

Ti-6Al-4V rod was machined with a diameter of 28 mm and a length of 150 mm. The head part was made to be conical to insert a material into a pair of calibers. The beta-transus temperature of this alloy was reported to be 1268 K [9]. The Ti-6Al-4V rod was solution-treated at 1323 K for 2 h followed by quenching in a water bath to induce a fine lath structure. This alloy was then soaked in a furnace at 1073 K for 1 h and caliber-rolled in the ambient atmosphere. The caliber-rolling process used in this work consists of six deformation passes. First, third, and fifth calibers are oval-shaped, while second, fourth, and sixth calibers are circular. The sample was inserted in a caliber, rotated by 90 degrees, and then immediately inserted in the next caliber for each deformation path, as illustrated in Figure 1. There was a single reheating process at 1073 K for 2 min after the fourth deformation pass. The sample was air-cooled after the sixth deformation pass. The total reduction of area was determined to be 85%.

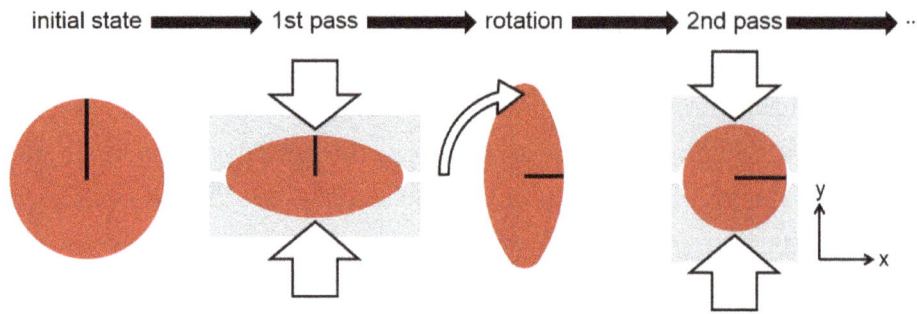

Figure 1. Schematic illustration of multi-pass caliber-rolling process.

Disc samples with a diameter of 3 mm were obtained from the caliber-rolled alloy, and then mechanically thinned with 240-grit SiC papers to a thickness less than 200 μm. Afterwards, they were jet-polished at 22 V and 265 K in a solution containing 40 mL of HClO₄, 240 mL of 2-butoxy ethanol, and 400 mL of methanol. Transmission electron microscope (TEM) was used to observe microstructures at 200 kV with JEM-2100F FE-TEM machine (JEOL, Tokyo, Japan). Grain size was measured from the image using the linear intercept method [23]. Meanwhile, other discs were mirror-polished with 1 μm alumina powder and 0.25 μm colloidal silica for electron backscatter diffraction (EBSD) analysis. The analysis was performed at 20 kV with Quanta 3D FEG machine (FEI, Hillsboro, OR, USA).

Tensile properties were measured using rod-type specimens whose gauge length and diameter were 6 and 3 mm, respectively. Instron 8862 machine (INSTRON, Norwood, MA, USA) was used for both room- and high-temperature tensile tests. Room-temperature tensile test was carried out at a strain rate of 5×10^{-3} s^{-1} with an extensometer to obtain reliable data. High-temperature tensile test was conducted in a halogen furnace at two temperature conditions (873 K and 1073 K) and two strain rates (5×10^{-3} s^{-1} and 5×10^{-4} s^{-1}). Each sample was heated for 10 min before commencing the test. Load-relaxation test (LRT) was conducted in a similar manner to the high-temperature tensile test, except that the samples were deformed up to a true strain of 0.2 and then hold to investigate the load relaxation behavior. The obtained LRT data were converted into stress-strain rate relationship based on the following equation [24]:

$$\sigma = P(L_0 + X - P/K)/A_0L_0 \tag{1a}$$

$$\dot{\varepsilon} = -(dP/dt)(L_0 + X - P/K) \tag{1b}$$

$$K^{-1} \approx C_m + L_0/A_0E \tag{1c}$$

where P is load, L_0 is the gauge length, A_0 is the cross-sectional area of gauge region, X is the displacement, C_m is the elastic compliance of testing machine, and E is Young's modulus of Ti-6Al-4V alloy. Strain-rate-jump test (SRJT) was performed at 1073 K, during which a strain rate changed from 5×10^{-4} s^{-1} to 5×10^{-3} s^{-1} at a true strain of 0.6. Strain-rate sensitivity (m) was determined as follows:

$$m = \partial \log \sigma / \partial \log \dot{\varepsilon} \tag{2}$$

3. Results

Figure 2 demonstrates the manufactured caliber-rolled Ti-6Al-4V bulk rod. The length and diameter of the rod was approximately 1200 and 10 mm, respectively. It is also noted that the length can be even increased by tailoring the dimension of initial material. Such a large dimension enables the caliber-rolled rod to be directly used in the industry. Indeed, the authors fabricated a dental implant fixture with this material, which exhibited satisfying mechanical strength and fatigue resistance both in ambient atmosphere and simulated body fluid [21].

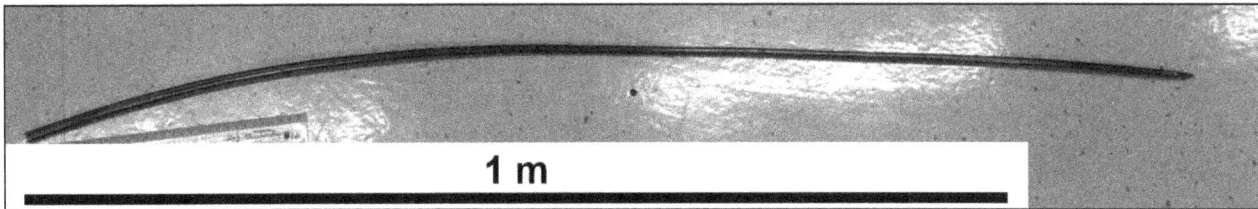

Figure 2. Ti-6Al-4V bulk rod manufactured by the multi-pass caliber-rolling process.

The grain structure of caliber-rolled rod was observed by TEM analysis as shown in Figure 3. It is obvious that the rod possessed the UFG structure with a grain size of 0.2 ± 0.05 μm. Such a strong grain refinement has thus far been accomplished by SPD processes. For example, Ko *et al.* [25] achieved a similar UFG structure with a mean grain size of ~0.3 μm after applying four-pass ECAP deformation at 873 K; however, the length of fabricated sample was much shorter (80 mm) compared to the present UFG rod.

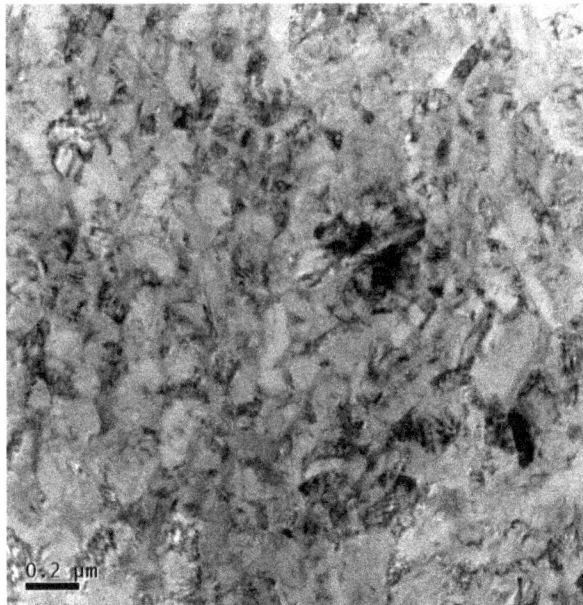

Figure 3. Transmission electron microscope (TEM) micrograph of caliber-rolled Ti-6Al-4V rod. The image was taken perpendicular to the rolling direction (RD).

Figure 4 presents the EBSD results for the investigated materials. The solution-treated microstructure consisted of martensitic laths as intended. Beta phase was not confirmed in this alloy, as reported in the literature [26,27]. Two types of martensitic laths were observed; coarse primary laths were formed first, and then fine secondary laths were generated between the primary laths [28]. The latter occupies the most area in the solution-treated alloy, whose lath thickness is less than 5 μm. The similar lath structure with no presence of beta phase was confirmed after the heating step prior to the caliber-rolling (*i.e.*, 1073 K for 1 h), although the thickness was increased to ~10 μm. The beta fraction increased to 6% after the caliber-rolling. Figure 5 shows two types of beta constituents; most of them exist as a form of nano-sized beta precipitations, while fragmented beta lamellae are also observed. The beta lamellae were formed parallel to the RD and the thickness was measured to be ~0.2 μm or less. Chao *et al.* [27] attributed this type of phase transformation to adiabatic heating and strain-induced transformation. A fraction of high-angle grain boundaries of 0.6–0.8 was confirmed in Ti-6Al-4V alloys groove-rolled at 923–1023 K [22]. Similar results are expected in the present material as both rolling processes have the similar deformation mechanism.

The confidence index (CI) of caliber-rolled rod was too low to provide meaningful microstructural information. Tirumalasetty *et al.* [29] ascribed the low CI in UFG alloys to distorted Kikuchi patterns in region with high dislocation density. Alternatively, the EBSD analysis was conducted after annealing the caliber-rolled alloy at 873 K for 1 h, followed by water-quenching; such a condition was reported to minimize recrystallization and maintain the texture in this material [30]. Figure 4c,d demonstrate two types of grains: fine and coarse-grain groups. The fine grains are closer to the original microstructure of caliber-rolled alloy because the clear image and high CI of coarse grains imply the formation of microstructure during the subsequent annealing process. It should be noted that the fine grains are equiaxed in both planes, supporting the conclusion from TEM observation that the caliber-rolling gave rise to the equiaxed UFG structure for the present material.

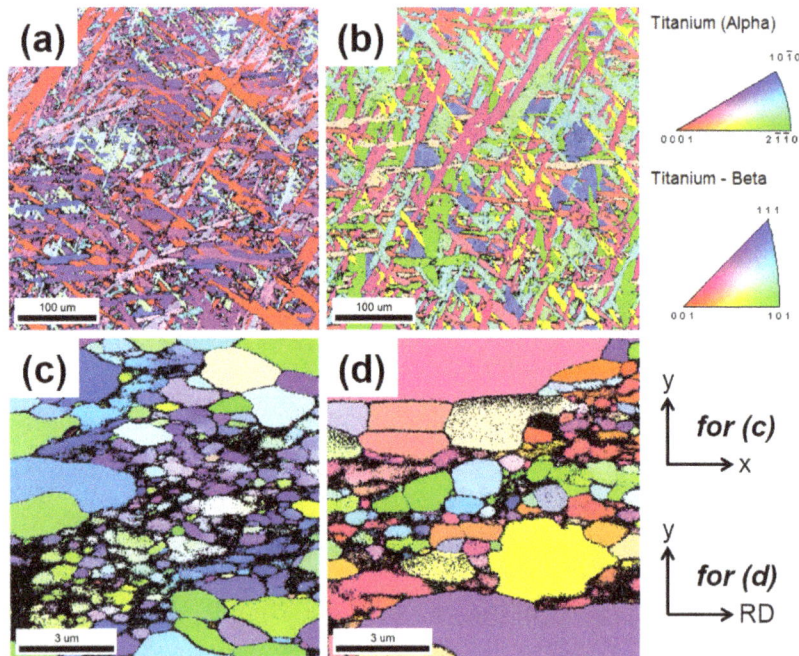

Figure 4. Electron backscatter diffraction (EBSD) pole figure map of the investigated Ti-6Al-4V rod: (**a**) solution-treated; (**b**) heat-treated prior to the caliber-rolling; and (**c,d**) caliber-rolled and annealed alloys. Figure 4c is a plane perpendicular to the RD, while Figure 4d is parallel to the RD. The black dots indicate the area of confidence index (CI) ≤ 0.1. The *x* and *y* axes are demonstrated in Figure 1.

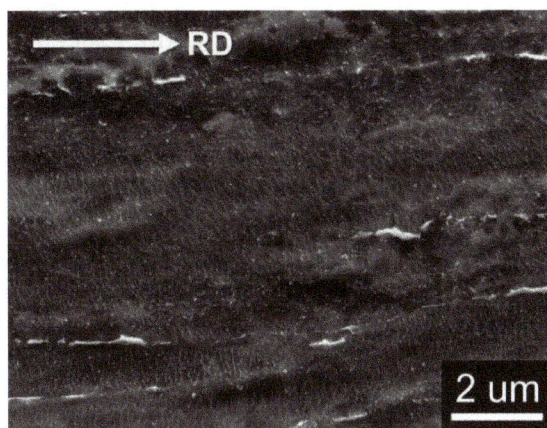

Figure 5. SEM micrograph of the caliber-rolled Ti-6Al-4V rod. The dark and bright areas indicate alpha and beta phases, respectively. The image was taken parallel to the RD marked as the arrow in the micrograph.

Figure 6 presents the texture of the caliber-rolled rod in the form of RD inverse pole figure. The alloy exhibited the fiber texture with the basal planes parallel to the RD. The fraction of beta phase is small enough to be neglected. Narayana Murty *et al.* [22] recently reported the similar texture in Ti-6Al-4V alloys groove-rolled at 873–1023 K, although their texture showed stronger orientation along <10-10> and <2-1-10> directions. According to the work [22], the relatively randomized texture observed in the present alloy may be attributed to higher deformation temperature related to martensite decomposition and phase transformation during the deformation process as well as the annealing treatment.

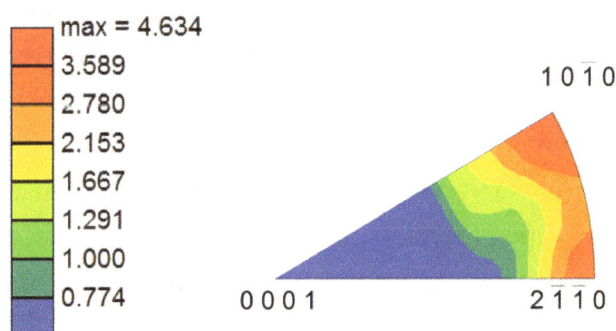

Figure 6. RD inverse pole figure for the caliber-rolled Ti-6Al-4V rod obtained by EBSD.

The manufactured UFG rod can be utilized in two ways in the industry. First, the material can be directly machined to be used for biomedical products, such as dental implant fixture, bone screw, bone plate, and micro-drill [1]. Room-temperature tensile properties are important in this case to ensure the resistance to fatigue fracture for biomedical uses. Second, the rod can be further processed at elevated temperature for automobile and aerospace industries, which requires the evaluation of high-temperature mechanical properties. Therefore, the tensile properties of the caliber-rolled Ti-6Al-4V alloy were investigated at both room and elevated temperatures.

Figure 7 shows room-temperature tensile properties of the caliber-rolled rod as well as UFG Ti-6Al-4V alloys in the literature [5–8]. The caliber-rolled rod provided the high yield stress (YS = 1345 MPa) and ultimate tensile stress (UTS = 1425 MPa) due to its UFG structure and resultant grain-boundary strengthening. The tensile properties of the manufactured alloy was compared with SPD-processed UFG

Ti-6Al-4V alloys in terms of the product of strength and elongation; such an approach has been widely used in the field of structural materials as strength increases in sacrifice of elongation in many cases [31]. The UFG bulk Ti-6Al-4V rod fabricated in this work exhibited the value of 18,525 MPa%, as shown in Figure 7b, which was comparable with most of UFG Ti-6Al-4V alloys in the literature.

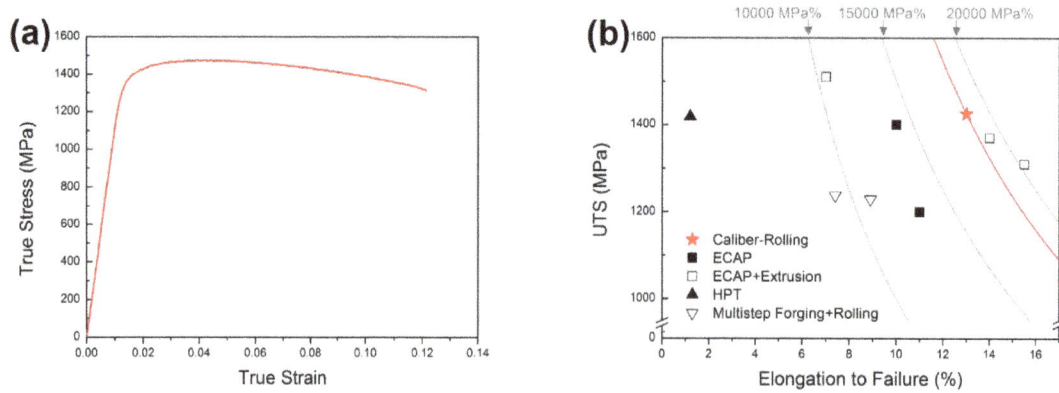

Figure 7. (a) True stress-strain curve of caliber-rolled Ti-6Al-4V rod at room temperature and **(b)** comparison of the tensile properties of various ultrafine-grained (UFG) Ti-6Al-4V alloys [5–8].

Figure 8a shows flow curves of the caliber-rolled alloy at elevated temperatures. Increasing deformation temperature or decreasing strain rate increased the ductility in sacrifice of strength. High-temperature deformation mechanisms at each condition were discussed in Section 4 on the basis of the internal-variable theory. It is of particular note that the caliber-rolled rod deformed at 1073 K and 5×10^{-4} s^{-1} exhibited the different characteristics than the others. The sample recorded a total elongation of 967% as shown in the inset of Figure 8a, whose flow curve showed a long plateau as generally found in superplastic materials. In addition, the SRJT results shown in Figure 8b provided a strain-rate sensitivity of 0.44 for the caliber-rolled Ti-6Al-4V at 1073 K. A material with a strain-rate sensitivity of 0.3–0.8 is considered to possess superplasticity [32]. All of these factors suggest the superplastic behavior of the present bulk rod at the high temperature, which will be useful in the related industries.

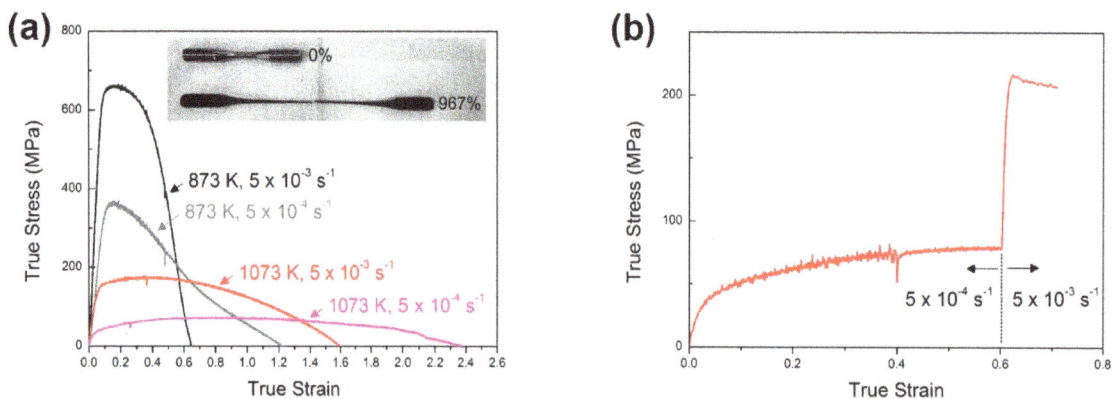

Figure 8. High-temperature tensile properties of caliber-rolled Ti-6Al-4V rod: **(a)** true stress-strain curve at elevated temperatures and **(b)** SRJT results obtained at 1073 K. The inset in Figure 8a compares the undeformed specimen and the sample deformed at 1073 K and 5×10^{-4} s^{-1}.

4. Discussion

The caliber-rolling process gave rise to the strong grain refinement by which the grain size became similar to those refined by the SPD processes. Such a microstructural evolution can be understood in terms of the dynamic globularization [33–36]. A groove formed at the phase/grain boundaries splits a platelet into several pieces by deepening along the boundary in the platelet. The broken-up lamellae are transformed into globular grains to stabilize the surface energy.

Two factors contributed to the effective dynamic globularization and resultant grain refinement in the caliber-rolled Ti-6Al-4V alloy. First, the present alloy consisted of fine martensitic laths prior to the caliber-rolling process. According to the literature [27,37,38], such a microstructure is beneficial for the grain refinement through dynamic globularization because an initial lamellar thickness directly affects a final grain size. Second, the globularizing fraction increases with increase in an applied strain following an Avrami-type equation in Ti-6Al-4V [39]. The authors have proven that the caliber-rolling imposes more than twice as high strain than a conventional rolling with the same reduction of area [30]. In this work, the equivalent strain was determined to be 0.7, 1.2, 1.9, 2.4, 3.3, and 4.0 after the one- to six-pass caliber-rolling in a two-phase titanium alloy. This is attributed to the redundant strain accumulating without the volume change of workpiece [40].

Superplastic behavior was confirmed in the manufactured alloy at 1073 K and 5×10^{-4} s^{-1}. The high-temperature deformation data were interpreted on the basis of internal-variable theory to investigate the deformation mechanisms. Ha and Chang [41] suggested this theory to measure the contribution of grain matrix deformation (GMD) and phase/grain-boundary sliding (P/GBS) to deformation behavior at elevated temperatures. In the internal-variable theory, stress is composed of internal stress of long-range dislocation interactions (σ^I) and friction stress of short-range dislocation-lattice interactions (σ^F). Strain rate consists of the rate of internal strain ($\dot{\imath}$), non-recoverable plastic strain ($\dot{\alpha}$), and P/GBS ($\dot{\eta}$) strain. Among these factors, the friction stress and internal strain rate are negligible under the present conditions, providing stress and strain rate as follows:

$$\sigma = \sigma^I + \sigma^F \approx \sigma^I \tag{3a}$$

$$\dot{\varepsilon} = \dot{\imath} + \dot{\alpha} + \dot{\eta} \approx \dot{\alpha} + \dot{\eta} \tag{3b}$$

The internal stress, non-recoverable plastic strain rate, and P/GBS strain rate at a deformation temperature of T are determined from the following relations:

$$(\sigma^*/\sigma^I) = \exp(\dot{\alpha}^*/\dot{\alpha})^p \tag{4a}$$

$$\dot{\alpha}^* = \nu^I (\sigma^*/G)^{n(I)} \exp(-Q^I/RT) \tag{4b}$$

$$(\dot{\eta}/\dot{\eta}_0) = [(\sigma - \Sigma_\eta)/\Sigma_\eta]^{1/M} \tag{4c}$$

$$\dot{\eta}_0 = \nu^\eta (\Sigma_\eta/\mu^\eta)^{n(\eta)} \exp(-Q^\eta/RT) \tag{4d}$$

Here, σ^* and Σ_η are stress for GMD and P/GBS, respectively. $\dot{\alpha}^*$ and $\dot{\eta}_0$ are their conjugate reference strain rates. ν^I and ν^η are the jump frequency for dislocations. Q^I and Q^η are the activation energy for GMD and P/GBS, respectively. R is the gas constant and other parameters are material constants [4].

Figure 9 shows LRT results and corresponding internal-variable analysis. The GMD curve deviated from the experimental data at lower strain rates of Figure 9a and almost entire range of Figure 9b. These deviations were corrected by the P/GBS curve. It is thus concluded that P/GBS was not activated at 873 K and a strain rate of 5×10^{-3} s^{-1}, whereas both mechanisms contributed to the high-temperature deformation in the other cases. The relative contribution of each mechanism was quantified on the basis of the LRT data. At 873 K and a strain rate of 5×10^{-4} s^{-1}, GMD mainly contributed to the deformation behavior (92% for GMD and 8% for P/GBS). Similar results were obtained at 1073 K and a strain rate of 5×10^{-3} s^{-1} (85% for GMD and 15% for P/GBS) from the extrapolated data in Figure 9b. These results explain why the superplasticity was not observed under the three conditions mentioned above.

On the other hand, the contribution of P/GBS significantly increased at 1073 K and a strain rate of 5×10^{-4} s^{-1} (55% for GMD and 45% for P/GBS). This is rationalized in light of the decreasing activation energy for deformation with increasing temperature and decreasing strain rate. The authors have reported that the activation energy decreased to that for interphase/grain-boundary diffusion under such conditions, resulting in the increasing contribution of P/GBS [4]. This conclusion is also supported by the fact that the friction stress for P/GBS decreased as the deformation temperature increased; the value of log Σ_η was -2.7 at 873 K and -5.7 at 1073 K. This is in good accord with the superplastic behavior observed at 1073 K and 5×10^{-4} s^{-1}.

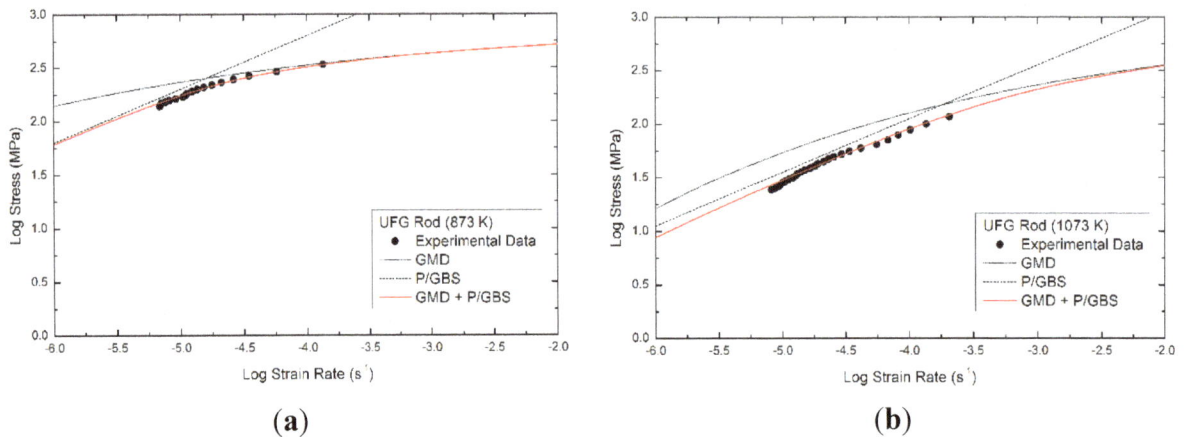

Figure 9. Internal-variable analysis of caliber-rolled Ti-6Al-4V rod at (**a**) 873 K and (**b**) 1073 K.

5. Conclusions

In this work, the multi-pass caliber-rolling process successfully manufactured Ti-6Al-4V bulk rod with the UFG microstructure. The dimension of the manufactured rod (1200 mm in length) was significantly larger than most UFG Ti-6Al-4V samples fabricated by SPD processes and large enough to be directly used in the industry. The length can be even increased by tailoring the dimension of initial material. The alloy consisted of equiaxed ultrafine grains with a mean size of 0.2 μm. Such an effective grain refinement originated from the fine lath structure in the initial material and redundant strain accumulating without the volume change of workpiece. A small amount of beta particles was formed during the caliber-rolling process due to adiabatic heating and strain-induced phase transformation. The caliber-rolled rod showed the fiber texture with the basal planes parallel to the RD. The grain refinement through the caliber-rolling affected the grain-boundary strengthening at room temperature and superplastic behavior at high temperature. The product of strength and elongation of the alloy at room temperature was calculated to be 18525 MPa%, which was similar to the values of most SPD-processed UFG Ti-6Al-4V alloys in the literature. The manufactured rod exhibited the superplastic behavior at 1073 K and 5×10^{-4} s^{-1}. The internal-variable analysis revealed the increasing P/GBS contribution to deformation under these conditions, while GMD controlled the deformation at the lower temperature and/or higher strain rate.

Acknowledgments

The authors gratefully acknowledge the financial support from The Boeing Company for the present research.

Author Contributions

T. Lee and C.S. Lee conceived and designed the experiments; T. Lee performed all experiments except for EBSD, analyzed the data, and wrote the paper; D.S. Shih and C.S. Lee guided the direction of the work and contributed reagents/materials/analysis tools; Y. Lee carried out the entire EBSD analyses; all authors participated in discussions.

Conflicts of Interest

The authors declare no conflict of interest.

References

1. Lee, Y.T. *Titanium*; Korea Metal Journal: Seoul, Korea, 2009.
2. Park, C.H.; Ko, Y.G.; Park, J.-W.; Lee, C.S. Enhanced superplasticity utilizing dynamic globularization of Ti-6Al-4V alloy. *Mater. Sci. Eng. A* **2008**, *496*, 150–158.
3. Semiatin, S.L.; Fagin, P.N.; Betten, J.F.; Zane, A.P.; Ghosh, A.K.; Sargent, G.A. Plastic flow and microstructure evolution during low-temperature superplasticity of ultrafine Ti-6Al-4V sheet material. *Metall. Mater. Trans. A* **2010**, *41*, 499–512.
4. Lee, T.; Kim, J.H.; Semiatin, S.L.; Lee, C.S. Internal-variable analysis of high-temperature deformation behavior of Ti-6Al-4V: A comparative study of the strain-rate-jump and load-relaxation tests. *Mater. Sci. Eng. A* **2013**, *562*, 180–189.
5. Mishra, R.S.; Stolyarov, V.V.; Echer, C.; Valiev, R.Z.; Mukherjee, A.K. Mechanical behavior and superplasticity of a severe plastic deformation processed nanocrystalline Ti-6Al-4V alloy. *Mater. Sci. Eng. A* **2001**, *298*, 44–50.
6. Salishchev, G.A.; Galeyev, R.M.; Valiakhmetov, O.R.; Safiullin, R.V.; Lutfullin, R.Y.; Senkov, O.N.; Froes, F.H.; Kaibyshev, O.A. Development of Ti-6Al-4V sheet with low temperature superplastic properties. *J. Mater. Process. Technol.* **2001**, *116*, 265–268.
7. Semenova, I.P.; Raab, G.I.; Saitova, L.R.; Valiev, R.Z. The effect of equal-channel angular pressing on the structure and mechanical behavior of Ti-6Al-4V alloy. *Mater. Sci. Eng. A* **2004**, *387–389*, 805–808.

8. Saitova, L.R.; Höppel, H.W.; Göken, M.; Semenova, I.P.; Valiev, R.Z. Cyclic deformation behavior and fatigue lives of ultrafine-grained Ti-6AL-4V ELI alloy for medical use. *Int. J. Fatigue* **2009**, *31*, 322–331.

9. Park, C.H.; Kim, J.H.; Yeom, J.-T.; Oh, C.-S.; Semiatin, S.L.; Lee, C.S. Formation of a submicrocrystalline structure in a two-phase titanium alloy without severe plastic deformation. *Scr. Mater.* **2013**, *68*, 996–999.

10. Azushima, A.; Kopp, R.; Korhonen, A.; Yang, D.Y.; Micari, F.; Lahoti, G.D.; Groche, P.; Yanagimoto, J.; Tsuji, N.; Rosochowski, A.; *et al.* Severe plastic deformation (SPD) processes for metals. *CIRP Ann. Manuf. Technol.* **2008**, *57*, 716–735.

11. Kimura, Y.; Inoue, T.; Yin, F.; Tsuzaki, K. Inverse temperature dependence of toughness in an ultrafine grain-structure steel. *Science* **2008**, *320*, 1057–1060.

12. Yin, F.; Hanamura, T.; Inoue, T.; Nagai, K. Fiber texture and substructural features in the caliber-rolled low-carbon steels. *Metall. Mater. Trans. A* **2004**, *35*, 665–677.

13. Torizuka, S.; Ohmori, A.; Narayana Murty, S.V.S.; Nagai, K. Effect of strain on the microstructure and mechanical properties of multi-pass warm caliber rolled low carbon steel. *Scr. Mater.* **2006**, *54*, 563–568.

14. Inoue, T.; Yin, F.; Kimura, Y. Strain distribution and microstructural evolution in multi-pass warm caliber rolling. *Mater. Sci. Eng. A* **2007**, *466*, 114–122.

15. Lee, T.; Park, C.H.; Lee, D.-L.; Lee, C.S. Enhancing tensile properties of ultrafine-grained medium-carbon steel utilizing fine carbides. *Mater. Sci. Eng. A* **2011**, *528*, 6558–6564.

16. Lee, T.; Koyama, M.; Tsuzaki, K.; Lee, Y.-H.; Lee, C.S. Tensile deformation behavior of Fe–Mn–C TWIP steel with ultrafine elongated grain structure. *Mater. Lett.* **2012**, *75*, 169–171.

17. Chun, Y.S.; Lee, J.; Bae, C.M.; Park, K.-T.; Lee, C.S. Caliber-rolled TWIP steel for high-strength wire rods with enhanced hydrogen-delayed fracture resistance. *Scr. Mater.* **2012**, *67*, 681–684.

18. Lee, T.; Park, C.H.; Lee, S.-Y.; Son, I.-H.; Lee, D.-L.; Lee, C.S. Mechanisms of tensile improvement in caliber-rolled high-carbon steel. *Met. Mater. Int.* **2012**, *18*, 391–396.

19. Doiphode, R.; Kulkarni, R.; Narayana Murty, S.V.S.; Prabhu, N.; Prasad Kashyap, B. Effect of severe caliber rolling on superplastic properties of Mg-3Al-1Zn (AZ31) alloy. *Mater. Sci. Forum* **2012**, *735*, 327–331.

20. Lee, T.; Heo, Y.-U.; Lee, C.S. Microstructure tailoring to enhance strength and ductility in Ti–13Nb–13Zr for biomedical applications. *Scr. Mater.* **2013**, *69*, 785–788.

21. Jung, H.S.; Lee, T.; Kwon, I.K.; Kim, H.S.; Hahn, S.K.; Lee, C.S. Surface modification of multi-pass caliber-rolled Ti alloy with dexamethasone-loaded graphene for dental applications. *ACS Appl. Mater. Interfaces* **2015**, *7*, 9598–9607.

22. Narayana Murty, S.V.S.; Nayan, N.; Kumar, P.; Narayanan, P.R.; Sharma, S.C.; George, K.M. Microstructure-texture-mechanical properties relationship in multi-pass warm rolled Ti-6Al-4V alloy. *Mater. Sci. Eng. A* **2014**, *589*, 174–181.

23. Dieter, G.E. *Mechanical Metallurgy*; Metric, S.I., Ed.; McGraw-Hill: London, UK, 1988.

24. Lee, D.; Hart, E. Stress relaxation and mechanical behavior of metals. *Metall. Mater. Trans. B* **1971**, *2*, 1245–1248.

25. Ko, Y.G.; Lee, C.S.; Shin, D.H.; Semiatin, S.L. Low-temperature superplasticity of ultra-fine-grained Ti-6Al-4V processed by equal-channel angular pressing. *Metall. Mater. Trans. A* **2006**, *37*, 381–391.

26. Matsumoto, H.; Bin, L.; Lee, S.-H.; Li, Y.; Ono, Y.; Chiba, A. Frequent occurrence of discontinuous dynamic recrystallization in Ti-6Al-4V Alloy with α' martensite starting microstructure. *Metall. Mater. Trans. A* **2013**, *44*, 3245–3260.

27. Chao, Q.; Hodgson, P.D.; Beladi, H. Ultrafine grain formation in a Ti-6Al-4V alloy by thermomechanical processing of a martensitic microstructure. *Metall. Mater. Trans. A* **2014**, *45*, 2659–2671.

28. Williams, J.C.; Taggart, R.; Polonis, D.H. The morphology and substructure of Ti-Cu martensite. *Metall. Mater. Trans. B* **1970**, *1*, 2265–2270.

29. Tirumalasetty, G.K.; van Huis, M.A.; Kwakernaak, C.; Sietsma, J.; Sloof, W.G.; Zandbergen, H.W. Unravelling the structural and chemical features influencing deformation-induced martensitic transformations in steels. *Scr. Mater.* **2014**, *71*, 29–32.

30. Lee, C.S.; Lee, T.; Kim, J.H.; Park, C.H. Fabrication of ultrafine-grained Ti-6Al-4V bulk sheet/rod for related industries and their mechanical characteristics. In Proceedings of the 8th Pacific Rim International Conference on Advanced Materials and Processing (PRICM 8), Waikoloa, HI, USA, 4–9 August 2013.

31. Bouaziz, O.; Allain, S.; Scott, C.P.; Cugy, P.; Barbier, D. High manganese austenitic twinning induced plasticity steels: A review of the microstructure properties relationships. *Curr. Opin. Solid State Mater. Sci.* **2011**, *15*, 141–168.

32. Reed-Hill, R.E.; Cribb, W.R.; Monteiro, S.N. Concerning the analysis of tensile stress-strain data using log dσ/dεp *versus* log σ diagrams. *Metall. Mater. Trans. B* **1973**, *4*, 2665–2667.

33. Seshacharyulu, T.; Medeiros, S.C.; Morgan, J.T.; Malas, J.C.; Frazier, W.G.; Prasad, Y.V.R.K. Hot deformation and microstructural damage mechanisms in extra-low interstitial (ELI) grade Ti-6Al-4V. *Mater. Sci. Eng. A* **2000**, *279*, 289–299.

34. Stefansson, N.; Semiatin, S.L. Mechanisms of globularization of Ti-6Al-4V during static heat treatment. *Metall. Mater. Trans. A* **2003**, *34*, 691–698.

35. Zherebtsov, S.; Murzinova, M.; Salishchev, G.; Semiatin, S.L. Spheroidization of the lamellar microstructure in Ti-6Al-4V alloy during warm deformation and annealing. *Acta Mater.* **2011**, *59*, 4138–4150.

36. Park, C.H.; Won, J.W.; Park, J.-W.; Semiatin, S.L.; Lee, C.S. Mechanisms and kinetics of static spheroidization of hot-worked Ti-6Al-2Sn-4Zr-2Mo-0.1Si with a lamellar microstructure. *Metall. Mater. Trans. A* **2012**, *43*, 977–985.

37. Park, C.H.; Park, J.-W.; Yeom, J.-T.; Chun, Y.S.; Lee, C.S. Enhanced mechanical compatibility of submicrocrystalline Ti-13Nb-13Zr alloy. *Mater. Sci. Eng. A* **2010**, *527*, 4914–4919.

38. Matsumoto, H.; Yoshida, K.; Lee, S.-H.; Ono, Y.; Chiba, A. Ti-6Al-4V alloy with an ultrafine-grained microstructure exhibiting low-temperature-high-strain-rate superplasticity. *Mater. Lett.* **2013**, *98*, 209–212.

39. Song, H.-W.; Zhang, S.-H.; Cheng, M. Dynamic globularization kinetics during hot working of a two phase titanium alloy with a colony alpha microstructure. *J. Alloys Compd.* **2009**, *480*, 922–927.

40. Maccagno, T.M.; Jonas, J.J.; Hodgson, P.D. Spreadsheet modelling of grain size evolution during rod rolling. *ISIJ Int.* **1996**, *36*, 720–728.

41. Ha, T.K.; Chang, Y.W. An internal variable theory of structural superplasticity. *Acta Mater.* **1998**, *46*, 2741–2749.

14

Processing and Properties of Aluminum and Magnesium Based Composites Containing Amorphous Reinforcement: A Review

Jayalakshmi Subramanian [1,2], **Sankaranarayanan Seetharaman** [1,*] **and Manoj Gupta** [1]

[1] Department of Mechanical Engineering, National University of Singapore,
9 Engineering Drive 1, Singapore 117576, Singapore;
E-Mails: jayalakshmi.subramanian@gmail.com (J.S.); mpegm@nus.edu.sg (M.G.)

[2] Department of Mechanical Engineering, Bannari Amman Institute of Technology,
Satyamangalam, Tamil Nadu 638401, India

* Author to whom correspondence should be addressed; E-Mail: seetharaman.s@nus.edu.sg

Academic Editor: Hugo F. Lopez

Abstract: This review deals with the processing and properties of novel lightweight metal matrix composites. Conventionally, hard and strong ceramic particles are used as reinforcement to fabricate metal matrix composites (MMCs). However, the poor mechanical properties associated with the interfacial de-cohesion and undesirable reactions at (ceramic) particle–(metallic) matrix interface represent major drawbacks. To overcome this limitation, metallic amorphous alloys (bulk metallic glass) have been recently identified as a promising alternative. Given the influential properties of amorphous metallic alloys, their incorporation is expected to positively influence the properties of light metal matrices when compared to conventional ceramic reinforcement. In view of this, a short account of the existing literature based on the processing and properties of Al- and Mg-matrix composites containing amorphous/bulk metallic glass (BMG) reinforcement is presented in this review.

Keywords: aluminum; magnesium; amorphous materials/Bulk Metallic Glass (BMG) reinforcement; processing methods; mechanical properties

1. Introduction

Metallic glasses are novel metallic materials that exhibit superior properties such as high strength (~2 GPa) and elastic strain limit (~2%) [1]. Recently, there have been some attempts to use these metallic glasses as reinforcement in metal matrix composites. Conventionally, metal-matrix composites are made by incorporating hard and strong ceramic reinforcement such as Al_2O_3 or SiC, in the form of fibers, flakes or particles [2]. However, the undesirable reactions and the interfacial de-cohesion at the (ceramic) particle–(metallic) matrix interface often result in poor mechanical properties [3]. In this regard, considering the metastable nature of the amorphous/metallic glass materials, they are considered viable for use as reinforcement in light metal matrices such as Al and Mg with relatively low melting points. Further, the reinforcement/matrix interface being metallic in nature is expected to negate the adverse effects experienced in conventional ceramic reinforced composites [4]. In this respect, the present article will give an in-detail review of the research efforts that have been undertaken so far on the development of amorphous/metallic glass reinforced light-metal matrix composites. In the first section, the preparation and properties of aluminum metal matrix composites containing amorphous reinforcement are presented and the second section deals with that of magnesium matrix composites.

2. Preparation and Characterization of Amorphous/Glass Reinforced Al-Metal Matrix Composites (MMCs)

2.1. Ni-Based Amorphous/Glassy Reinforcements

Lee *et al*. fabricated amorphous/glass reinforced Al-composite for the first time, by reinforcing 20 vol% of Ni-Nb-Ta amorphous alloy (in the form of ribbons) in Al-matrix (Al-356 alloy) using the melt infiltration technique [4]. In this study, the Al-6.5Si-0.25Mg (wt%) alloy was selected as matrix material and the excellent castability of the selected alloy ensured fabrication of defect-free composites by the infiltration process. Here, the Ni-Nb-Ta based $Ni_{39.2}Nb_{20.6}Ta_{40.2}$ (wt%) alloy with excellent thermal stability against crystallization (crystallization onset temperature = 721 °C, which is higher than the liquidus temperature of the A356 alloy = 613 °C) was carefully chosen as the reinforcing material, so as to retain the amorphous structure of the reinforcement. In the first stage, amorphous ribbons (thickness: ~30 μm, width: ~1 mm) of the above mentioned Ni-Nb-Ta alloy were fabricated using the melt spinning process and then cold-pressed (pressure: 16 MPa) into cylindrical preforms (diameter: 9 mm and height: 15 mm) for composite making. In the next stage, the melt infiltration process involved the heating and pressure infiltration of the molten A356 alloy into the prepared amorphous preform. Upon successful synthesis, the structural, thermal, and mechanical properties of the bulk composites were investigated. The results of optical microscopic analysis (Figure 1a) showed the absence of any macro scale defects and showed a homogenous distribution of amorphous reinforcement. Also, no new additional phases were reported in this study which was attributed to the thermal and structural stability of the amorphous reinforcement. Using X-ray diffraction analysis, the crystallographic properties of the composite samples were studied in comparison to that of the amorphous ribbon and the base Al-alloy. The reported X-ray diffractograms are shown in Figure 1b. They show the retention of the amorphous structure of the Ni-Nb-Ta alloy even after the infiltration process which confirmed the suitability of the infiltration process to fabricate light metal matrix composites containing amorphous particles of high thermal stability. Mechanical property measurements were carried out under indentation and compression loads and the reported properties (Table 1) revealed an increment in strength by amorphous reinforcement addition.

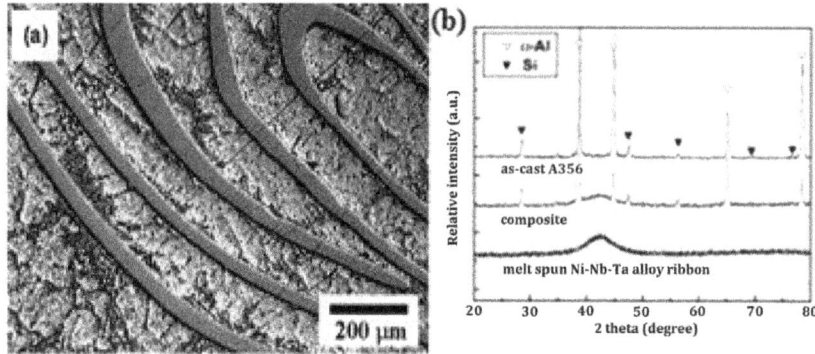

Figure 1. (a) Optical micrographs of Ni-Nb-Ta metallic glass ribbon reinforced A356 alloy based composite; **(b)** X-ray diffractograms of Ni-Nb-Ta metallic glass ribbon reinforced Al-matrix composite in comparison with the melt spun Ni-Nb-Ta amorphous alloy and as-cast A356 alloy (reprinted from Lee *et al.* 2004 [4], with permission from © Elsevier).

Table 1. Properties of bulk metallic glass (BMG) reinforced Al-MMCs.

Matrix	Reinforcement	Amount	Processing Condition	Hardness (H_v)	Compressive Yield Strength (MPa)	Ultimate Compressive Strength (MPa)	Strain at Fracture (%)	Reference
Al 356	Ni-Nb-Ta	20 vol%	Melt infiltration	-	163	320	16	[4]
Pure Al	$Ni_{70}Nb_{30}$	30 wt%	Powder Metallurgy (Compaction + Sintering)	-	111	146	-	[5]
Pure Al	$Ni_{70}Nb_{30}$	30 wt%	Mechanical Alloying + *Sintering*	-	94	-	-	[6]
			Mechanical Alloying + *Hot Press*	-	106	-	-	
			Mechanical Alloying + *Hot Extrusion*	-	134	-	-	
Pure Al	$Ni_{60}Nb_{40}$	5 vol%	Mechanical Alloying + Cold Compaction Microwave Sintering + Hot Extrusion	74.5	114 (50) *	300 ** (60) *	>50 ** (16.8) *	[7]
		15 vol%		103.3	125 (75) *	333 ** (85) *	>50 ** (18.0) *	
		25 vol%		125.2	155 (102) *	375 ** (120) *	>50 ** (9.5) *	
Pure Al	Zr-Ti-Nb-Cu-Ni-Al	40 vol%	Mechanical Alloying + Hot Extrusion	-	-	200	-	[8]
		60 vol%		-	-	250	-	
Al 520	$Cu_{54}Zr_{36}Ti_{10}$	15 vol%	Mechanical Alloying + High Frequency Induction Sintering	-	580	840	14	[9]

Table 1. *Cont.*

Matrix	Reinforcement	Amount	Processing Condition	Hardness (H$_v$)	Compressive Yield Strength (MPa)	Ultimate Compressive Strength (MPa)	Strain at Fracture (%)	Reference
Al 6061	Fe-Co based	15 vol%	Powder Metallurgy + High Frequency Induction Sintering	-	167	570	13	[10]
Al 2024	Fe-Nb-Ge-P-C-B	7.2 vol%	Powder Metallurgy + Hot Extrusion	-	403	660	12	[11]
Al 2024	Fe$_{49.9}$Co$_{35.1}$Nb$_{7.7}$B$_{4.5}$Si$_{2.8}$	10 vol%	Gas Atomization + Hot Pressing + Hot Extrusion	-	(179) *	(297) *	(7) *	[12]
		20 vol%			(200) *	(320) *	(6.7) *	
		30 vol%			(225) *	(340) *	(5.1) *	
		40 vol%			(229) *	(363) *	(4.7) *	
Al	Mg$_{65}$Cu$_{20}$Zn$_5$Y$_{10}$	10 vol%	Ball Milling + Hot Pressing	-	203	247	25	[13]
		30 vol%			221	323	5.8	

* Number in brackets represents the tensile properties; ** Tests stopped at 50% strain due to equipment limitation.

Following the work of Lee *et al.* [4], Yu *et al.* [5] fabricated Ni$_{70}$Nb$_{30}$ metallic glass particle-reinforced Al-metal matrix composite using the powder metallurgy route by sintering below the melting temperature of Al. In this work, the Ni$_{70}$Nb$_{30}$ glass particles were prepared by ball milling pure Ni and Nb elemental powders for 20 h. As it is known that most of the metallic glasses have a crystallization temperature lower than the melting point of Al alloys (~933 K), the selection of sintering temperature and time plays an important role in the solid state powder metallurgy processing of bulk metallic glass (BMG) particle reinforced light metal matrix composites. In this study, to avoid the crystallization of amorphous reinforcement, the Al-30 wt% Ni$_{70}$Nb$_{30}$ glass particles reinforced composite was compacted at room temperature and sintered at 773 K for 2 h. Microstructural features showed the retention of the amorphous state of the glassy particles after sintering, with no interfacial products. Investigation on the mechanical properties showed an increase in compressive yield and ultimate tensile strengths when compared to pure Al, in addition to a 69% increase in the Young's modulus value. Eckert *et al.* [6] also used Ni$_{60}$Nb$_{40}$ (at%) amorphous powder reinforcement. The amorphous Ni$_{60}$Nb$_{40}$ powder (30 wt%) prepared by mechanical alloying was reinforced in pure Al by mixing, sintering, hot pressing, and hot extrusion. Sintering was performed at ~823 K which is below the recrystallization temperature of Ni$_{60}$Nb$_{40}$ amorphous powder. Results of X-ray diffraction analysis conducted on the composite specimen confirmed the retention of the amorphous structure of Ni$_{60}$Nb$_{40}$ (at%) reinforcement and only the crystalline peaks of Al were present. The results of indentation and compression tests (Table 1) revealed superior mechanical properties (in all conditions), which was attributed to the inherent superior strength of the Ni$_{60}$Nb$_{40}$ (at%) amorphous alloy (strength ~2GPa and hardness ~800 Hv).

Recently, Jayalakshmi *et al.* [7] prepared Al-matrix composites containing different amounts of Ni$_{60}$Nb$_{40}$ (at%) amorphous reinforcement using rapid microwave sintering assisted powder metallurgy technique. The amorphous Ni$_{60}$Nb$_{40}$ alloy reinforcement required for the study was initially prepared by ball milling the elemental Ni and Nb metal powder in a planetary ball milling machine for 87 h. The ball milling parameters include milling speed of 200 rpm, and ball to powder ratio of 3:1. In the next stage, the required amount (5, 15, 25 vol%) of the prepared amorphous reinforcement was blended with pure Al powder in a planetary ball milling machine (without balls) for 1 h at 200 rpm. The blended composite powder mixture was then cold compacted into a cylindrical billet of diameter 36 mm and height 50 mm. The compacted billets were then sintered using the microwave sintering approach which employed the combined action of microwave and a microwave couple external heating source to rapidly heat the composite materials (to 823 K, which is less than the crystallization temperature of $Ni_{60}Nb_{40}$ amorphous alloy) in a short period of time (12 min). The sintered billets were then soaked at 673 K for 1 h and hot extruded at 623 K into cylindrical rods of diameter 8 mm on which the structural and mechanical property characterization were performed. The results of structural characterization (SEM microscopy and X-ray analyses) revealed the retention of the amorphous structure and the uniform distribution of the reinforcement without any interfacial reaction products (Figure 2). Structural dilatation and change in the aspect ratio of reinforcement in 25 vol% amorphous particle reinforced composite was also observed. The observed features were attributed to the local stress variations and the temperature gradients within the composite arising *in-situ* during hot extrusion, due to reduced inter-particle spacing.

Figure 2. *Cont.*

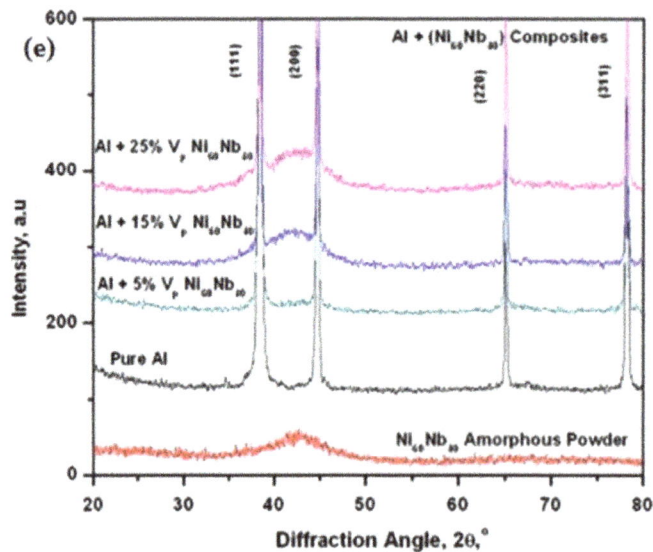

Figure 2. Results of microstructural (**a–d**) analyses conducted on Al-matrix composites reinforced with Ni60Nb40 metallic glass powder, Results of X-ray diffraction analyses (**e**) conducted on Al-matrix composites reinforced with Ni60Nb40 metallic glass powder (reprinted from Jayalakshmi *et al.* 2014 [7], with permission from © Elsevier).

Electrical resistivity measurements were also conducted on pure Al and its composite samples. The results showed higher resistivity values for composites due to the disordered structure of the amorphous phase. However, these values were found to be less than that of the amorphous Ni-Nb alloys. The reported mechanical properties under indentation, tension, and compression loads are listed in Table 1. When compared to pure Al, the micro-hardness value of 25 vol% composite was increased by ~130%. Compression test results showed ~45%–100% enhancement in compression yield strength without fracture. The reported flow curves are shown in Figure 3. Further, it is also worth mentioning that the long extrusions obtained in this study were sufficient to investigate the tensile properties (gauge length 25 mm, diameter: 5 mm) and the report on the tensile behavior of amorphous alloy/glass-reinforced composites was the first of its kind. It showed that the increment in strengths (both yield strength and ultimate strength) did not follow an increasing trend with reinforcement volume fraction. It showed that a minimum critical volume fraction of reinforcement was required for strength enhancement. A maximum increase in strength by ~60% was obtained in 25 vol% composite. Unlike compression test, the tensile test results showed ductility reduction (however not drastic as ceramic particle reinforced Al-MMCs) due to amorphous particle reinforcement addition.

The microscopic analysis of tension test failed samples (Figure 4) showed prominent ductile features and good interfacial bonding between the matrix and reinforcement. Particle breakage was also reported in 25 vol% composite.

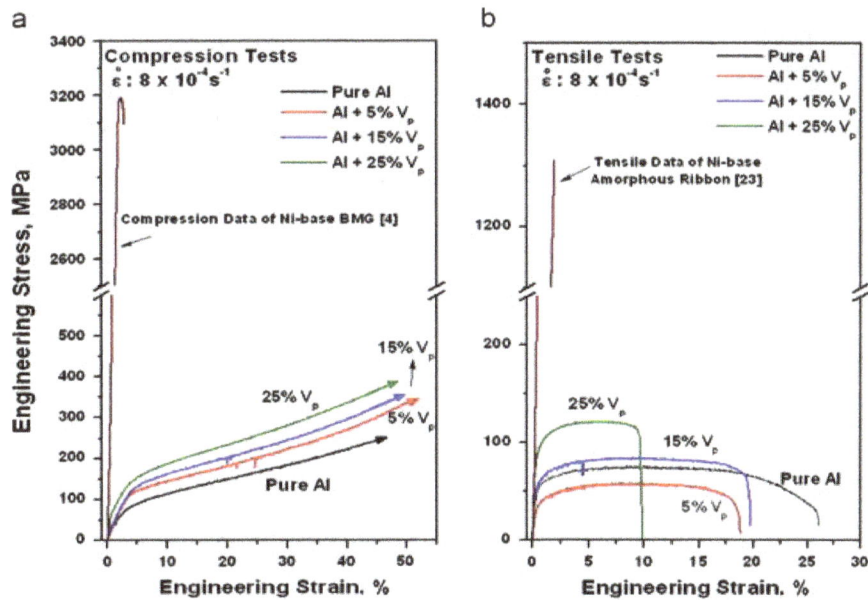

Figure 3. Stress-strain curves of Al-Ni$_{60}$Nb$_{40}$ composites under (**a**) compression and (**b**) tension loads (reprinted from Jayalakshmi *et al.* 2014 [7], with permission from © Elsevier).

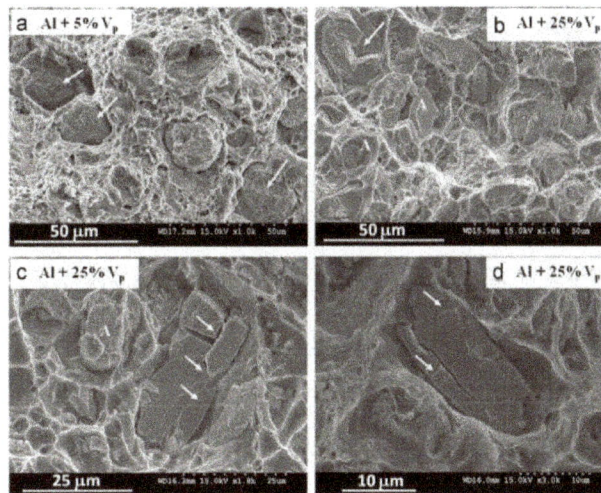

Figure 4. (**a,b**) Low magnification and (**c,d**) high magnification fractographs of Al-Ni$_{60}$Nb$_{40}$ composites fractured under tensile loads. White arrows indicate better interfacial bonding between particle and matrix with no particle debonding in (**a** & **b**), small sized particles remaining intact marked as 'A' in (**b** & **c**), high aspect ratio particles undergoing multiple cracking and fracture in (**c**) and vein patterns caused by strain localization in amorphous particles in (**d**) (reprinted from Jayalakshmi *et al.* 2014 [7], with permission from © Elsevier).

2.2. Zr-Based Amorphous/Glassy Reinforcements

Scudino *et al.* [8] reported the synthesis, structure and compressive behavior of Zr-based metallic glass particle reinforced Al-matrix composites. The Zr$_{57}$Ti$_8$Nb$_{2.5}$Cu$_{13.9}$Ni$_{11.1}$Al$_{7.5}$ (at%) BMG reinforcement required for the study was prepared by ball-milling the elemental Zr, Ti, Nb, Cu, Ni, and

Al powder for 120 h under argon atmosphere, with a ball to powder ratio of 13:1 and milling speed of 150 rpm. For making the composites, 40 and 60 vol% of the prepared amorphous powder was mixed with pure Al powder and the powder mixture was consolidated by hot pressing followed by hot extrusion. An extrusion ratio of 6:1 was employed and the extrusion was performed under argon atmosphere at a temperature 673 K, which is slightly lower than the crystallization temperature range (716–757 K) of the metallic glass reinforcement. The structural properties of the developed Al/Zr-BMG composites were investigated using X-ray diffraction and electron microscopy. The results of X-ray diffraction studies indicated the presence of only pure Al-crystalline peaks and the retention of the reinforcements' amorphous structure which confirmed that the crystallization of amorphous reinforcement was completely avoided during the hot pressing and extrusion. The SEM micrographs (Figure 5) showed no porosity and uniform distribution of the reinforcement particles. Particle clustering was also seen in the high volume fraction of amorphous reinforcement addition.

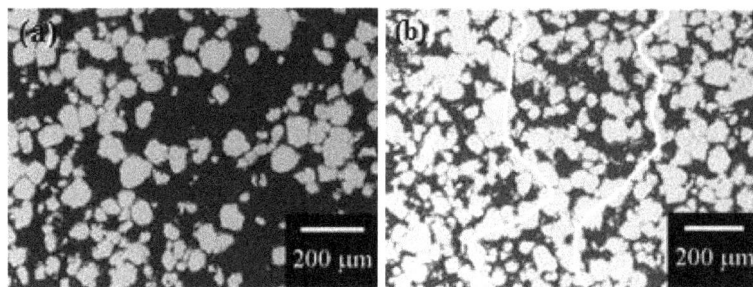

Figure 5. Scanning electron microscopy (SEM) micrographs of (**a**) 40 vol% and (**b**) 60 vol% Zr-based glassy particles reinforced Al-matrix composites. White lines in (**b**) represent the continuous network of particles, (reprinted from Scudino *et al.* 2009 [8], with permission from © Elsevier).

The results of compressive properties reported in comparison to that of the melt spun $Zr_{57}Ti_8Nb_{2.5}Cu_{13.9}Ni_{11.1}Al_{7.5}$ ribbon and pure Al showed that the composite reinforced with 60% reinforcement exhibited superior strength (however less than the amorphous Zr-based ribbon) and work softening behavior. The microscopic features of the fractured samples investigated using scanning electron microscopy (SEM) (Figure 6) showed cracks in the reinforcing particles which occurred parallel to the compression direction. However, the Al matrix underwent large plastic deformation and displayed dimple rupture, which are both indicative of ductile fracture.

Figure 6. Fractographs of Zr-based metallic glass reinforced Al composites. (**a**) Low magnification image with loading direction indicated by arrow; (**b**) particle breakage (reprinted from Scudino *et al.* 2009 [8], with permission from © Elsevier).

2.3. Cu-Based Amorphous/Glassy Reinforcements

Cu-based glassy material reinforced in Al-A520 alloy were developed by Dudina *et al.* [9] using high frequency induction sintering. In this study, the $Cu_{54}Zr_{36}Ti_{10}$ (at%) metallic glass ribbons were prepared bythe arc melting/melt spinning process. The prepared amorphous alloy ribbons were then cut and ball-milled for 24 h. The cut ribbons were then mixed together with the Al-matrix alloy in a vibratory mill for 8 h. The composite powder mixture was processed using high frequency induction sintering at a temperature just below the crystallization temperature of the amorphous alloy. The processed composite materials were subjected to X-ray, electron microscopy and compression analysis. While the results of X-ray diffraction analysis indicated no additional phases and transformation of the amorphous phase, electron microscopic observations revealed uniform distribution of the amorphous phase with no preferred orientation. The compressive stress-strain curves as shown in Figure 7 (properties listed in Table 1) revealed promising strength enhancement by efficient load transfer and grain refinement. This was accompanied by a slight reduction in fracture strain. The fracture morphology of the compression test failed specimen indicated no metallic glass particle de-bonding from the matrix (Figure 7b,c).

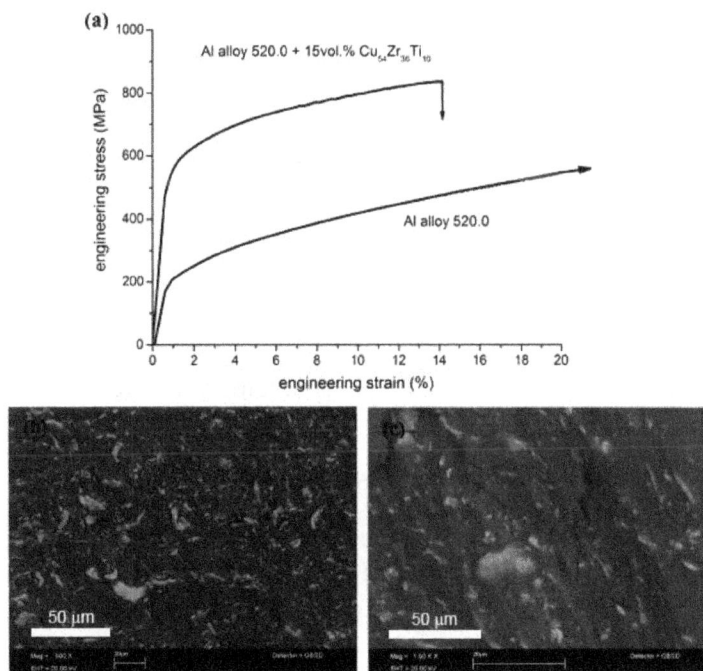

Figure 7. (**a**) Compression stress-strain curves and (**b,c**) fractographic evidences of 15 vol% $Cu_{54}Zr_{36}Ti_{10}$ glassy particle reinforced A520 alloy composite (reprinted from Dudina *et al.* 2010 [9], with permission from © Elsevier).

2.4. Fe-Based Amorphous/Glassy Reinforcements

Fujii *et al.* [10] fabricated Fe-based ($Fe_{72}B_{14.4}Si_{9.6}Nb_4$) metallic glass particle reinforced Al-based composite materials using the friction stir processing method. The selection of the Fe-based metallic glass particle as reinforcement in this study was based on the fact that the glass transition temperature is relatively higher than the welding temperature (673–723 K) of the Al alloys. For the base matrix material, pure Al plates (1050-H24) with dimension $300 \times 70 \times 5$ mm^3 were used. Structural and mechanical property measurements were performed and the results revealed coarsening of Al-grains due to the dispersion of Fe-based BMG particles. Further, it was also reported that dispersion of Fe-BMG particles had little effect on the hardness, although the results of hardness measurements showed improvement. This was attributed to the formation of Al$_{13}$Fe$_4$ precipitates in the stir zone due to the reaction between pure Al and Fe based metallic glass.

Aljerf et al. [14] prepared [(Fe$_{0.5}$Co$_{0.5}$)$_{75}$B$_{20}$Si$_5$]$_{96}$Nb$_4$ amorphous alloy reinforced Al6061 alloy composites using the high frequency induction sintering assisted powder metallurgy method. In this study, Fe-Co based amorphous alloy ribbons of the above mentioned composition were first ball-milled with Al-alloy and the Al/Fe-BMG composite mixture was then densified using induction sintering at 828 K. Similar to earlier studies, DSC and X-ray diffraction analyses were used to comprehend the crystallographic nature of the amorphous reinforcement and the results confirmed the retention of the amorphous structure in the composites (Figure 8). The flow curves under compression (Figure 9) clearly highlight the enhancement in strength properties due to amorphous reinforcement addition. The reported mechanical properties are listed in Table 1.

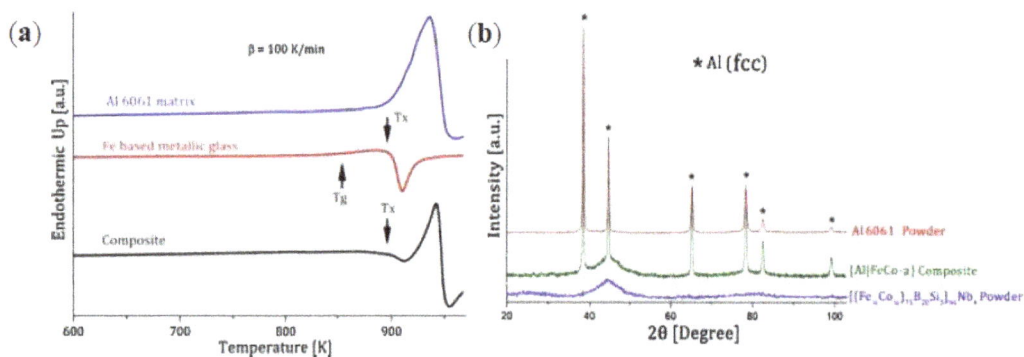

Figure 8. (a) DSC thermograms and (b) X ray diffractograms of Al matrix composite reinforced with Fe-based metallic glass compared to Al 6061 alloy and the Fe-based metallic glass powder (reprinted from Aljerf et al. 2012 [14], with permission from © Elsevier).

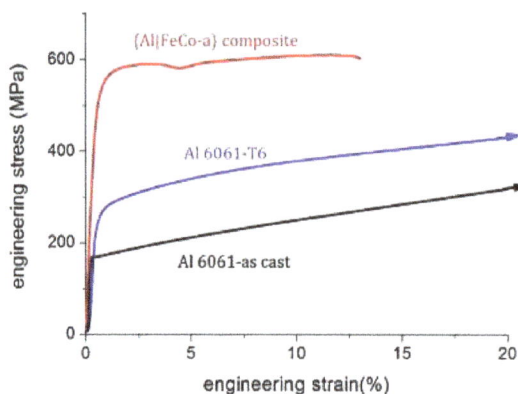

Figure 9. Compression flow curves of Al matrix composite reinforced with Fe-based metallic glass compared to as cast and heat treated Al 6061 alloy (reprinted from Aljerf et al. 2012 [14], with permission from © Elsevier).

Zheng *et al.* [11] produced Fe-based BMG particle reinforced 2024 Al-alloy composite using the powder metallurgy method. The Al 2024 alloy matrix and the amorphous $Fe_{73}Nb_5Ge_2P_{10}C_6B_4$ reinforcement materials required, were initially prepared using gas and water atomization respectively. The prepared powders were then ball milled under an argon atmosphere to fabricate the composite powder. Milling parameters include ball to powder ratio of 10:1, sun-disk rotation speed of 280 rpm, planetary-disk rotation speed of 480 rpm, process control agent of stearic acid and argon protective atmosphere. The composite powder mixture was then consolidated and sintered at 823 K, 400 MPa in a 20 mm diameter stainless steel die using induction heating for 30 min. The sintered billet was hot extruded at 823 K at an extrusion ratio of 10:1. Structural characterization showed the distribution of refined amorphous particles and the nanostructure Al-2024 matrix with clear interface between the matrix and amorphous reinforcement (Figure 10). Compression property measurements conducted on the composite samples showed enhanced yield and fracture strength resulting from the nanostructure of the Al-matrix and the uniform distribution of the amorphous reinforcement particles. In a recent study, Marko *et al.* [12] reinforced Al-2024 alloy with different volume fractions of gas atomized $Fe_{49.9}Co_{35.1}Nb_{7.7}B_{4.5}Si_{2.8}$ glassy particles using the powder metallurgy method by hot pressing (for 10 min at 673 K, 500 MPa) followed by hot extrusion at the same temperature. Given that the crystallization temperature of the glass reinforcement is ~873 K, the sintering temperature was far below the crystallization temperature and avoided crystallization of the glassy particles during sintering. This is yet another work that after ref. [7] investigated the room temperature tensile properties which indicated a 27% and 20% increase in yield and ultimate strengths respectively (listed in Table 1), when compared to the unreinforced alloy matrix. The ductility of the glass particle reinforced composite (range between 5% and 10%) was lower than that of the unreinforced alloy. Nevertheless, the glass particle reinforced composites have a definite advantage over the conventional ceramic reinforced composites, as the conventional composites usually have negligible/no ductility.

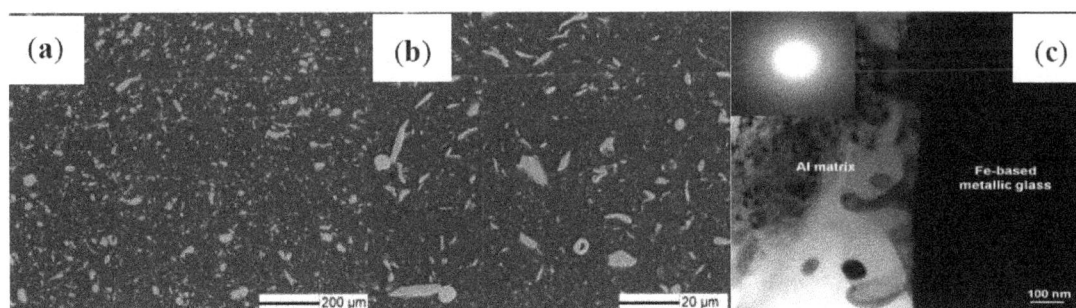

Figure 10. SEM micrographs of Al 2024 alloy reinforced with Fe based metallic glass. (**a**) Low magnification and (**b**) high magnification; (**c**) TEM image showing the interfacial characteristics (reprinted from Zheng *et al.* 2014 [11], with permission from © Elsevier).

2.5. Mg-Based Amorphous/Glassy Reinforcements

Wang *et al.* [13] reinforced $Mg_{65}Cu_{20}Zn_5Y_{10}$ (at%) glassy particles in pure Al matrix. Injection cast $Mg_{65}Cu_{20}Zn_5Y_{10}$ metallic glass rods were ball-milled to produce metallic glass powder particles. 10 and 30 vol% of the particles were mixed with pure Al powder and ball-milled to form composite powder (for 10 h with a ball-to-powder ration of 10:1). Uniaxial hot pressing was used to compact the composite powder in a vacuum at 453 K (falls in the super-cooled liquid region of the Mg-glassy particles) and 700 MPa. Uniform distribution of the glassy particles and good interfacial characteristics were observed. The room temperature compressive properties showed a yield strength increment by a factor of ~3 (203 MPa) in the 10 vol% composite (when compared to the unreinforced Al), along with a compressive deformation of ~25% (Table 1). Theoretical strength estimation using modified shear lag (MSL) model was used to explain the strengthening effect which showed good agreement with the experimental data. It was concluded that dislocation strengthening was the dominant strengthening mechanism in the developed composites.

3. Preparation and Characterization of Amorphous/Glass Reinforced Mg-MMCs

Dudina *et al*. [15] initiated the work on reinforcing amorphous/metallic glass in magnesium matrix. The effect of 15 vol%. Vitraloy 6, *i.e.*, $Zr_{57}Nb_5Cu_{15.4}Ni_{12.6}Al_{10}$ amorphous alloy reinforcement in AZ31 Mg-alloy was investigated. The composite materials were prepared by using high frequency induction sintering under pressure. In the first stage, Zr-based metallic glass ribbons (prepared by induction melting followed by melt spinning) were cut into pieces and ball-milled in a low speed vibratory mill to prepare amorphous powder particles. The milling was performed for 15 h under an argon atmosphere. In the next stage, the ball milled Zr-based BMG powder (15 vol%) was mixed and milled with the cut ribbons of Mg-alloy. The milled composite powder mixture was then consolidated by using high frequency induction sintering under pressure (50 MPa) at a temperature of 713 K for 120 s. Similar to Al/BMG metal matrix composites, the selection of the sintering temperature was based on the crystallization temperature of Vitraloy 6 metallic glass reinforcement. The consolidated bulk Mg/Zr-BMG composite materials were then characterized for their structural and mechanical properties. Structural investigations involved X-ray diffraction and electron microscopy analyses. The results (Figure 11) revealed the retention of the reinforcement's amorphous structure and the uniform dispersion of the amorphous reinforcement without any interfacial reaction products or pores. This confirms the sound consolidation of the powder materials.

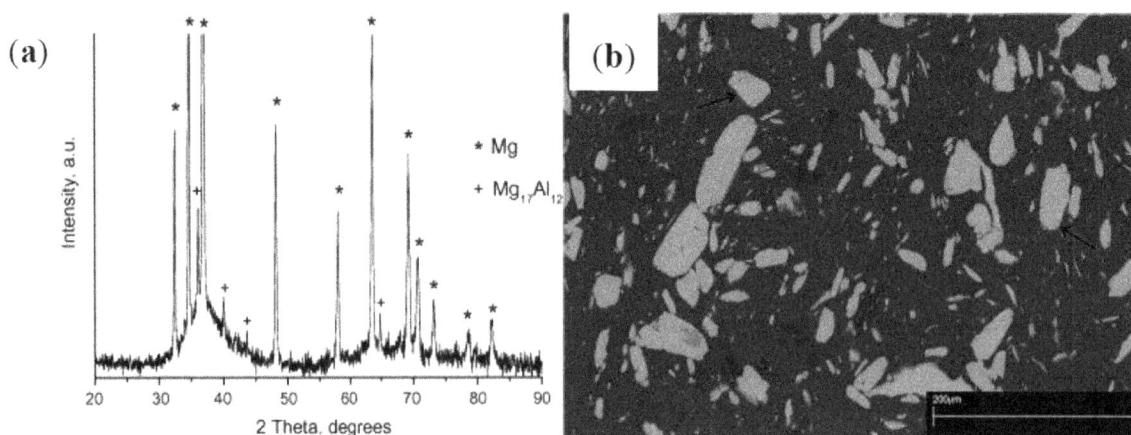

Figure 11. (a) X-ray diffractograms and (b) SEM micrograph of Zr-based glassy particle reinforced AZ91 Mg-alloy composite prepared by high frequency induction heating under pressure. Metallic glass particles are marked with black arrows in (b) (reprinted from Dudina *et al*. 2009 [15], with permission from © Elsevier).

Mechanical property measurements under indentation and compression loads were performed on samples 4 mm long and 2×2 mm^2 cross section. The reported results (Table 2) indicated a significant enhancement in strength properties due to the efficient load transfer and dislocation strengthening.

Table 2. Properties of BMG reinforced Mg-MMCs.

Matrix	Reinforcement	Amount	Processing Condition	Hardness (H$_v$)	Compressive Yield Strength (MPa)	Ultimate Compressive Strength (MPa)	Strain at Fracture (%)	Reference
AZ 91	Zr-Nb-Cu-Ni-Al	15 vol%	Ball Milling + High Frequency Induction Sintering	123	325	542	10.5	[15]
Pure Mg	Ni-Nb	3 vol%	Ball Milling + Microwave Sintering + Hot Extrusion	62	85	283	17.6	[16]
		5 vol%		84	135	320	18.4	
		10 vol%		95	90	322	17.2	
Pure Mg	Ni-Ti	3 vol%	Ball Milling + Microwave Sintering + Hot Extrusion	49	67 (94) *	291 (144) *	15.9 (8.8) *	[17]
		5 vol%		62	89 (127) *	368 (183) *	15.1 (6.5) *	
		10 vol%		66	102 (148) *	417 (178) *	14.9 (2.0) *	

* Number in brackets represents the tensile properties.

Pure Mg-based composite materials reinforced with varying volume fractions of Ni$_{60}$Nb$_{40}$ (at%) metallic glass particles were synthesized by Jayalakshmi *et al.* [16] using the blend-press-sinter based powder metallurgy method. In the first stage, the amorphous Ni$_{60}$Nb$_{40}$ reinforcement required for the study was prepared by ball-milling. The processing details were similar to that mentioned in Jayalakshmi *et al.* [7]. In the next stage, the required amount (3, 5, 10 vol%) of Ni$_{60}$Nb$_{40}$ BMG reinforcement was mixed with pure Mg powder and the mixture was blended in a planetary ball mill for 1 h. This was followed by the densification of the blended composite powder using uniaxial cold compaction and rapid microwave sintering. A sintering time of 12 min 30 s was selected for microwave sintering so as to reach the sintering temperature of 823 K, which is below the crystallization temperature of the Ni$_{60}$Nb$_{40}$ reinforcement. The sintered billets were then hot extruded at 623 K into cylindrical rods of 8 mm diameter on which the structural and mechanical property investigations were performed. The results of structural characterization by X-ray analysis, optical and electron microscopy revealed matrix grain refinement (Figure 12), uniform distribution of reinforcement at low volume fraction and agglomeration at high volume fractions (Figure 13), absence of interfacial reaction production (Figure 13d) and the retention of the reinforcement's amorphous structure at all volume fractions (Figure 14a). The results of X-ray analyses also indicated the influence of the amorphous reinforcement in changing the dominant crystallographic orientation of the Mg-matrix (Figure 14b). It was reported that the basal planes of the composites were not aligned entirely parallel to the extrusion as generally would be in the case of extruded magnesium materials.

The results of micro-hardness and compression tests (listed in Table 2) showed a remarkable increment with increasing amount of amorphous reinforcement. This behavior was attributed to the inherent high hardness, strength and elastic strain limit of the amorphous reinforcement.

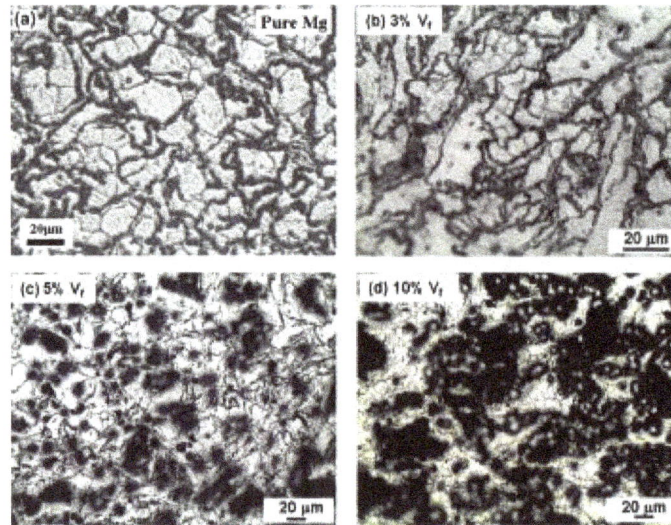

Figure 12. (**a**) Grain characteristics of pure Mg, and (**b–d**) its composite reinforced with different amounts of $Ni_{60}Nb_{40}$ amorphous reinforcement (reprinted from Jayalakshmi *et al*. 2014 [16], with permission from © Elsevier).

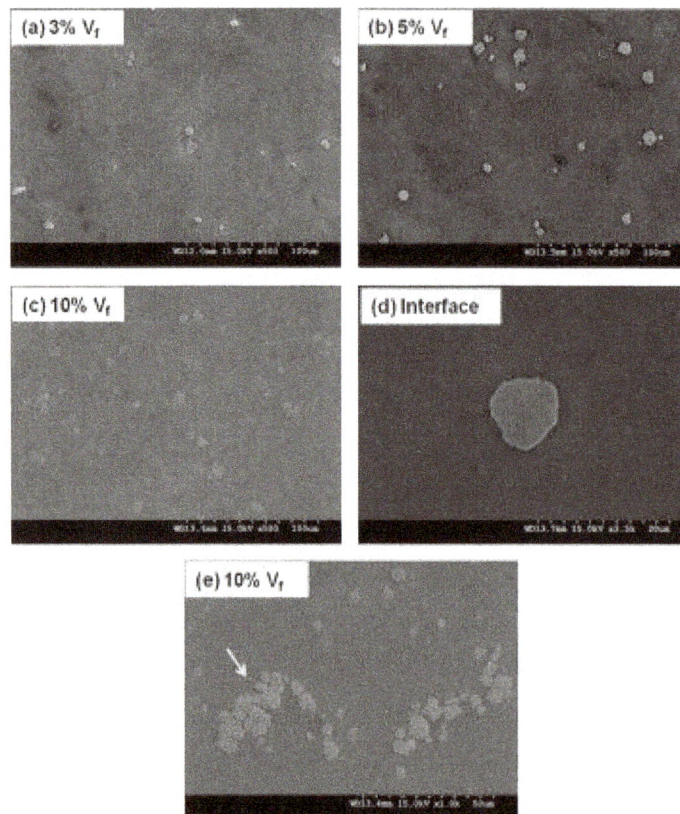

Figure 13. SEM images showing the distribution of $Ni_{60}Nb_{40}$ amorphous particles in Mg matrix composites. White arrow in (e) represents clustering of particles. (reprinted from Jayalakshmi *et al*. 2014 [16], with permission from © Elsevier).

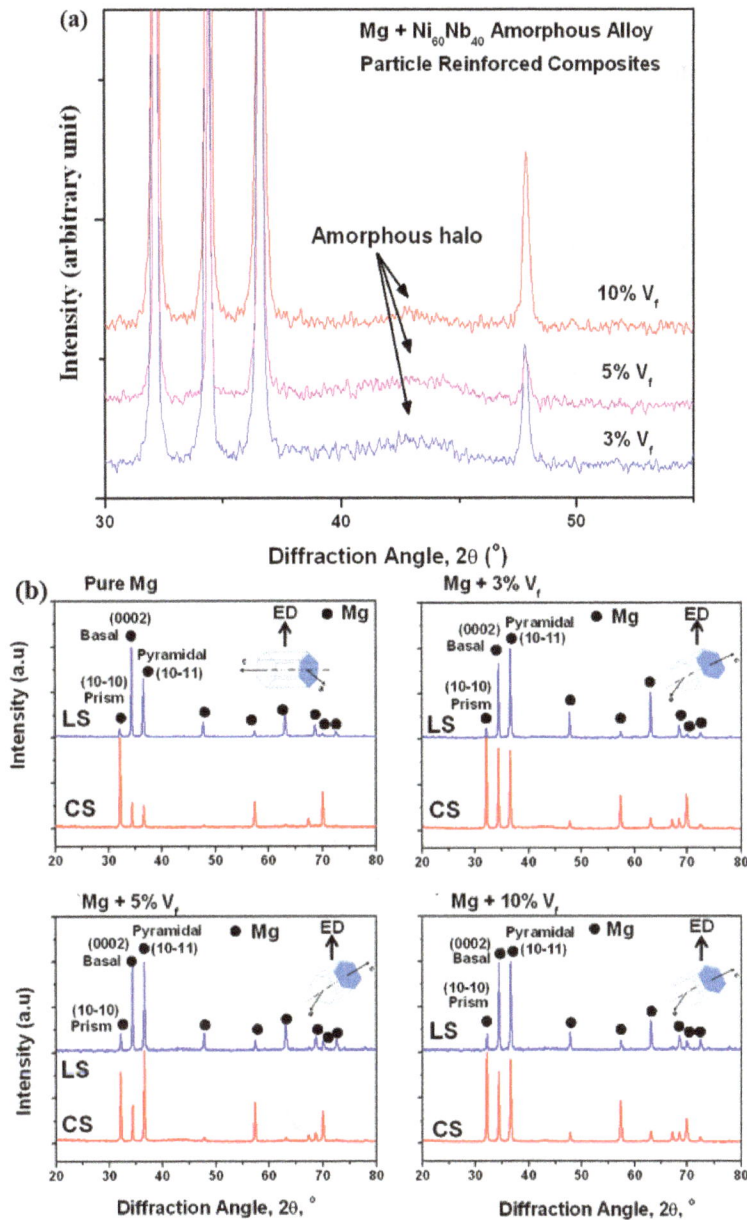

Figure 14. (a) XRD diffractograms of Mg-$Ni_{60}Nb_{40}$ composites showing the retention of amorphous structure at all volume fractions and (b) change in crystal orientation due to amorphous reinforcement addition. (reprinted from Jayalakshmi *et al*. 2014 [16], with permission from © Elsevier).

Recently, pure Mg reinforced with varying volume fractions of $Ni_{50}Ti_{50}$ (at%) were fabricated using the microwave sintering assisted powder metallurgy technique [17]. The $Ni_{50}Ti_{50}$ amorphous reinforcement used in this study was prepared by ball milling the Ni and Ti powder particles for 55 h in a planetary ball milling machine. The ball milling parameters were similar to those used in reference [16]. For composite fabrication, the required amounts (3, 6 and 10 vol%) of the prepared $Ni_{50}Ti_{50}$ amorphous reinforcement were mixed with pure Mg powder and the composite mixture was then cold compacted in an isostatic press. The compacted billet was then densified using microwave sintering. The sintered composite materials were then hot extruded into cylindrical rods and the samples cut from the cylindrical rods were subjected to X-ray, electron microscopy and mechanical property characterization. The SEM observations indicated a fair distribution of amorphous reinforcements with a clear matrix/reinforcement interface (Figure 15).

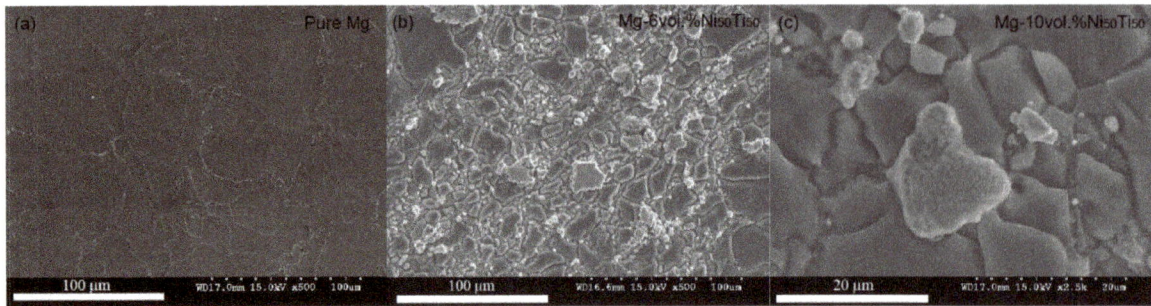

Figure 15. SEM micrographs showing the microstructural characteristics of: (**a**) pure Mg and its composite containing different amounts of $Ni_{50}Ti_{50}$ amorphous reinforcement (**b,c**). (reprinted from Sankaranarayanan *et al*. 2015 [17], with permission from © Elsevier).

The reported mechanical properties are listed in Table 2 which indicated superior strength properties due to amorphous $Ni_{50}Ti_{50}$ reinforcement addition (Figure 16). The tensile properties of such amorphous particle reinforced magnesium composites were reported for the first time in the case of $Ni_{50}Ti_{50}$ amorphous particles reinforced magnesium composites. The results showed ~98% increase in tensile yield strength and ~50% increase in the ultimate strength. While the tensile ductility was adversely affected, the reported properties were comparable or superior to that of ceramic particle reinforced Mg-MMCs.

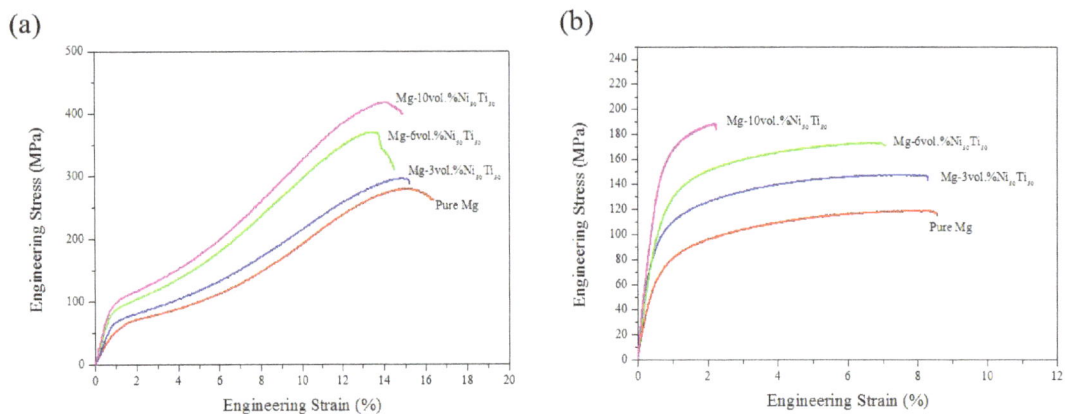

Figure 16. Engineering stress-strain curves of the developed Mg-composites under (**a**) compressive loading (**b**) tensile loading (reprinted from Sankaranarayanan *et al*. 2015 [17], with permission from © Elsevier).

The microscopic features of tension test failed samples of $Ni_{50}Ti_{50}$ particle reinforced Mg-composites were studied in comparison to pure Mg as shown in Figure 17. It revealed dominant quasi-cleavage fracture in both pure Mg and its composites. Sparse particle segregation (debonding) features were also reported in $Mg/Ni_{50}Ti_{50}$ composites.

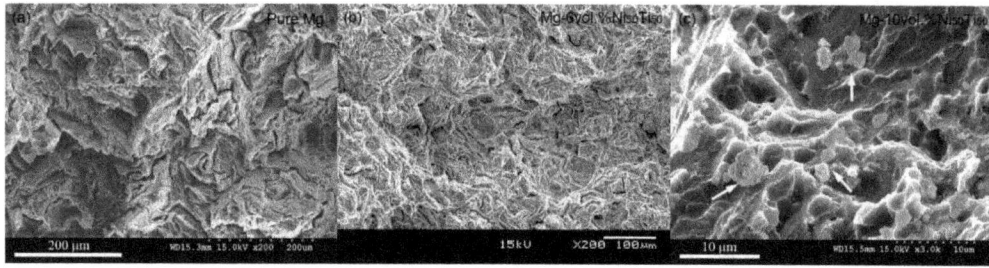

Figure 17. Representative tensile fractographs showing typical cleavage mode fracture in: (**a**) pure Mg and (**b**) Mg-6 vol% $Ni_{50}Ti_{50}$ and (**c**) prominent particle segregation in Mg-10 vol% $Ni_{50}Ti_{50}$. White arrows represent fine amorphous particles. (reprinted from Sanakranarayanan *et al.* 2015 [17], with permission from © Elsevier).

4. Influence of Processing Method on the Mechanical Properties

Figure 18 compares the compressive properties of some of the amorphous alloy/metallic glass reinforced composites produced by different methods (as explained in the preceding paragraphs). Considering that in the graph (*y*-axis) the strength is normalized, the mechanical properties of the composites seem to be dominated by the processing method. Those composites produced by high frequency induction sintering exhibit the highest strength properties, irrespective of the matrix or reinforcement volume fraction. The high frequency induction sintering method is a rapid sintering method with high-temperature exposure and application of pressure over a short duration of time. Amongst the strength of the three composites produced by the high frequency induction sintering method, composites with Mg- and Fe-based glassy particles exhibit the highest strength, which could be due to the higher thermal stability, and enhanced glass formation characteristics of the alloy that may impart a better structural stability during processing, which would in turn improve the mechanical properties. The next best properties are observed in those composites produced by another unconventional rapid sintering method, *i.e.*, microwave sintering, followed by hot extrusion. The method can also produce composites in larger dimensions such that the tensile behavior of the composites can also be studied. Composites produced by this method also show enhanced mechanical properties. These results highlight the fact that rapid sintering is essential to achieve an improved strengthening effect for composites reinforced with amorphous alloy/metallic glass reinforcements, as they can minimize interaction time with the matrix, promote rapid sintering, reduce porosity/defects, and retain an amorphous structure, resulting in an efficient load bearing capacity of the inherently high strength amorphous alloy/metallic glass reinforcements. Overall, the mechanical properties of the developed new composites are promising and should be designed so as to compensate for the disadvantages faced by conventional MMCs and at the same time utilize the superior properties of the glassy materials. A synergistic effect of superior properties of the matrix and reinforcement is expected to give rise to novel and advanced hybrid materials.

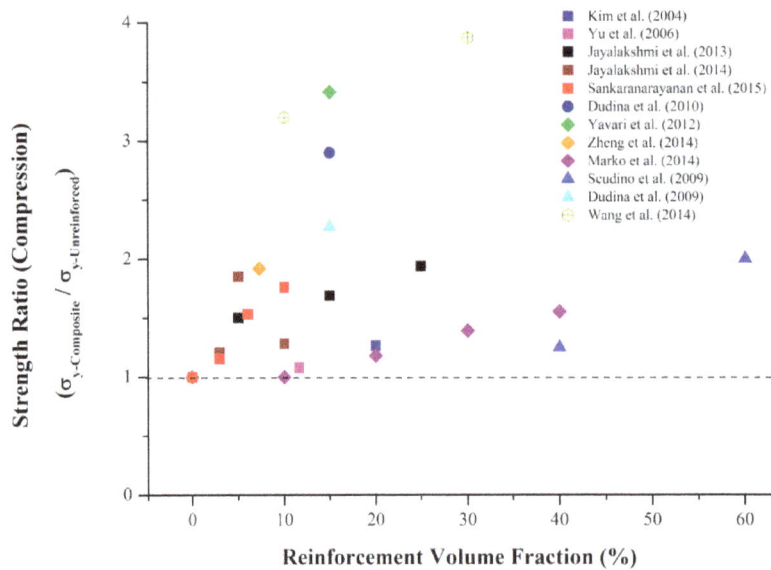

Figure 18. Compressive properties of amorphous alloy/metallic glass reinforced composites produced by different methods.

5. Conclusions

In most of the studies reviewed in this article, the incorporation of metallic glass reinforcement into light metal matrices (Al or Mg) enhanced the mechanical properties. The reported mechanical property enhancement was attributed to the inherent superior mechanical properties of the amorphous/metallic glass reinforcement. Further, the absence of interfacial reaction products (unlike conventional light metal matrix composites) also contributes towards the improvement of the composite properties. The method of processing the composite seems to play an influential role in defining the resulting properties. Composites produced by rapid processing methods such as high frequency induction sintering and rapid microwave sintering showed superior properties to those produced by conventional methods. Hence, by critically designing the nature and incorporation of metallic glass reinforcement with superior properties, the limitations faced by conventional metal matrix composite materials can be overcome.

Acknowledgments

The authors wish to acknowledge the funding support given by Ministry of Education, Singapore under grant "Singapore Ministry of Education Academic Research Fund—Tier 1" and WBS No: R-265-000-493-112, for carrying out this project.

Author Contributions

All authors contributed to the paper. Jayalakshmi Subramanian and Sankaranarayanan Seetharaman collected the data and prepared the manuscript. Manoj Gupta designed the scope of the paper. All authors discussed the conclusions and reviewed the manuscript.

Conflicts of Interest

The authors declare no conflict of interest.

References

1. Wang, W.-H.; Dong, C.; Shek, C. Bulk metallic glasses. *Mater. Sci. Eng. R Rep.* **2004**, *44*, 45–89.

2. Kainer, K.U. *Metal Matrix Composites: Custom-made Materials for Automotive and Aerospace Engineering*; John Wiley & Sons: Hoboken, NJ, USA, 2006.

3. Suresh, S.; Mortensen, A.; Needleman, A. *Fundamentals of Metal-matrix Composites*; Butterworth-Heinneman: Stoneham, MA, USA, 1993; p. 400.

4. Lee, M.; Kim, J.-H.; Park, J.; Kim, J.; Kim, W.; Kim, D. Fabrication of Ni-Nb-Ta metallic glass reinforced Al-based alloy matrix composites by infiltration casting process. *Scr. Mater.* **2004**, *50*, 1367–1371.

5. Yu, P.; Kim, K.; Das, J.; Baier, F.; Xu, W.; Eckert, J. Fabrication and mechanical properties of Ni-Nb metallic glass particle-reinforced Al-based metal matrix composite. *Scr. Mater.* **2006**, *54*, 1445–1450.

6. Eckert, J.; Calin, M.; Yu, P.; Zhang, L.; Scudino, S.; Duhamel, C. Al-based alloys containing amorphous and nanostructured phases. *Rev. Adv. Mater. Sci* **2008**, *18*, 169–172.

7. Jayalakshmi, S.; Gupta, S.; Sankaranarayanan, S.; Sahu, S.; Gupta, M. Structural and mechanical properties of $Ni_{60}Nb_{40}$ amorphous alloy particle reinforced Al-based composites produced by microwave-assisted rapid sintering. *Mater. Sci. Eng. A* **2013**, *581*, 119–127.

8. Scudino, S.; Liu, G.; Prashanth, K.; Bartusch, B.; Surreddi, K.; Murty, B.; Eckert, J. Mechanical properties of Al-based metal matrix composites reinforced with Zr-based glassy particles produced by powder metallurgy. *Acta Mater.* **2009**, *57*, 2029–2039.

9. Dudina, D.; Georgarakis, K.; Aljerf, M.; Li, Y.; Braccini, M.; Yavari, A.; Inoue, A. Cu-based metallic glass particle additions to significantly improve overall compressive properties of an Al alloy. *Compos. Part A Appl. Sci. Manuf.* **2010**, *41*, 1551–1557.

10. Aljerf, M.; Georgarakis, K.; Louzguine-Luzgin, D.; Le Moulec, A.; Inoue, A.; Yavari, A. Strong and light metal matrix composites with metallic glass particulate reinforcement. *Mater. Sci. Eng. A* **2012**, *532*, 325–330.

11. Zheng, R.; Yang, H.; Liu, T.; Ameyama, K.; Ma, C. Microstructure and mechanical properties of aluminum alloy matrix composites reinforced with Fe-based metallic glass particles. *Mater. Des.* **2014**, *53*, 512–518.

12. Markó, D.; Prashanth, K.; Scudino, S.; Wang, Z.; Ellendt, N.; Uhlenwinkel, V.; Eckert, J. Al-based metal matrix composites reinforced with $Fe_{49.9}Co_{35.1}Nb_{7.7}B_{4.5}Si_{2.8}$ glassy powder: Mechanical behavior under tensile loading. *J. Alloys Compd.* **2014**, *615*, S382–S385.

13. Wang, Z.; Tan, J.; Sun, B.; Scudino, S.; Prashanth, K.; Zhang, W.; Li, Y.; Eckert, J. Fabrication and mechanical properties of Al-based metal matrix composites reinforced with $Mg_{65}Cu_{20}Zn_5Y_{10}$ metallic glass particles. *Mater. Sci. Eng. A* **2014**, *600*, 53–58.

14. Fujii, H.; Sun, Y.; Inada, K.; Ji, Y.; Yokoyama, Y.; Kimura, H.; Inoue, A. Fabrication of Fe-based metallic glass particle reinforced al-based composite materials by friction stir processing. *Mater. Trans.* **2011**, *52*, 1634–1640.

15. Dudina, D.; Georgarakis, K.; Li, Y.; Aljerf, M.; LeMoulec, A.; Yavari, A.; Inoue, A. A magnesium alloy matrix composite reinforced with metallic glass. *Compos. Sci. Technol.* **2009**, *69*, 2734–2736.

16. Jayalakshmi, S.; Sahu, S.; Sankaranarayanan, S.; Gupta, S.; Gupta, M. Development of novel Mg-Ni$_{60}$Nb$_{40}$ amorphous particle reinforced composites with enhanced hardness and compressive response. *Mater. Des.* **2014**, *53*, 849–855.

17. Sankaranarayanan, S.; Shankar, H.; Jayalakshmi, S.; Nguyen, Q.; Gupta, M. Development of high performance magnesium composites using Ni$_{50}$Ti$_{50}$ metallic glass reinforcement and microwave sintering approach. *J. Alloys Compd.* **2015**, *627*, 8.

Permissions

List of Contributors

Ahmad Mostafa
Department of Mechanical and Industrial Engineering, Concordia University, 1455 de Maisonneuve Blvd. West, Montreal, QC H3G 1M8, Canada

Mamoun Medraj
Department of Mechanical and Industrial Engineering, Concordia University, 1455 de Maisonneuve Blvd. West, Montreal, QC H3G 1M8, Canada
Department of Mechanical and Materials Engineering, Masdar Institute, Masdar City, P. O. Box 54224, Abu Dhabi, UAE

Liang Song
School of Materials Science and Engineering, Heilongjiang University of Science and Technology, Harbin 150022, China
School of Materials Science and Engineering, Harbin University of Science and Technology, Harbin 150080, China

Erjun Guo
School of Materials Science and Engineering, Harbin University of Science and Technology, Harbin 150080, China

Liping Wang
School of Materials Science and Engineering, Harbin University of Science and Technology, Harbin 150080, China

Dongrong Liu
School of Materials Science and Engineering, Harbin University of Science and Technology, Harbin 150080, China

Nicolas Eustathopoulos
SIMAP, University Grenoble Alpes-CNRS, F-38000 Grenoble, France

Susanne Enghardt
Department of Prosthetic Dentistry, Faculty of Medicine, Technische Universität Dresden, Fetscherstr, 74 D-01307 Dresden, Germany

Gert Richter
Department of Prosthetic Dentistry, Faculty of Medicine, Technische Universität Dresden, Fetscherstr, 74 D-01307 Dresden, Germany

Edgar Richter
Forschungszentrum Dresden-Rossendorf e.V., Institute of Ion Beam Physics and Materials Research, Bautzner Landstr, 400 D-01328 Dresden, Germany

Bernd Reitemeier
Department of Prosthetic Dentistry, Faculty of Medicine, Technische Universität Dresden, Fetscherstr, 74 D-01307 Dresden, Germany

Michael H. Walter
Department of Prosthetic Dentistry, Faculty of Medicine, Technische Universität Dresden, Fetscherstr, 74 D-01307 Dresden, Germany

Di Tie
School of Metal and Metallurgy, Northeastern University, Wenhua Rd. No.3, Shenyang 110003, China

Ren-guo Guan
School of Metal and Metallurgy, Northeastern University, Wenhua Rd. No.3, Shenyang 110003, China

Ning Guo
Postgraduate Academy, Shenyang Ligong University, Nanpingzhong Rd. No.6, Shenyang 110000, China

Zhouyang Zhao
School of Metal and Metallurgy, Northeastern University, Wenhua Rd. No.3, Shenyang 110003, China

Ning Su
School of Metal and Metallurgy, Northeastern University, Wenhua Rd. No.3, Shenyang 110003, China

Jing Li
CNPC Institute, Taiyanggongnan St. NO.23, Beijing 100010, China

Yang Zhang
School of Metal and Metallurgy, Northeastern University, Wenhua Rd. No.3, Shenyang 110003, China
Postgraduate Academy, Shenyang Ligong University, Nanpingzhong Rd. No.6, Shenyang 110000, China
CNPC Institute, Taiyanggongnan St. NO.23, Beijing 100010, China

Mahmoud Ebrahimi
Department of Mechanical Engineering, Iran University of Science and Technology, Tehran 16846-13114, Iran

Faramarz Djavanroodi
Department of Mechanical Engineering, Iran University of Science and Technology, Tehran 16846-13114, Iran
Department of Mechanical Engineering, Prince Mohammad Bin Fahd University, Al Khobar 31952, Saudi Arabia

Sobhan Alah Nazari Tiji
Department of Mechanical Engineering, Iran University of Science and Technology, Tehran 16846-13114, Iran

Hamed Gholipour
Department of Mechanical Engineering, Iran University of Science and Technology, Tehran 16846-13114, Iran

Ceren Gode
School of Denizli Vocational Technology, Program of Machine, Pamukkale University, Denizli 20100, Turkey

Suqing Zhang, Tijun Chen
Key Laboratory of Advanced Processing and Recycling of Nonferrous Metals, Lanzhou University of Technology, Lanzhou 730050, China

Faliang Cheng and Pubo Li
Key Laboratory of Advanced Processing and Recycling of Nonferrous Metals, Lanzhou University of Technology, Lanzhou 730050, China

Halil Ibrahim Kurt
Technical Sciences, University of Gaziantep, 27310 Gaziantep, Turkey

Murat Oduncuoglu
Technical Sciences, University of Gaziantep, 27310 Gaziantep, Turkey

Özge Balcı
Particulate Materials Laboratories (PML), Department of Metallurgical and Materials Engineering, İstanbul Technical University, 34469 İstanbul
Institute for Complex Materials, IFW Dresden, 270116 Dresden, Germany

Konda Gokuldoss Prashanth
Institute for Complex Materials, IFW Dresden, 270116 Dresden, Germany
R&D Engineer, Additive manufacturing Center, Sandvik AB, 81181 Sandviken, Sweden

Sergio Scudino
Institute for Complex Materials, IFW Dresden, 270116 Dresden, Germany

Duygu Ağaoğulları
Particulate Materials Laboratories (PML), Department of Metallurgical and Materials Engineering, İstanbul Technical University, 34469 İstanbul

İsmail Duman
Particulate Materials Laboratories (PML), Department of Metallurgical and Materials Engineering, İstanbul Technical University, 34469 İstanbul

M. Lütfi Öveçoğlu
Particulate Materials Laboratories (PML), Department of Metallurgical and Materials Engineering, İstanbul Technical University, 34469 İstanbul

Volker Uhlenwinkel
Institut für Werkstofftechnik, Universität Bremen, D-28359 Bremen, Germany

Jürgen Eckert
Institute for Complex Materials, IFW Dresden, 270116 Dresden, Germany
TU Dresden, Institut für Werkstoffwissenschaft, D-01062 Dresden, Germany

Nicolas Lippitz
Institute for Materials, TU Braunschweig, Langer Kamp 8, Braunschweig D-38106, Germany

Joachim Rösler
Institute for Materials, TU Braunschweig, Langer Kamp 8, Braunschweig D-38106, Germany

Wei Zhong
School of Mechanical Engineering, Jiangsu University of Science and Technology, Zhenjiang 212003, China
Wuxi Pneumatic Technology Research Institute, Wuxi 214072, China

Xin Li
State Key Laboratory of Fluid Power Transmission and Control, Zhejiang University, Hangzhou 310027, China

Guoliang Tao
State Key Laboratory of Fluid Power Transmission and Control, Zhejiang University, Hangzhou 310027, China

Toshiharu Kagawa
Precision and Intelligence Laboratory, Tokyo Institute of Technology, Yokohama 226-8503, Japan

Igor Altenberger
Central Laboratory, Research & Development, Wieland-Werke AG, Graf-Arco-Str. 36, 89079 Ulm, Germany

Hans-Achim Kuhn
Central Laboratory, Research & Development, Wieland-Werke AG, Graf-Arco-Str. 36, 89079 Ulm, Germany

Mozhgan Gholami
Institute of Materials Science and Engineering, TU Clausthal, Agricola-Str. 6, 38678 Clausthal- Zellerfeld, Germany

Mansour Mhaede
Institute of Materials Science and Engineering, TU Clausthal, Agricola-Str. 6, 38678 Clausthal-Zellerfeld, Germany

Lothar Wagner
Institute of Materials Science and Engineering, TU Clausthal, Agricola-Str. 6, 38678 Clausthal-Zellerfeld, Germany

Taekyung Lee
Department of Mechanical Engineering, Northwestern University, Evanston, IL 60208, USA

Donald S. Shih
Boeing Research & Technology, St. Louis, MO 63166, USA

Yongmoon Lee
Graduate Institute of Ferrous Technology, Pohang University of Science and Technology (POSTECH), Pohang 790-784, Korea

Chong Soo Lee
Graduate Institute of Ferrous Technology, Pohang University of Science and Technology (POSTECH), Pohang 790-784, Korea

Jayalakshmi Subramanian
Department of Mechanical Engineering, National University of Singapore, 9 Engineering Drive 1, Singapore 117576, Singapore
Department of Mechanical Engineering, Bannari Amman Institute of Technology, Satyamangalam, Tamil Nadu 638401, India

Sankaranarayanan Seetharaman
Department of Mechanical Engineering, National University of Singapore, 9 Engineering Drive 1, Singapore 117576, Singapore

Manoj Gupta
Department of Mechanical Engineering, National University of Singapore, 9 Engineering Drive 1, Singapore 117576, Singapore

www.ingramcontent.com/pod-product-compliance
Lightning Source LLC
Chambersburg PA
CBHW050437200326

41458CB00014B/4974